Principles of Real Analysis

Moo Hong, Joo

KyoWooSa

Principles of Real Analysis

Moo Hong, Joo

Department of Mathematics
Kyungwon University
Sŏngnam 461-701, KOREA

148-19 Jaekidong, Dongdaemoonku,
Seoul 132-062, Korea

Phone : 82-2-925-2861
Fax : 82-2-925-2860
e-mail : kw2861@chollian.net
http://www.kyowoo.co.kr

ISBN 978-89-8172-716-1 (93410)
Printed in the Republic of korea.

Preface

The subject of real analysis is one of the fundamental areas of mathematics, and is the foundation for the study of many advanced topics, not only in mathematics, but also in engineering and the physical sciences. A through understanding of the concepts of real analysis has also become increas -ingly important for the study of advanced topics in economics and the social science topics such as measure theory and integration are fundamental in mathematics, and physics as well as engineering, economics and many other areas.

This book is written as a text for a first course in abstract analysis, and preliminary versions of the text have been used both in courses for under graduates and in courses for graduate students.

In outline the course consists of a study of the familiar concepts of real analysis, and to provides familiarity with the abstract approach to analysis which is valuable in many areas of applied mathematics and essential to the study of advanced analysis.

I hope that many readers will use exercises to borden their horizons and see how this material fits in the general scheme of things more through under -standing of the theorems and their proofs.

Every reader should persue them, although only the most ambitious will try to work them all out.

Preface

Contents

Chapter 1

INTRODUCTION

In this introductory chaper of the book, we will give an elementary account of sets, functions and relations, with the primary purpose of introducing the notation to be used in the sequel. To save the reader unnecessary effort, this topic will be developed at as low a level is possible and with minimal coverage.

1.1 Set Theory

From the logician's point of view, mathematics is the theory of sets and its consequences. For the analyst, sets and concepts immediately definable from sets are essential tools. and manipulation of sets is an operation he must carry out continually. Accordingly we begin with two sections on sets and functions, containing few proofs, and intended largely to fix notation and terminology and to form a review for the reader in need of one. Section 3 and 4, on the axiom of choice and infinite arithmetic, are more serious: they contain detailed proofs and are recommended for close study by readers unfamiliar with their contents.

As a starting point we describe what we mean by a *set*. We will not attempt to define the term **set**, but we demonstrate the use of the term by examples. The term **set** is used (roughly) to describe any collection of objects. If a set consists of a finite number of objects, we may denote the set by listing its objects. If a set consists of infinitely many objects such a listing is impossible. In this case we may describe the set by naming a property common to all the objects in the set. In general, to describe the set of all objects having a particular property P, we write

$$\{x \mid x \text{ has property } P\}$$

If x is an object in a set A, we write

$$x \in A$$

and say that x is an **element**(or a **point**) of A. If x is not an element of the set A, we write

$$x \notin A$$

We shall say two sets A and B are *equal* and write $A = B$ if A and B have the same elements. Thus $A = B$ if whenever $x \in A$, then $x \in B$, and whenever $x \in B$, then $x \in A$.

1.1 Definition *If A and B are sets, the* ***union*** *of A and B is the set*

$$A \cup B = \{x | x \in A \textit{ or } x \in B\}$$

The ***intersection*** *of A and B is the set*

$$A \cap B = \{x | x \in A \textit{ and } x \in B\}$$

We may extend the definitions of union and intersection to more than two sets as follows. If $\mathscr{A}$ is a collection of sets, we define

$$\cup\mathscr{A} = \{x \mid x \in A \text{ for some } A \in \mathscr{A}\}$$
$$\cap\mathscr{A} = \{x \mid x \in A \text{ for every } A \in \mathscr{A}\}$$

In case the family $\mathscr{A}$ is indexed by the positive integers, $\mathscr{A} = \{A_1, A_2, A_3, \cdots\}$, we write

$$\cup\mathscr{A} = \cup_{n=1}^{\infty} A_n \quad \cap\mathscr{A} = \cap_{n=1}^{\infty} A_n$$

For example, if

$$\mathscr{A} = \{A_1, A_2, A_3, \cdots\}$$

where $A_1 = \{1\}, A_2 = \{1,2\}, \cdots, A_n = \{1, 2, \cdots, n\}, \cdots$
then

$$\cap\mathscr{A} = \bigcap_{n=1}^{\infty} A_n = \{1\} \quad \cup\mathscr{A} = \bigcup_{n=1}^{\infty} A_n$$

is the set of positive integers.
Example If $A_n = \{x : x \text{ is a real number}, |x| < \frac{1}{n}\}, n = 1, 2, 3, \cdots$ then $\cap_{n=1}^{\infty} A_n = \{0\}$.

1.2 Definition *The **empty set** is the set with no elements and is denoted $\emptyset$.*

1.3 Definition *If every element of the set A is an element of the set B, we write $A \subset B$ and say that A is **contained** in B or that A is a **subset** of B.*

We write $B \supset A$ if $A \subset B$ and say that B ***contains*** A.

We note that according to Definition 1.3, $A \subset B$ allows the possibility that $A = B$. If follows immediately from our definitions that $A = B$ if and only if $A \subset B$ and $B \subset A$. We call A a ***proper subset*** of B if $A \subset B$ and $A \neq B$.

1.4 Definition *If A and B are sets, the **difference** of A and B is the set*

$$A \backslash B = \{x \mid x \in A \text{ and } x \notin B\}$$

If we are working with sets all of which are subsets of some particular set U, we sometimes say that U is the *universe* in which we are working. For example, if we are working with sets of integers, we could agree that the universe U is the set $\mathbb{Z}$ of all integers.

1.5 Definition *If we are working in a fixed universe U, and $A \subset U$, we write*

$$A' = U \backslash A$$

*The set $U\backslash A$ is called the **complement** of A* relative to U *(or just simply the **complement** of A if the universe U is understood).*

For example, if U is the set of real numbers, we would state that the complement of the set of rational numbers is the set of irrational numbers.

1.6 Theorem *For arbitrary sets A,B,C, and X, the following laws are valid :*

(1.6.1) *The commutative laws:*

$$A \cup B = B \cup A,$$
$$A \cap B = B \cap A.$$

(1.6.2) *The associative laws:*

$$A \cup (B \cup C) = (A \cup B) \cup C,$$
$$A \cap (B \cap C) = (A \cap B) \cap C.$$

(1.6.3) *The distributive laws:*

$$A \cap (B \cup C) = (A \cap B) \cup (A \cap C),$$
$$A \cup (B \cap C) = (A \cup B) \cap (A \cup C).$$

We now prove two theorems which are known as **De Morgan's laws**. These theorems can often be used to convert a statement about unions into a statement about intersections and vice versa.

1.7 Theorem *If A and B are subsets of a nuivers U, then*

$$(A \cup B)' = A' \cap B' \quad (A \cap B)' = A' \cup B'$$

(De Morgan's formulae)

or

$$X \backslash (A \cup B) = (X \backslash A) \cap (X \backslash B),$$
$$X \backslash (A \cap B) = (X \backslash A) \cup (X \backslash B).$$

Proof. We prove only the first equation leaving the second as an exercise.

Let $x \in (A \cup B)'$. Then $x \notin A \cup B$, and hence $x \notin A$ and $x \notin B$. Thus $x \in A'$ and $x \in B'$ which implies $x \in A' \cap B'$. Therefore,

$$(A \cup B)' \subset A' \cap B'$$

Next suppose $x \in A' \cap B'$. Then $x \in A'$ and $x \in B'$, and hence $x \notin A$ and $x \notin B$. Thus $x \notin A \cup B$, which implies $x \in (A \cup B)'$. Therefore,

$$A' \cap B' \subset (A \cup B)'$$

which completes the proof. □

For a set A, the family of all subsets of A is a well-defined family of sets which is known as the **power set of** A and is denoted by $\wp(A)$. For example, if $A = \{1, 2\}$, then $\wp(A) = \{\emptyset, \{1\}, \{2\}, \{1, 2\}\}$.

1.8 Theorem *Let A, B, C be any sets. Then we have:*

(i)	$A \cup A = A$;	(i')	$A \cap A = A$;
(ii)	$A \cup \emptyset = A$;	(ii')	$A \cap \emptyset = \emptyset$;
(iii)	$A \subset A \cup B$;	(iii')	$A \cap B \subset A$;
(iv)	$A \subset B$ *if and only if* $A \cup B = B$;	(iv')	$A \subset B$ *if and only if* $A \cap B = A$.

The proof of this theorem is very simple and is left to the reader.

1.9 Exercise Describe in words the sets $\cup_{n=1}^{\infty}(\cap_{k=n}^{\infty} A_k)$ and $\cap_{n=1}^{\infty}(\cup_{k=n}^{\infty} A_k)$ where $\{A_1, A_2, \cdots, A_k, \cdots\}$ is any family of sets indexed by $\mathbb{N}$. Also prove that the first set is a subset of the second.

In a similar way one may prove the following theorem.

1.10 Theorem *Let $\mathcal{A}$ be a collection of subsets of universe U, and let*

$$\mathcal{A}' = \{A' \mid A \in \mathcal{A}\}$$

then

$$(\cup\mathcal{A})' = \cap\mathcal{A}' \quad (\cap\mathcal{A})' = \cup\mathcal{A}'$$

1.11 Exercise Simplify as much as possible:
(a) $(A \cup (B \cap (C \cup W')))'$;
(b) $((X' \cup Y) \cap (X \cup Y'))$;
(c) $(A \cap B \cap C) \cup (A' \cap B \cap C) \cup (A \cap B' \cap C) \cup (A \cap B \cap C')$
$\cup (A \cap B' \cap C') \cup (A' \cap B \cap C') \cup (A' \cap B' \cap C)$.

1.12 Definition *If $A \cap B = \emptyset$, then A and B are said to be* ***disjoint****. If $\mathcal{A}$ is a family of sets such that each pair of distinct members of $\mathcal{A}$ are disjoint, then $\mathcal{A}$ is said to be* ***pairwise disjoint****. Thus an indexed family $\{A_i\}_{i\in I}$ is pairwise disjoint if $A_\iota \cap A_\eta = \emptyset$ whenever $\iota \neq \eta$*

1.13 Definition *For sets A and B, the* ***symmetric difference of A and B*** *is the set $\{x : x \in A \text{ or } x \in B \text{ and } x \notin A \cup B\}$, and we write $A \triangle B$ for this set. Note that $A \triangle B$ is the set consisting of those points which are in exactly one of A and B, and that it may also be defined by $A \triangle B = (A \cap B') \cup (A' \cap B)$. The symmetric difference is sketched in Fig. 1.1*

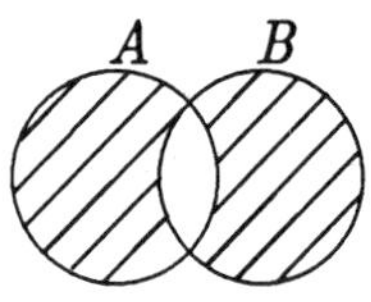

$A \triangle B$

Figure 1.1

1.14 Exercise [PORETSKY] Given two sets X and Y, prove that $X = \emptyset$ if and only if $Y = X \triangle Y$.

1.15 Exercise Describe in words the sets $\bigcup_{n=1}^{\infty}\left(\bigcap_{k=n}^{\infty} A_k\right)$ and $\bigcap_{n=1}^{\infty}\left(\bigcup_{k=n}^{\infty} A_k\right)$ where $\{A_1, A_2, \cdots, A_k, \cdots\}$ is any family of sets indexed by $\mathbb{N}$. Also prove that the first set is a subset of the second.

1.16 Exercise Prove:
(a) $A \triangle (B \triangle C) = (A \triangle B) \triangle C$;
(b) $A \cap (B \triangle C) = (A \cap B) \triangle (A \cap C)$;
(c) $A \triangle A = \emptyset$;
(d) $\emptyset \triangle A = A$.

1.17 Exercise Let $\{A_i\}_{i\in I}$ and $\{B_i\}_{i\in I}$ be nonvoid families of sets. Prove that

(i) $$\left(\bigcup_i A_i\right) \triangle \left(\bigcup_i B_i\right) \subset \bigcup_i (A_i \triangle B_i).$$

Prove by an example that the inclusion may be proper. Can you assert anything about (i) if the $\cup$'s are changed to $\cap$'s ?

1.18 Exercise For any sets A , B , and C, prove that

$$A \triangle B \subset (A \triangle C) \cup (B \triangle C) ,$$

and show by an example that the inclusion may be proper.

1.19 Exercise Let $\{M_n\}_{n=1}^{\infty}$ and $\{N_n\}_{n=1}^{\infty}$ be families of sets such that the sets N_n are pairwise disjoint. Define $Q_1 = M_1$ and $Q_n = M_n \cap (M_1 \cup \cdots \cup M_{n-1})'$ for $n = 2, 3, \cdots$. Prove that $N_n \triangle Q_n \subset \bigcup_{k=1}^{n} (N_k \triangle M_k) \quad (n = 1, 2, \cdots)$.

1.20 Exercise Consider an alphabet with a finite number of letters, say a, where $a > 1$. A *word* in this alphabet is a finite sequence of letters, not necessarily distinct. Two words are equal if and only if they have the same number of letters and if the letters are same and same order. Consider all words of length l, where $l > 1$. How many words of length l have at least two repetitions of a fixed letter ? How many have three such repetitions ? In how many words of length l do there occur two specified distinct letters?

PROBLEMS [1.1]

1. Let U be a universe. Let A be a subset of U and let $\mathcal{A}$ be a collection of subsets of U. Prove:

   (a) $A \cap (\cup\mathcal{A}) = \cup\{A \cap X \mid X \in \mathcal{A}\}$

   (b) $A \cup (\cap\mathcal{A}) = \cap\{A \cup X \mid X \in \mathcal{A}\}$

2. Show that("$\Leftrightarrow$" denotes "if and only if")

(a) $A \times B = \emptyset \Leftrightarrow A = \emptyset$ or $B = \emptyset$

(b) If $C \times D \neq \emptyset$, then $C \times D \subset A \times B \Leftrightarrow C \subset A$ and $D \subset B$.

(c) $(A \times B) \cup (C \times B) = (A \cup C) \times B$

(d) $(A \times B) \cap (C \times D) = (A \cap C) \times (B \cap D)$

3. Prove the DeMorgan laws.

4. The **symmetric difference** of two sets A and B in $\wp(X)$, denoted by $A \triangle B$, is defined by

$$A \triangle B = (A \sim B) \cup (B \sim A)$$

Show that: $A \triangle B = (A \cup B) \sim (A \cap B) = B \triangle A$

5. If $\Re$ is a ring of sets in $\wp(X)$, show that if $A, B \in \Re$, then

(a) $A \cap B \in \Re$

(b) $A \triangle B \in \Re$

(c) $\emptyset \in \Re$

6. If $\Re$ is a ring of sets in $\wp(X)$, show that $\Re$ is an algebra if and only if $\sim A \in \Re$ whenever $A \in \Re$.

7. If symmetric difference (Problem 4) is taken as addition and intersection is taken as multiplication, show that a ring of sets is a commutative ring in the algebraic sense. What is the additive inverse of an element of this ring?

1.2 Relations and Functions

In this section we take up the concepts of relation and function. familiar in several forms from elementary analysis. We adopt the currently popular point of view that relations and functions are indistingulshable from their graphs, *i.e.*, they are sets of ordered pairs. As in the case of sets, we content ourselves with a highly informal discussion of the subject.

2.1 Definition *If X and Y are sets, the* ***Cartesian product*** *of X and Y denoted $X \times Y$ is the set*

$$X \times Y = \{(x, y) \mid x \in X \text{ and } y \in Y\}$$

We write $(x, y) = (u, v)$ if and only if $x = u$ and $y = v$. Thus $(1,2) \neq (2,1)$ while $\{1,2\} = \{2,1\}$

2.2 Definition *A* ***relation*** *is any set of ordered pairs. Thus a relation in any set which is a subset of the Cartesian product of two sets. Observe that $\emptyset$ is a relation.*

2.3 Definition *Let f be any relation. We define the* ***domain of*** *f to be the set $D_f = \{x : (x, y) \in f$ for some $y\}$ and we define the* ***range of*** *f to be the set $R_f = \{y : (x, y) \in f$ for some $x\}$. The symbol f^{-1} denotes the* ***inverse*** *of $f : f^{-1} = \{(y, x) : (x, y) \in f\}$.*

2.4 Definition *Let f be a relation and let A be a set. We define the* ***image of A under f*** *to be the set*

$$f(A) = \{y : (x, y) \in f \text{ for some } x \in A\}.$$

Observe that $f(A) \neq \emptyset$ if and only if $A \cap \text{dom } f \neq \emptyset$. The **inverse image** of A under f is the set $f^{-1}(A)$.

2.5 Definition *Let f and g be relations. We define the* ***composition*** *[* ***product, iterate*** *are also used] f and g to be the relation $g \circ f = \{(x, z) :$ for some $y, (x, y) \in f$ and $(y, z) \in g\}$.*

The composition of f and g may be void. In fact, $g \circ f \neq \emptyset$ if and only if $(\text{rng } f) \cap (\text{dom } g) \neq \emptyset$.

2.6 Definition *Let f and g be relations such that $f \subset g$. Then we say that g is an* ***extension of*** *f and that f is a* ***restriction of*** *g.*

We now discuss some special kinds of relations that are needed in the sequel. Wherever convenient, we will use the conventional notation x/y to mean that $(x, y) \in f$.

Another important special type of relation is called an equivalence relation.

2.7 Definition *A relation R in A is an* ***equivalence relation*** *if:*

E_1.	aRa *for all* $a \in A$;	[*reflexive*]
E_2.	aRb *implies* bRa;	[*symmetric*]
E_3.	aRb *and* bRc *imply* aRc.	[*transitive*]

Notice that both an equivalence relation and a partial order have the reflexive and transitive properties. However, their second properties differ. Property E_2 is called the symmetry property.

Equality, as we use it in the sense of logical identity, is an example of an equivalence relation. Congruence and similarity of triangles in the plane are also equivalence relations. In the set of real numbers the relation R, defined by the rule : aRb if $a - b$ is an integral multiple of some fixed number c, is an equivalence relation.

Closely associated with the concept of an equivalence relation is that of a partition of a set. A **partition** of a set A is a pairwise disjoint collection of nonempty subsets of A whose union is A. We now show that an equivalence relation in A defines a partition of A and, conversely, a partition of A yields an equivalence relation in A.

2.8 Definition *A relation R in A is called a* ***partial order in*** A *if*

P_1.	aRa *for all* $a \in A$;	[*reflexive*]
P_2.	aRb *and* bRa *imply* $a = b$;	[*antisymmetric*]
P_3.	aRb *and* bRc *imply* aRc.	[*transitive*]

A nonempty set A with a partial order defined in it is called a ***partially ordered set*** Frequently the notation $\prec$ is used for a partial order and $a \prec b$ is read "a precedes b". Note that any nonempty subset of a partially ordered set is itself a partially ordered set with respect to the partial order which is the restriction of the original partial order.

Some examples of partially ordered sets follow.

(1) The set $\mathbb{R}$ of all real numbers is partially ordered by the relation $\prec$. In this case P_1, P_2 and P_3 become:

$a \prec a$ for all $a \in \mathbb{R}$;
$a \prec b$ and $b \prec a$ imply $a = b$;
$a \prec b$ and $b \prec c$ imply $a \prec c$.

(2) The power set $\wp(X)$ of some fixed set X is partially ordered by $\subset$. This is called the partial ordering of $\wp(X)$ by inclusion.

(3) Let $\mathbb{N}$ be the set of positive integers and let $a|b$ denote that a divides b. Then $\mathbb{N}$ is partially ordered by this relation.

(4) Let A be the set of all real-valued functions defined on a set X and let $f \prec g$ denote that $f(x) \prec g(x)$ for all $x \in X$. This relation $\prec$ is partial order in A.

(5) Let A be the set of all functions f with domain $D_f \subset X$ and range $R_f \subset Y$. Let $g \prec f$ denote that f is an extension of g, that is, $g \subset f$. Then A is a partially ordered set and is said to be partially ordered by extension. Note that this example is closely related to example (2).

If $\prec$ is a linear ordering such that

(6) $\emptyset \neq A \subset P$ implies there exists an element $a \in A$ such that $a \prec x$ for each $x \in A$ [a is the **smallest element of A**], then $\prec$ is called a **well ordering on** P.

A **partially ordered set** is an ordered pair $(P, \prec)$ where P is a set and $\prec$ is a partial ordering on P. If $\prec$ is a linear ordering, $(P, \prec)$ is called a **linearly** [**simply**, **completely**, **totally**] **ordered set**. If $\prec$ is a well ordering, then $(P, \prec)$ is called a **well-ordering set**.

2.9 Definition *Two element a and b of a partially ordered set A are **comparable** if $a \prec b$ or $b \prec a$.*

2.10 Definition *A partially ordered set is said to be **totally** (**completely**, **linearly**) **ordered** or to be a **chain** if every two elements of the set are comparable.*

2.11 Example (a) Let $\mathfrak{F}$ be any family of sets. Then set inclusion $\subset$ is a partial ordering on $\mathfrak{F}$ and $(\mathfrak{F}, \subset)$ is a partially ordered set.

For short we say that $\mathfrak{F}$ is partially ordered by $\subset$. The reader should note that, depending on $\mathfrak{F}$, this relation may fail to be a linear ordering ; for example, $\mathfrak{F} = \wp(\{0, 1\})$.

(b) Let P be the set of all nonnegative rational numbers:

$$P = \{x : x \in Q, x \geq 0\},$$

and let $\leq$ be the usual ordering on P. Then $\leq$ is a linear ordering on P and P has a smallest element 0, but P, with this ordering, is not a well-ordered set, since there are nonvoid subsets of P containing no smallest element. For example, let $A = \{x \in P : x \neq 0\}$. Then $\frac{x}{2} \in A$ whenever $x \in A$, so A contains no smallest element.

(c) The set $\mathbb{N}$ of all positive integers with its usual ordering is a linearly ordered set. It is also a well-ordered set. This last assertion is equivalent to PEANO'S axiom of mathematical-induction.

The set of all real numbers, $\mathbb{R}$, partially ordered by $\leq$ is a chain. In example (3) the subset $\{1, n, n^2, n^3, \cdots\}$, where n is any positive integer, is a chain.

If B is a subset of a partially ordered set A, then an element c in A is said to be an **upper bound** of B if $b \prec c$ for all $b \in B$. An element c in A is said to be the **least upper bound** of B if c is an upper bound of B and, if d is any other upper bound of B, then $c \prec d$, An element a_0 in the partially ordered set A is said to be a **maximal element** of A if $a \in A$ and $a_0 \prec a$ imply that $a = a_0$. In example (5) any function defined on X is a maximal element of A.

In many proofs a property of partially ordered sets, called Zorn's Lemma, seems to be more suitable to use than the Axiom of Choice. Though we will not do so, it can be shown that Zorn's Lemma is equivalent to the Axiom of Choice. We take Zorn's Lemma as another assumption.

2.12 Zorn's Lemma *If A is a partially ordered set in which each chain has an upper bound, then A has a maximal element.*

2.13 Definition *A relation f is said to be* ***single-valued*** *if $(x, y) \in f$ and $(x, z) \in f$ imply $y = z$. If f and f^{-1} are both single-valued. then f is called a* **one-to-one** *relation. The definitions of* ***many-to-one, one-to-many*** *and* ***many-to-many*** *relations are analogous.*

Single-valued relations play such an important role in analysis that we make the following definition.

2.14 Definition *A single-valued relation is called a* ***function*** *[* ***mapping, transformation, operation, correspondence, application*** *].*

2.15 Definition *Let X and Y be sets. A* ***function*** *from X into Y is a subset f of $X \times Y$ satisfying*

(i) If (x, y) and (x, y') belong to f, then $y = y'$.
(ii) If $x \in X$, then $(x, y) \in f$ for some $y \in Y$.

If f is a function from X into Y, we will write $f : X \to Y$.

The crucial property of a function from X into Y is that with **each** element x in X there is associated a **unique** element y in Y.

2.16 Example The sine function, $\{(x, \sin x) : x \in \mathbb{R}\}$ is many-to-one. The inverse of this function, $\{(\sin x, x) : x \in \mathbb{R}\}$, is a one-to-many relation. The relation $\{(x, y) : x, y \in \mathbb{R}, x^2 + y^2 = 1\}$ is a many-to-many relation. The function $\left\{(x, \tan x) : x \in \mathbb{R}, -\frac{\pi}{2} < x < \frac{\pi}{2}\right\}$ is a one-to-one function.

2.17 Definition *Let f be a function from X into Y. Let $A \subset X$ and $B \subset Y$.*

1. *X is called the* ***domain of*** *f. Y is called the* ***codomain of*** *f.*
2. *If $(x, y) \in f$, we write $y = f(x)$ and call y the (* ***direct*** *)* ***image*** *of x under f.*

3. *The* **range** *of f is the set*

$$\{f(x) \mid x \in X\}$$

4. *The* **image** *of A under f is the set $f(A) = \{f(x) \mid x \in A\}$.*

5. *The* **inverse image** *of B under f is the set*

$$f^{-1}(B) = \{x \mid f(x) \in B\}$$

6. *f is* **onto** *Y if $f(X) = Y$.*

7. *f is* **one to one** *if $f(x) = f(x')$ always implies that $x = x'$ for $x, x' \in X$.*

If f is a one-to-one function from X into Y, we may define the **inverse function** to f, denoted f^{-1}, from the range of f onto X by the rule

$$(y, x) \in f^{-1} \text{ if and only if } (x, y) \in f$$

Notice that the *function* f^{-1} is defined only if f is one to one but that $f^{-1}(B)$ is defined for an arbitrary function f and for all sets $B \subset Y$.

If $g : X \to Y$ and $f : Y \to Z$, we define the **composition** $f \circ g : X \to Z$ by the rule

$$(f \circ g)(x) = f(g(x)) \quad \text{for each } x \in X$$

If $f : X \to Y$ and $A \subset X$, we define the function

$$f|_A = \{(x, y) \in f \mid x \in A\}$$

$f|_A$ is called the **restriction** of f to A, and f is said to an **extension** if $f|_A$ to X.

While there is only one restriction of a given function $f : X \to Y$ to a given subset A of X, the extensions of a given function $g : A \to Y$ over a set X which contains A are usually numerous.

For example, let y denoted an arbitrary point in Y; then the function $\iota_y : X \to Y$ defined by

$$\iota_y(x) = \begin{cases} g(x), & (\text{ if } x \in A) \\ y, & (\text{ if } x \in X \setminus A) \end{cases}$$

is an extension of the given function $g : A \to Y$ over the set X.

The definition of the function $\iota_y : X \to Y$ given above is special case of the construction of **combined functions**. Let F denote a given family of subsets of a set X. Assume that F **covers** X; that is to say, X is equal to the union of the sets in F. Assume that, for each $A \in F$, there has been given a function $f_A : A \to Y$. Thus, we obtain a family

$$\Phi = \{f_x | A \in F\}$$

of functions **indexed** by the members of the family F. The family Φ fo functions is said to be **combinable** iff, for any two sets $A, B \in F$, the functions $f_A : A \to Y$ and $f_B : B \to Y$ **agree** on the intersection $A \cap B$, that is,

$$f_A|_{A\cap B} = f_B|_{A\cap B}.$$

If the family Φ of functions is combinable, then Φ defines uniquely a function $f : X \to Y$ given by taking $f(x) = f_A(x)$ if $x \in A \in F$. This function f will be called the **combined function** of the family Φ of functions.

2.18 Proposition *If $f : X \to Y$ and $\{A_s : s \in S\}$ is a collection of sets in $\wp(X)$, then*

$$f\left(\bigcup_{s \subset S} A_s\right) = \bigcup_{s \subset S} f(A_s)$$

and

$$f\left(\bigcap_{s \in S} A_s\right) \subset \bigcap_{s \in S} f(A_s).$$

However, if f is a one-to-one function from X into Y, then we have the nicer result:

2.19 Proposition *If f is a one-to-one function from X into Y and $\{A_s : s \in S\}$ is a collection of sets in $\wp(X)$, then*

$$f\left(\bigcap_{s\in S} A_s\right) = \bigcap_{s\in S} f(A_s)$$

To obtain corresponding results for the function f we again need to require that f be one-to-one and, in this case f must also be onto.

2.20 Proposition *If f is a one-to-one function from X onto Y and $A, C \in \wp(X)$, then*

$$f(\sim A) = \sim (f(A))$$

and

$$f(A \sim C) = f(A) \sim f(C).$$

Proof. The proof is left to the reader. □

We consider now the behavior of f and f^{-1} with respect to complements and differences

2.21 Proposition *If $f : X \to Y$ and $B, C \in \wp(Y)$, then*

$$f^{-1}(\sim B) = \sim f^{-1}(B)$$

and

$$f^{-1}(B \sim C) = f^{-1}(B) \sim f^{-1}(C)$$

Proof. The proof is left to the reader(Problem 6). □

2.22 Proposition *If $f : X \to Y$ and $\{B_s : s \in S\}$ is a collection of sets in $\wp(Y)$, then*

$$f^{-1}\left(\bigcup_{s\in S} B_s\right) = \bigcup_{s\in S} f^{-1}(B_s)$$

and

$$f^{-1}\left(\bigcap_{s\in S} B_s\right) = \bigcap_{s\in S} f^{-1}(B_s)$$

Proof. The proof is left to the reader. □

2.23 Definition *Let X be any set and E any subset of X. The function ξ_E with domain X and range contained in $\{0,1\}$ such that*

$$\xi_E = \begin{cases} 1 & \text{if } x \in E, \\ 0 & \text{if } x \in X \cap E'. \end{cases}$$

is called the ***characteristic function of E****.*

It will always be clear from the context what the domain of ξ_E is. Characteristic functions are very useful in analysis, and will be encountered frequently throughout this text. One particular characteristic function is used so much that it has a special symbol. The **diagonal** D **of** $X \times X$ is defined as $D = \{(x,x) : x \in X\}$. The value of the characteristic function of D at (x,y) is written δ_{xy} and is called KRONECKER's $\delta-$ symbol. Thus $\delta_{xy} = 1$ if $x = y$ and $\delta_{xy} = 0$ if $x \neq y$; here x and y are arbitrary points in X.

2.24 Exercise Prove that $f \circ (g \circ h) = (f \circ g) \circ h$ for all relations f,g, and h.

2.25 Exercise Show that the equality $f(f^{-1}(B) \cap A) = B \cap f(A)$ fails for every relation f that is not a function.

2.26 Exercise For (a,b) and (c,d) in $\mathbb{N} \times \mathbb{N}$, define $(a,b) \leq (c,d)$ if either:$a < c$, or a=c and $b \leq d$. Prove that, with this relation, $\mathbb{N} \times \mathbb{N}$ is a well-ordered set.

2.27 Exercise Let n be a positive integer and let $P_n = \{k \in N : k \text{ is a divisor of } n\}$. For $a, b \in P_n$ define $a \preceq b$ to mean that a is a divisor of b, *i.e.*, $a|b$.

(a) Prove that P_n, with $\preceq$ is a partially ordered set.

(b) Find necessary and sufficient conditions on n that P_n be a linearly ordered set.

2.28 Exercise Let X be a set with a binary operation p defined on it, *i.e.*, p is a function from $X \times X$ into X. Write $p(x, y) = xy$. Suppose that this operation satisfies

(i) $x(yz) = (xy)z$,
(ii) $xy = yx$,
(iii) $xx = x$,

for all x, y, z in X. Define $\leq$ on X by $x \leq y$ if and only if $xy = y$. Prove that: (a) X is a partially ordered set; (b) each pair of elėments of X has a least upper bound, *i.e.*, if $x, y \in X$, then there exist $z \in X$ such that $x \leq z$, $y \leq z$, and if $x \leq w$, $y \leq w$, then $z \leq w$.

2.29 Exercise Let f be a function from X into Y. Suppose that there is a function g from Y to X such that $f \circ g(y) = y$ for all $y \in Y$ and $g \circ f(x) = x$ for all $x \in X$. Prove that f is a one-to-one function from X onto Y and that $g = f^{-1}$.

2.30 Theorem *If $h = g \circ f$ denotes the composition of the functions $f : X \to Y$ and $g : Y \to Z$, then the following statements are true :*

i. If h is surjective, then so is g.

ii. If h is injective, then so is f.

Proof. To establish *i*), let us assume that h is surjective. By definition, we have $h(X) = Z$. and

$$Z = h(X) = g[f(X)] \subset g(Y) \subset Z.$$

This implies that $g(Y) = Z$ and hence g surjective. This proves *ii*).

To establish *ii*), let us assume that h is injective. Consider any two points a and b of X with $f(a) = f(b)$. Then we have

$$h(a) = g[f(a)] = g[f(b)] = h(b).$$

Since h is injective, it follows that $a = b$. This proves *ii*). □

To conclude the present section, we will give a few more examples of functions of special kinds.

2.31 Example A function $f : \mathbb{N} \to X$ from the set $\mathbb{N}$ of natural numbers into a given set X is called a *sequence* (of points) in X. For each $n \in \mathbb{N}$, the image $x_n = f(n)$ is called the nth *term* of the sequence f. Customarily, the sequence f is written in the form

$$f = \{x_1, x_2, \cdots, x_n, \cdots\}.$$

In particular, if X is the set $\mathbb{R}$ of real numbers, then f is called a sequence of real numbers; if X is the set $\mathbb{Z}$ of intergers, then f is called a sequence of integers.

P R O B L E M S [1.2]

1. Let f and g be functions defined on $\mathbb{R}$, the set of all renumbers, by the rules $f(x) = 3x + 2$ and $g(x) = x^2$.

   (a) Show that f is one-to-one and g is not one-to-one.

   (b) Determine the invers relations f^{-1} and g^{-1} and show that f^{-1} is a function but that g^{-1} is not a function.

   (c) Show that $g \circ f = f \circ g$.

2. If $f : X \to Y$, $g : Y \to Z$ and f and g are ono-to-one, show that $g \circ f$ is a one-to-one function from X into Z and that

$$(g \circ f)^{-1} = f^{-1} \circ g^{-1}.$$

3. Let (A_n) be a decreasing sequence of sets; that is, $A_n \supset A_{n+1}$ for each $n \in \mathbb{N}$, the set of positive integers. Define $B_0 = \bigcap_{n=1}^{\infty} A_n$ and $B_n = A_n \sim A_{n+1}$, $n \in \mathbb{N}$. Show that $\{B_k : k \in \{0\} \cup \mathbb{N}\}$ is a pairwise disjoint collection of sets and that
$$A_1 = \bigcup_{k=0}^{\infty} B_k.$$

4. Prove Propositon 2.22

5. If $f : X \to Y$ and $\{A_s : s \in S\}$ is a collection of sets in $\wp(X)$, prove
$$f\left(\bigcup_{s\in S} A_s\right) = \bigcup_{s\in S} f(A_s).$$

6. If $f : X \to Y$ and $B, C \in \wp(Y)$, show that
$$f^{-1}(\sim B) = \sim (f^{-1}(B))$$
and
$$f^{-1}(B \sim C) = f^{-1}(B) \sim f^{-1}(C).$$

7. If $f : X \to Y$ and $B \in \wp(Y)$, show that $f(f^{-1}(B)) \subset B$. Given an example where $f(f^{-1}(B)) \neq B$. What conditions on f insure that $f(f^{-1}(B)) = B$ for all $B \in \wp(Y)$?

8. If $f : X \to Y$ and $A \in \wp(X)$, show that $A \subset f^{-1}(f(A))$. Given an example where $A \neq f^{-1}(f(A))$. What conditions on f insures that $A = f^{-1}(f(A))$ for all $A \in \wp(X)$?

9. If $f : X \to Y$ and $A \in \wp(X)$, show that $\sim f(A) \subset f(\sim A)$ if and if f is onto Y.

1.3 The Axiom of Choice

In the study of algebra, analysis, and topology one frequently encounters situations in which the tools of elementary set theory are too weak to permit constructions, proofs, or even definitions that one may need. In the early 1900's the German mathematician ERNST ZERMELO propounded an innocent appering but actually very strong axiom, called the axiom of choice [*Auswahlpostulat*], which has many important consequence, and which has also excited vigorous controversy. In this section we take up the axiom of choice, establish the equivalence of four other assertions with it, and point out two important applications. Other applications of the axiom of choice will appear throughout the book.

3.1 Definition *Let $\{A_i\}_{i\in I}$ be any family of sets. The* ***Cartesian product*** *of this family, written $\times_{i\in I}A_i$, is the set of all functions x having domain I such that $x_i = x(i) \in A_i$ for each $i \in I$. Each such function x is called a* ***choice function*** *for the family $\{A_i\}_{i\in I}$. For $x \in \times_{i\in I}A_i$ and $i \in I$, the value $x_i \in A_i$ is known as the ith* ***coordinate of x*** .

One may ask if there are any choice functions for a given family of sets. Of course if $I = \emptyset$, then the void function $\emptyset$ is a choice function for any family indexed by I. If $I \neq \emptyset$ and $A_i = \emptyset$ for some $i \in I$, then $\times_{i\in I}A_i = \emptyset$. These two special cases are of little interest. In general the question cannot be answered on the basis of the usual axioms of set theory. We will use the following axiom.

3.2 Axiom of choice. *The Cartesian product of any nonvoid family of nonvoid sets is a nonvoid set, i.e., if $\{A_i\}_{i\in I}$ is a family of sets such that $I \neq \emptyset$ and $A_i \neq \emptyset$ for each $i \in I$, then there exist at least one choice function for the family $\{A_i\}_{i\in I}$.*

P.J.COHEN has recently proved that this axiom is independent of the other axioms of set theroy [Proc. Nat. Acad. Sci. U. S. A. 50, 1143-1148(1963); 51, 105-110 (1964)].

3.3 Definition *Let A and I be sets. We defined A^I to be the Cartesian product $\times_{i\in I}A_i$, where $A_i = A$ for each $i \in I$. Thus A^I is the set of all functions f such that* $\operatorname{dom} f = I$ *and* $\operatorname{rng} f \subset A$. *If, for some $n \in \mathbb{N}$, I is the set $\{1, \cdots, n\}$, then we write $A^I = A^n$. Some authors write A^∞ for $A^{\mathbb{N}}$.*

A typical member of A^n is, to be sure, a function and as such is a set of n ordered pairs. We follow conventional notation, however, and list the values of such a function as an ordered n-tuple. Thus $A^n = \{(a_1, \cdots, a_n) : a_k \in A \text{ for } k = 1, \cdots, n\}$. Similary $A^{\mathbb{N}} = \{(a_1, a_2, \cdots) : a_k \in A \text{ for } k \in \mathbb{N}\}$. The set $\mathbb{R}^n$ is called **Euclidean n-space** and K^n is called **unitary n-space**.

3.4 Exercise Let $A = \{0, 1\}$. Then $A^{\mathbb{N}}$ is the set of all sequences $a = (a_1, a_2, \cdots, a_n, \cdots)$ where each a_n is 0 or 1. In many ways this set resembles CANTOR'S ternary set $P = [0,1] \cap \left(\left(\frac{1}{3}, \frac{2}{3}\right) \cup \left(\frac{1}{9}, \frac{2}{9}\right) \cup \left(\frac{7}{9}, \frac{8}{9}\right) \cup \cdots\right)$. The mapping φ defined by $\varphi(a) = 2\sum_{n=1}^{\infty} \frac{a_n}{3^n}$ is a one-to-one mapping from $A^{\mathbb{N}}$ onto P. Anticipating future developments, we remark that $A^{\mathbb{N}}$ can be made a metric space by introducing the metric ρ, where $\rho(a, b) = \frac{1}{n}$ if $a_1 = b_1, a_2 = b_2, \cdots, a_{n-1} = b_{n-1}$, and $a_n \neq b_n$; and $\rho(a, b) = 0$ if $a = b$. Under this metric on $A^{\mathbb{N}}$, φ and φ^{-1} are both continuous. The set $A^{\mathbb{N}}$ becomes an Abelian group under the operation $+$ defined by $(a + b)_n = a_n + b_n (\text{mod } 2)$ for $n \in \mathbb{N}$. [There are many other ways to make $A^{\mathbb{N}}$ into an Abelian group.]

3.5 Definition *Let $(P, \leq)$ be any partially ordered set and let $A \subset P$. An element $u \in P$ is called an* ***upper bound for*** *A if $x \leq u$ for each $x \in A$. An element $m \in P$ is called a* ***maximal element of*** *P if $x \in P$ and $m \leq x$ implies $m = x$. Similary we define* ***lower bound*** *and* ***minimal element.***[1] ***A chain in*** *P is any subset C of P such that C is linearly ordered under the given order relation $\leq$ on P.*

This terminology of partially ordered sets will often be applied to an arbitrary family of sets. When this is done, it should be understood that the family is being regarded as a partially ordered set under the relation $\subset$ of set inclusion. Thus a maximal member of $\mathscr{A}$ is set $M \in \mathscr{A}$ such that M is proper subset of no other

[1]We agree that every element of P is both an upper bound and a lower bound for the void set $\emptyset$; but naturally $\emptyset$ *contains* neither a maximal nor a minimal element.

member of $\mathscr{A}$ and a chain of sets is a family $\mathscr{B}$ of sets such that $A \subset B$ or $B \subset A$ whenever $A, B \in \mathscr{B}$.

3.6 Definition *Let $\mathcal{F}$ be a family of sets. Then $\mathcal{F}$ is said to be a* ***family of finite character*** *if for each set A we have $A \in \mathcal{F}$ if and only if each finite*[2] *subset of A is in $\mathcal{F}$.*

We shell need the following technical fact.

3.7 Lemma . *Let $\mathcal{F}$ be a family of finite character and let $\mathscr{B}$ be a chain in $\mathcal{F}$. Then $\cup\mathscr{B} \in \mathcal{F}$.*

Proof. It suffices to show that each finite subset of $\cup\mathscr{B}$ is in $\mathcal{F}$. Let $F = \{x_1, \cdots, x_n\} \subset \cup\mathscr{B}$. Then there exist sets $B_1, \cdots, B_n$ in $\mathscr{B}$ such that $x_j \in B_j (j = 1, \cdots, n)$. Since $\mathscr{B}$ is a chain there is a $j_0 \in \{1, \cdots, n\}$ such that $B_j \subset B_{j_0}$ for each $j = 1, \cdots, n$. Then $F \subset B_{j_0} \in \mathcal{F}$. But $\mathcal{F}$ is of finite character, and so $F \in \mathcal{F}$. □

There are many problems in set theory, algebra, and analysis to which the axiom of choice in the from(3.2) is not immediately applicable, but to which one or another equivalent axiom is applicable at once. We next list four such statement. The names "lemma" and "theorem" are attached to them only for historical, as they are all equivalent to Axiom(3.2).

3.8 Tukey's Lemma *Every nonvoid family of finite character has a maximal member.*

3.9 Hausdorff Maximality Principle *Every nonvoid partially ordered set contains a maximal chain.*

3.10 Zorn's Lemma *Every nonvoid partially ordered set in which each chain has an upper bound has a maximal element.*

3.11 Well-ordering Thorem [ZERMELO] *Every set can be well ordered;*i.e., *if S is a set, then there exists some well-ordering $\leq$ on S.*

[2]A set F is said to be **finite** if either $F = \emptyset$ or there exist $n \in \mathbb{N}$ and a one-to-one function from $\{1, 2, \cdots, n\}$ onto F.

3.12 Theorem *The following five propositions are pairwise equivalent:*

(*i*) *The axiom of choice;*

(*ii*) *Tukey's lemma;*

(*iii*) *The Hausdorff maximality principle;*

(*iv*) *Zorn's lemma;*

(*v*) *The well-ordering theorem.*

Proof. The proof is left to the reader. □

PROBLEMS [1.3]

1. Prove theorem 3.12

2. Let X and Y be arbitrary sets with given partial orders in X and Y, both denoted by the same symbol $<$. A function $f : X \to Y$ is said to **preserve** the partial order $<$ (or **increasing**) iff, for any two members a and b of X, $a < b$ implies $f(a) < f(b)$. A function $f : X \to Y$ is said to **reverse** the partial order $<$ (or **decreasing**) iff, for any two members a and b of X, $a < b$ implies $f(b) < f(a)$. Consider the following functions $f : \mathbb{R} \to \mathbb{R}$ of the set real numbers. Determine those which preserve the usual partial order $<$ in $\mathbb{R}$ and those which reverse it.

   (a) $f(x) = 3x + 1$ for each $x \in \mathbb{R}$
   (b) $f(x) = 1 - x$ for each $x \in \mathbb{R}$
   (c) $f(x) = x^2$ for each $x \in \mathbb{R}$
   (d) $f(x) = x^3$ for each $x \in \mathbb{R}$
   (e) $f(x) = e^x$ for each $x \in \mathbb{R}$
   (f) $f(x) = e^{-x}$ for each $x \in \mathbb{R}$

(g) $f(x) = -e^{-x}$ for each $x \in \mathbb{R}$

3. Prove that the Axiom of Choice is equivalent to the following statement: For every family $\mathcal{F}$ of subsets of a set X which covers X, there exists a subset C of X such that, for every member E of $\mathcal{F}$, the intersection $C \cap E$ is a singleton. Such a subset C of X will be called a *cross-section* of the family $\mathcal{F}$.

4. Let A be the set of all functions f with domain $D_f \subset X$ and range $R_f \subset Y$ and let A be partially ordered by extension; that is, $g \prec f$ if and only if $g < f$. If $\{g_s : s \in S\}$ is a chain in A, show that the function g such that $D_g = \bigcup_{s \in S} D_{g_s}$ and $g(x) = g_s(x)$ for all $x \in D_{g_x}$ is an upper bound for this chain in A.

Chapter 2

REAL NUMBER SYSTEM

A way of introducing the real number system, commonly used in introductory courses in analysis, is to give a set of axioms for the system. the real numbers are undefined objects with operations of addition and multiplication and an order relation which have properties specified by the given axioms. Also, one of the axioms states that the system is complete. This axiom may be given in various forms; for example, the least upper bound axiom states that a nonempty set of real numbers which is bounded above has a least upper bound.

Another approach to the real numbers is to introduce the natural numbers axiomatically and then to define successively the integers, rationals, and reals. This appoach has the advantage of a relatively simple initial axiom system. However, the path to the real numbers is quite long and tedious [12]. The development as far as the rational number system can be found in many books on modern abstract algebra.

A rational number is taken to be an equivalence class of ordered pairs of integers with second element nonzero where the equivalence is defined by: $(a, b)\ R\ (c, d)$ if $ad = bc$. Letting $[(a, b)]$

denote the equivalence class containing (a, b), if $r = [(a, b)]$ and $s = [(c, d)]$ one defines $r + s = [(ad + bc, bd)]$ and $rs = [(ac, bd)]$. Also, if b and d are positive, then

$$r < s \text{ if } ad < bc.$$

To obtain the representation of a rational number as a quotient of two integers identify the rational number $[(a, 1)]$ with the integer a. Note that $[(1, b)]$ is the multiplicative inverse of $[(b, 1)]$ since

$$[(b, 1)][(1, b)] = [(b, b)] = [(1, 1)] = 1.$$

Then, for any rational number $r = [(a, b)]$, we have

$$r = [(a, b)] = [(a, 1)][(1, b)] = \frac{[(a, 1)]}{[(b, 1)]} = \frac{a}{b}.$$

In this chapter we will develop the real number system from the rational number system. This development may provide insights not gained by an axiomatic introduction of the real numbers. Also, the methods used in this development will be used later in a more general context.

2.1 Ordered Rings and Fields

The basic properties of the rational numbers are described by the statement that the rational number system is an archimedean-ordered field. These terms will be defined in this section. First we define a related algebraic structure called a ring.

1.1 Definition *A **ring** is a set R with two functions from $R \times R$ into R, called addition and multiplication and denoted by the usual symbolism, which satisfy the following conditions:*

A_1. $(a + b) + c = a + (b + c)$ *for all* $a, b, c \in R$. *< associative >*

A_2. $a + b = b + a$ *for all* $a, b \in R$. *< commutative >*

A_3. *There is an element* 0 *in* R *such that*

$a + 0 = a$ *for all* $a \in R$. < *additive identity* >

A_4. *For each* $a \in R$ *there is an element* $-a \in R$ *such that*

$a + (-a) = 0$. < *additive inverse* >

M_2. $(ab)c = a(bc)$ *for all* $a, b, c \in R$. < *associative* >

D. $a(b + c) = ab + ac$ *and* $(b + c)a = ba + ca$

for all $a, b, c \in R$. < *distributive* >

Usually the ring is denoted by R, the same letter used to denote its set of elements. Any function from $R \times R$ into R is called a **binary operation** on R.

A ring R is called a **commutative ring** if it satisfies:

M_2. $ab = ba$ for all $a, b \in R$.

A ring R is said to be a **ring with unit** if it satisfies:

M_3. there is an element $1 \neq 0$ in R such that $a \cdot 1 = a = 1 \cdot a$ for all $a \in R$.

It is easy to show that the additive identity 0 and unit 1 (if it exists) are unique. Also, each element $a \in R$ has a unique additive inverse $-a$. In a ring R the operation of subtraction is defined as follows:

$$a - b = a + (-b).$$

The following results are easily proved for a ring R:

(1.2) $a + b = a + c$ implies $b = c$ < *cancellation* >

(1.3) $-(-a) = a$ and $-(a + b) = -a + (-b)$

(1.4) $a \cdot 0 = 0 = 0 \cdot a$ for all $a \in R$

(1.5) $a(-b) = -(ab) = (-a)b$ and $(-a)(-b) = ab$.

If R has a unit, we also have:

(1.6) $-a = (-1)a$

The integers and the rational numbers provide examples of a commutative ring with unit. The rational numbers have an additional property not shared with the integers: each nonzero element has a multiplicative inverse.

1.7 Definition *A* ***field*** $\mathbf{F}$ *is a commutative ring with unit which satisfies:*

M_4. *For each* $a \neq 0$ *in* $\mathbf{F}$ *there is an element* $a^{-1} \in \mathbf{F}$ *such that* $a \cdot a^{-1} = 1$.

The rational number system is a field. In a field $\mathbf{F}$ each nonzero element b has a unique multiplicative inverse b^{-1} and operation of division is defined as follows:

$$\frac{a}{b} = ab^{-1} \text{ where } b \neq 0.$$

Notice that 2·4 implies that 0 cannot have a multiplicative inverse.

Since a field is a commutative ring with unit, the results $1 \cdot 2$–$1 \cdot 6$ hold in a field $\mathbf{F}$ we also have:

(1.8) $ab = ac$ and $a \neq 0$ imply $b = c$ < *cancellation* >

(1.9) $(a^{-1})^{-1} = a$ and $(ab)^{-1} = a^{-1}b^{-1}$

(1.10) $\dfrac{a}{b} + \dfrac{c}{d} = \dfrac{ad + bc}{bd}$ and $\dfrac{a}{b} \cdot \dfrac{c}{d} = \dfrac{ac}{bd}$.

In the ring of integers there is a subset called the set of positive integers and there is a relation "less than" which can be defined in terms of this subset: $a < b$ if $b - a$ is positive. By recognizing the essential properties of the set of positive integers it is possible to generalize these notions so that they apply to other rings.

1.11 Definition *A ring R is said to be an* ***ordered ring*** *if there exists a subset P of R such that*

(1) $0 \notin P$;

(2) *If* $a \neq 0$, *then one and only one of the elements* a *or* $-a$ *is in* P;

(3) *If* $a, b \in P$, *then* $a + b \in P$ *and* $ab \in P$.

The set P is called the ***positive cone*** *of R and the elements of P are said to be* ***positive****.*

In an ordered ring the relation "less than" is defined as follows:

$$a < b \quad \text{if} \quad b - a \in P$$

Observe that a is positive if and only if $0 < a$. If $a < 0$, then is said to be **negative**. The relation "less then" (or "great than") has the following properties in any ordered ring R:

(1.12) For any elements a and b in R one and only one of the following relations holds: $a < b$, $a = b$, $b < a$.

(1.13) $a < b$ and $b < c$ imply $a < c$

(1.14) $a < b$ implies $a + c < b + c$ for all $c \in R$.

(1.15) $a < b$ and $0 < c$ imply $ac < bc$

(1.16) $a < b$ and $c < 0$ imply $ac > bc$

(1.17) $a \neq 0$ implies $a^2 > 0$.

If R has a unit 1, then 1.17 implies that $1 > 0$.

We may assume that any ordered ring with unit contains the integers by identifying the ring elements $0, 1, -1, 1 + 1$, etc., with the integers $0, 1, -1, 2$, etc. An ordered ring R with unit is said to be an **archimedean-ordered ring** if for each pair of positive elements a and b in R there is a positive integer n such that $b < na$.

A field **F** is said to be (archimedean-)ordered if as a ring it is (archimedean-)ordered. In an ordered field the "less then" relation has the additional properties:

(1.18) $0 < a < b$ implies $0 < b^{-1} < a^{-1}$

(1.19) $a < b < 0$ implies $b^{-1} < a^{-1} < 0$.

The **absolute value** of an element a, denoted by $|a|$, is defined in an ordered ring as follows:

$$|a| = \begin{cases} a & \text{if } a \geq 0 \\ -a & \text{if } a < 0 \end{cases}$$

The basic properties of the absolute value of elements of an ordered rign are:

(1) $|a| \geq 0$; $|a| = 0$ if and only if $a = 0$

(2) $|ab| = |a||b|$

(3) $|a + b| \leq |a| + |b|$ $\qquad$ $<$ *Triangle Inequality* $>$

These properties follow readily from the definition. In the proof of the Triangle Inequality the fact that $a \leq |a|$ and $-a \leq |a|$ can be helpful.

PROBLEMS [2.1]

1. Prove: 1.2-1.6
2. Prove: 1.8-1.10
3. Prove: 1.12-1.17
4. Prove: 1.18-1.19
5. Prove the basic properties of absolute value.
6. Prove that $||a| - |b|| \leq |a - b|$.
7. If $\mathcal{F}$ is the set of all functions from a set X into a ring R, taking the usual definitions of addition and multiplication of functions. show that $\mathcal{F}$ is a ring.
8. A subset S of a ring R is called a **subring** if it is itself a ring under the operations in R restricted to S. Give minimal requirements for a subset of a ring to be a subring.
9. A ring R is isomorphic to a ring R' if there is a one-to-one mapping φ from R onto R' such that

$$\varphi(a + b) = \varphi(a) + \varphi(b) \text{ and } \varphi(ab) = \varphi(a)\varphi(b).$$

If φ is an isomorphism of the ring R onto the ring R', show that $\varphi(0) = 0$, $\varphi(-a) = -\varphi(a)$, $\varphi(1) = 1$.

10. If R is the set of polynomials with rational coefficents, show that R is a ring when we use the usual definitions of addition and multiplication of polynomals. Let P be the set of polynomials in R with leading coefficient positive. Show P is a positive cone of R.

11. Show that an ordered field must be infinite.

12. Let a and b be elements of an ordered field $\mathbf{F}$ with $a < b$. Show that there exists a $c \in \mathbf{F}$ such that $a < c < b$.

2.2 Cauchy Sequences

In the development of the real number system from the rational number system the concept of a Cauchy sequence will play a central rôle. First we recall the definitions of a sequence and a convergent sequence. A sequence of elements of a field $\mathbf{F}$ is a function from the set $\mathbb{N}$ of positive integers into $\mathbf{F}$.

2.1 Definition *A sequence (a_n) of elements of an ordered field $\mathbf{F}$ is said to be **convergent** if there exists an element $a \in \mathbf{F}$ such that for each positive element ε in $\mathbf{F}$ there exists a positive integer n_0 such that*

$$|a_n - a| < \varepsilon \textit{ whenever } n \geq n_0.$$

In this case we say that (a_n) converges to a and write $\lim a_n = a$.

As an example we show that that the sequence $(1/n)$ of rational numbers converges to 0 in the rational field. Let ε be a positive rational number. Since the rational field is archimedean-ordered there exists a positive integer $n_0 > 1/\varepsilon$. Then

$$\left|\frac{1}{n} - 0\right| = \frac{1}{n} < \varepsilon \text{ whenever } n \geq n_0.$$

2.2 Definition *A sequence $< a_n >$ of elements of an ordered field $\mathbf{F}$ is called a **Cauchy sequence** if for each $\varepsilon > 0$ in $\mathbf{F}$ there exists a positive integer n_0 such that*

$$|a_n - a_m| < \varepsilon \textit{ whenever } m, n \geq n_0$$

The definition of a Cauchy sequence differs from that of a convergent sequence in that terms of a Cauchy sequence get close to each other whereas the terms of a convergent sequence get close to some fixed element in the ordered field. These may seem to be equivalent but such is not the case in every ordered field. A convergent sequence is always a Cauchy sequence but in some ordered fields a Cauchy sequence may fail to converge.

2.3 Proposition. *A convergent sequence in an ordered field* **F** *is a* Cauchy sequence *in* **F**.

Proof. Let (a_n) be a sequence of elements in **F** such that $\lim a_n = a$. Then for each $\varepsilon > 0$ in **F** there exists a positive integer n_0 such that

$$|a_n - a| < \varepsilon/2 \text{ whenever } n \geq n_0.$$

Thus

$$|a_n - a_m| < |a_n - a| + |a - a_m| < \varepsilon \text{ whenever } m, n \geq n_0. \qquad \square$$

To show that the converse of this proposition does not hold, consider the sequence of rational numbers:

$$1, 1.4, 1.41, 1.414, 1.4142, \cdots$$

obtained by the familiar algorithm for "finding" the square root of 2. This is a Cauchy sequence but it does not converge to any element in the rational field.

In some ordered fields it is true that every Cauchy sequence converges.

2.4 Definition *An ordered field* **F** *is siad to be* **complete** *if every Cauchy sequence of elements in* **F** *converges to an element in* **F**.

The discussion above shows that the rational field is not complete. We show later that the real field is complete. In fact, the extension of the rational field to the real field is basically the extension of an incomplete ordered field to a complete ordered field.

We now prove some properties of Cauchy sequence in an arbitrary ordered field. A sequence (a_n) of elements in an ordered field $\mathbf{F}$ is said to be **bounded** if there exists an element M in $\mathbf{F}$ such that $|a_n| \leq M$ for all n.

2.5 Propostion. *A Cauchy sequence (a_n) of elements in an ordered field $\mathbf{F}$ is bounded.*

Proof. Corresponding to the unit 1 in $\mathbf{F}$ there exists a positive integer n_0 such that

$$|a_n - a_m| < 1 \text{ whenever } m, n \geq n_0.$$

Thus, whenever $n \geq n_0$,

$$|a_n| - |a_{n_0}| \leq |a_n - a_{n_0}| < 1;$$

that is,

$$|a_n| < |a_{n_0}| + 1.$$

If we let $M = \max\{|a_{n_0}| + 1, |a_k| : k = 1, \cdots, n_0 - 1\}$, then

$$|a_n| \leq M \text{ for all } n. \qquad \square$$

If (a_n) is a sequence of elements in an ordered field $\mathbf{F}$ and (n_k) is an increasing sequence of positive integers, i.e., $n_k < n_{k+1}$, then $(a_{n_k}) = (a_n) \circ (n_k)$ is called a **subsequence** of (a_n). For example, $(1/2k)$ is a subsequence of $(1/n)$. In this example $a_n = (1/n)$ and $n_k = 2k$.

2.6 Propostion *If a Cauchy sequence (a_n) of elements in an ordered field $\mathbf{F}$ has a subsequence (a_{n_k}) which converges to a, then (a_n) converges to a.*

Proof. Take $\varepsilon > 0$ in $\mathbf{F}$. There exists a positive integer n_0 such that

$$|a_n - a_m| < \varepsilon/2 \quad \text{whenever } m, n \geq n_0.$$

Also, there exists a positive integer k_0 such that

$$|a_{n_k} - a| < \varepsilon/2 \quad \text{whenever } k \geq k_0.$$

If we take $k \geq k_0$ such that $n_k \geq n_0$, then

$$|a_n - a| \leq |a_n - a_{n_k}| + |a_{n_k} - a| < \varepsilon \text{ whenever } n \geq n_0. \qquad \square$$

Consider now Cauchy sequences of rational numbers. Let $\mathfrak{J}$ denote the set of all Cauchy sequences of rational numbers. We can define binary operations of addition and multiplication in $\mathfrak{J}$ so that $\mathfrak{J}$ with these operations is a commutative ring with unit. If (a_n) and (b_n) are in $\mathfrak{J}$, define

$$(a_n) + (b_n) = (a_n + b_n)$$
$$(a_n)(b_n) = (a_n b_n).$$

The fact that these definitions yield binary operations in $\mathfrak{J}$, that is, $(a_n + b_n)$ and $(a_n b_n)$ are in $\mathfrak{J}$ whenever (a_n) and (b_n) are in $\mathfrak{J}$, follows readily from the inequalities:

$$|(a_n + b_n) - (a_m + b_m)| \leq |a_n - a_m| + |b_n - b_m|$$

and

$$\begin{aligned} |a_n b_n - a_m b_m| &\leq |a_n b_n - a_m b_n| + |a_m b_n - a_m b_m| \\ &\leq M_2|a_n - a_m| + M_1|b_n - b_m| \end{aligned}$$

where $|a_n| \leq M_1$ and $|b_n| \leq M_2$ for all n.

It is a routine matter to check that $\mathfrak{J}$ with these operations satisfies the conditions for a commutative ring with unit. For example, the verification of A_2 is:

$$(a_n) + (b_n) = (a_n + b_n) = (b_n + a_n) = (b_n) + (a_n).$$

If we let $(\bar{a})$ denote the Cauchy sequence all of whose terms are the rational number a, then the element 0 in $\mathfrak{J}$ is the Cauchy sequence $(\bar{0})$. The additive inverse of (a_n) is $-(a_n) = (-a_n)$. And, the unit element 1 in $\mathfrak{J}$ is the Cauchy sequence $(\bar{1})$.

$\mathfrak{J}$ is not a field however since a nonzero element in $\mathfrak{J}$ may fail to have a multiplicative inverse. A sequence (b_n) such that $\lim b_n = 0$ is called a **null sequence**. The following proposition shows that any null sequence in $\mathfrak{J}$, e.g., $(1/n)$, cannot have a multiplicative inverse in $\mathfrak{J}$ and, hence, $\mathfrak{J}$ is not a field.

2.7 Proposition *If* (a_n) *and* (b_n) *are in* $\mathfrak{J}$ *and* $\lim b_n = 0$*, then* $\lim(a_n b_n) = 0$

Proof. Since any Cauchy sequence is bounded. there exists a positive rational number M such that $|a_n| \leq M$ for all n. Let ε be any positive rational number. Then there is a positive integer n_0 such that

$$|b_n| < \varepsilon/M \text{ whenever } n \geq n_0$$

and, hence,

$$|a_n b_n| < \varepsilon \text{ whenever } n \geq n_0. \qquad \square$$

PROBLEMS [2.2]

1. If (a_n) and (b_n) are sequences of elements in an ordered field $\mathbf{F}$ such that $\lim a_n = a$ and $\lim b_n = b$, show that

$$\lim(a_n + b_n) = a + b \text{ and } \lim(a_n b_n) = ab.$$

2. If in an ordered field $\mathbf{F}$ the sequence (a_n) converges to a, show that any subsequence (a_{n_k}) of (a_n) converges to a.

3. Prove that $\mathfrak{J}$ is a commutative ring with unit.

4. Show that the sequence $(\bar{a})$ converges to a.

5. If $(b_n) \in \mathfrak{J}$, $b_n \neq 0$ for all n, and (b_n) dose not converge to zero, show that (b_n) has a multiplicative inverse in $\mathfrak{J}$.

6. If (a_n) is a Cauchy sequence of integers in the rational field, does (a_n) converge ?

7. In an archimedean-ordered field $\mathbf{F}$ show that

$$\lim \frac{1}{2^n} = \lim \frac{1}{n} = 0.$$

2.3 The Real Numbers

Let $\mathfrak{J}$ be the set of all Cauchy sequences of rational numbers with

$$(a_n) + (b_n) = (a_n + b_n) \text{ and } (a_n)(b_n) = (a_n b_n),$$

and let $\mathcal{N}$ be the subset of $\mathfrak{J}$ consisting of the null sequences. Define a relation, denoted by R, in $\mathfrak{J}$ as follows:

$$(a_n)R(b_n) \text{ if } (a_n) - (b_n) = (a_n - b_n) \in \mathcal{N}.$$

It is easy to check that this relation is an equivalence relation and, hence, determines a partition of $\mathfrak{J}$ into equivalence classes which we denote by $[(a_n)]$.

3.1 Definition *The real number system $\mathbb{R}$ is the quotient set $\mathfrak{J}/R$ with the operations of addition and multiplication defined as follows:*

$$[(a_n)] + [(b_n)] = [(a_n) + (b_n)] \text{ and } [(a_n)][(b_n)] = [(a_n)(b_n)].$$

As usual when an operation on equivalence classes is defined in terms of representatives of the equivalence classes one must verify that it really is an operation on the equivalence classes; that is, that it is independent of the particular representative used. Take $(a_n)R(a'_n)$ and $(b_n)R(b'_n)$. Then $(a_n - a'_n)$ and $(b_n - b'_n)$ are null sequences and, hence,

$$((a_n + b_n) - (a'_n + b'_n)) = ((a_n - a'_n) + (b_n - b'_n))$$

is a null sequence; that is, $((a_n) + (b_n))R((a'_n) + (b'_n))$. Also,

$$(a_n b_n - a'_n b'_n) = (b_n(a_n - a'_n) + a'_n(b_n - b'_n))$$

is a null sequence; that is, $((a_n)(b_n))R((a'_n)(b'_n))$

In the proof that $\mathbb{R}$ is a field we will need the following proposition.

3.2 Proposition. *If $(a_n) \in \mathfrak{J} \sim \mathcal{N}$, then there exists a positive rational ε and a positive integer n_0 such that either $a_n > \varepsilon$ for all $n \geq n_0$ or $a_n < -\varepsilon$ for all $n \geq n_0$.*

Proof. Since $(a_n) \notin \mathcal{N}$ there exists a positive rational ε such that for each n there exists an $m \geq n$ such that $|a_m| \geq 2\varepsilon$. Since $(a_n) \in \mathfrak{J}$ there exists a positive integer n_0 such that

$$|a_n - a_m| < \varepsilon \text{ whenever } m, n \geq n_0.$$

Take $m \geq n_0$ such that $|a_m| \geq 2\varepsilon$, say $a_m \geq 2\varepsilon$. Then, for all $n \geq n_0$,

$$a_n > a_m - \varepsilon \geq \varepsilon.$$

If $a_m \leq -2\varepsilon$, then $a_n < -\varepsilon$ whenever $n \geq n_0$. □

3.3 Theorem *The real number system* $\mathbb{R}$ *is a field.*

Proof. The fact that $\mathbb{R}$ is a commutative ring with unit follows easily from the fact that $\mathfrak{J}$ is a commutative ring with unit. For example, the verification of A_2 is :

$$[(a_n)] + [(b_n)] = [(a_n) + (b_n)] = [(b_n) + (a_n)] = [(b_n)] + [(a_n)].$$

The element 0 in $\mathbb{R}$ is $[(\overline{0})] = \mathcal{N}$. Each $[(a_n)]$ in $\mathbb{R}$ has $[-(a_n)]$ as its additive inverse. The unit or multiplicative identity 1 in $\mathbb{R}$ is $[(\overline{1})]$.

To complete the proof that $\mathbb{R}$ is a field we must verify M_4: if $[(a_n)] \neq 0$, then there exists an element $[(b_n)]$ in $\mathbb{R}$ such that $[(a_n)][(b_n)] = [(\overline{1})] = 1$. If $[(a_n)] \neq 0$, then $(a_n) \in \mathfrak{J} \sim \mathcal{N}$ and, hence, by Proposition 3.2 there exists a positive rational ε and a positive integer n_0 such that $|a_n| > \varepsilon$ whenever $n \geq n_0$. Define (b_n) as follows:

$$b_n = \begin{cases} 0 & \text{for } n < n_0 \\ a_n^{-1} & \text{for } n \geq n_0 \end{cases}.$$

To show that $(b_n) \in \mathfrak{J}$ take a positive rational number δ. Then for $m, n \geq n_0$

$$|b_n - b_m| = |a_n^{-1} - a_m^{-1}| = |a_n^{-1}a_m^{-1}||a_m - a_n| < \varepsilon^{-2}|a_m - a_n|.$$

Since $(a_n) \in \mathfrak{J}$ there exists a positive integer $m_0 \geq n_0$ such that

$$|a_m - a_n| < \varepsilon^2\delta \text{ whenever } m, n \geq m_0$$

and, hence,

$$|b_n - b_m| < \delta \text{ whenever } m, n \geq m_0.$$

Finally, to show that $[(a_n)][(b_n)] = 1$ we must show that $(a_n b_n)$ $R(\overline{1})$ or, equivalently, $(a_n b_n - 1) \in \mathcal{N}$. Clearly, for any positive rational number δ,

$$|a_n b_n - 1| = 0 < \delta \text{ whenever } n \geq n_0.$$

Thus, $[(a_n)][(b_n)] = 1$ and M_4 is verified. □

We now make $\mathbb{R}$ an ordered field by the introduction of a positive cone which we will denote by $\mathbb{R}^+$. Proposition 3.2 states that if $[(a_n)] \neq 0$, then, for some positive rational ε and some positive integer n_0, $a_n > \varepsilon$ for all $n \geq n_0$ or $a_n < -\varepsilon$ for all $n \geq n_0$. We define the positive cone $\mathbb{R}^+$ to be those real numbers $[(a_n)]$ such that, for some positive rational ε and some positive integer $n_0, a_n > \varepsilon$ for all $n \geq n_0$. It is not difficult to show that this definition is independent of the particular representative (a_n).

That $\mathbb{R}^+$ qualifies as a positive cone of a field will now be shown:

(1) Since $0 = [(\overline{0})], 0 \notin \mathbb{R}^+$.

(2) Proposition 3.2 implies that if $[(a_n)] \neq 0$, then one and only one of the elements $[(a_n)]$ or $-[(a_n)]$ is in $\mathbb{R}^+$.

(3) If $[(a_n)]$ and $[(b_n)]$ are in $\mathbb{R}^+$, then for some positive rationals rationals ε and δ and for some positive integers n_0 and m_0 we have:

$$a_n > \varepsilon \text{ whenever } n \geq n_0$$
$$b_n > \varepsilon \text{ whenever } n \geq m_0$$

Therefore, if $n \geq \max\{n_0, m_0\}$,

$$a_n + b_n > \varepsilon + \delta > 0 \text{ and } a_n b_n > \varepsilon\delta > 0.$$

Thus, $[(a_n)] + [(b_n)] \in \mathbb{R}^+$ and $[(a_n)][(b_n)] \in \mathbb{R}^+$.

In the ordered field $\mathbb{R}$ we define the "less then" relation in the standard way:

$$[(a_n)] < [(b_n)] \quad \text{if } [(b_n)] - [(a_n)] \in \mathbb{R}^+$$

that is, $[(a_n)] < [(b_n)]$ if for some positive rational ε and some positive integer n_0, $b_n - a_n > \varepsilon$ whenever $n \geq n_0$.

Identifying the real numbers $[(\overline{0})]$, $[(\overline{1})]$, $[(-\overline{1})]$, $[(\overline{2})]$, etc., with the integers $0, 1, -1, 2$, etc., we can show that $\mathbb{R}$ is an archimedean-ordered field.

3.4 Theorem *The real number system* $\mathbb{R}$ *is an archmedean-ordered field.*

Proof. Let $[(a_n)]$ and $[(b_n)]$ be positive real numbers. We wish to show that there is a positive integer m such that $m[(a_n)] > [(b_n)]$. There exists a positive rational ε and a positive integer n_0 such that

$$a_n > \varepsilon \text{ whenever } n \geq n_0.$$

Since $(b_n) \in \mathfrak{I}$, it is bounded; that is, there exists a positive rational M such that

$$b_n \leq M \text{ for all } n.$$

Using the fact that the rational field is archimedean-ordered, we know that there exists a positive integer m such that

$$m\varepsilon > M.$$

Then

$$ma_n - b_n > m\varepsilon - M > 0 \text{ whenever } n \geq n_0.$$

Therefore,

$$[(b_n)] < [(ma_n)] = [(\overline{m})][(a_n)] = m[(a_n)]. \qquad \square$$

In our discussion we have identified the integer m with the rational number m and then with the real number $[(\overline{m})]$. We would like now to show that any rational number a can be identified with the real number $[(\overline{a})]$. In order to justify this identification we must show that elements a in the rational field behave the same as the corresponding elements $[(\overline{a})]$ in the real field. By "behave the same" we mean with respect to the operations of addition and multiplication and also with respect to the order. In more formal language we wish to show that the rational field is isomorphic to a subset of the real field.

3.5 Definition *If* $\mathbf{F}$ *and* $\mathbf{F}'$ *are fields. then* $\mathbf{F}$ *is* ***isomorphic*** *to a subset* $\mathbf{S}'$ *of* $\mathbf{F}'$ *if there is a one-to-one function* f *from* $\mathbf{F}$ *onto* $\mathbf{S}'$ *such that for all* a *and* b *in* $\mathbf{F}$

$$f(a+b) = f(a) + f(b) \text{ and } f(ab) = f(a)f(b).$$

3.6 Theorem *The rational field is isomorphic to a subset of the real field.*

Proof. For any rational number a, let $f(a) = [(\overline{a})]$. This is clearly a one-to-one function from the rational field $\mathbf{F}$ onto the subset $\mathbf{S}' = \{[(\overline{a})] : a \text{ rational }\}$ of R. For, if $f(a) = f(b)$, that is, $[(\overline{a})] = [(\overline{b})]$, then $(\overline{a-b})$ is a null sequence and hence $a = b$. Also

$$f(a+b) = [(\overline{a+b})] = [(\overline{a})] + [(\overline{b})] = f(a) + f(b)$$

and

$$f(ab) = [(\overline{ab})] = [(\overline{a})][(\overline{b})] = f(a)f(b). \qquad \square$$

The isomorphism f also preserves the order relation. If a is a positive rational number, then $f(a) = [(\overline{a})]$ is a positive real number. This implies that if $a < b$ then $[(\overline{a})] < [(\overline{b})]$.

Although a rational number may be considered to be a real number, it will sometimes be convenient to use a notational device to indicate whenther a rational number is being considered as an element of the rational field or the real field. When we wish to emphasize that the rational number a is being considered as an element of the real field we will write a^*. Note that $a^* = [(\overline{a})]$.

3.7 Theorem *Between any two distinct real numbers there is a rational number.*

Proof. If x and y are real numbers such that $x < y$, we wish to show that there is a rational number c such that

$$x < c^* < y.$$

Let $x = [(a_n)]$ and $y = [(b_n)]$. Then $[(a_n)] < [(b_n)]$ implies that for some positive rational ε and some positive integer n_0

$$b_n - a_n > \varepsilon \text{ whenever } n \geq n_0.$$

Since (a_n) and (b_n) are in $\mathfrak{J}$ there exists a positive integer $m_0 \geq n_0$ such that

$$|a_n - a_m| < \varepsilon/4 \text{ and } |b_n - b_m| < \varepsilon/4 \text{ whenever } m, n \geq m_0.$$

Let $c = \frac{1}{2}(a_{m_0} + b_{m_0})$. Then, for all $n \geq m_0$,

$$c - a_n > \frac{1}{2}(a_{m_0} + b_{m_0}) - a_{m_0} - \varepsilon/4 = \frac{1}{2}(b_{m_0} - a_{m_0}) - \varepsilon/4 > \varepsilon/4$$

and

$$b_n - c > b_{m_0} - \varepsilon/4 - \frac{1}{2}(a_{m_0} + b_{m_0}) = \frac{1}{2}(b_{m_0} - a_{m_0}) - \varepsilon/4 > \varepsilon/4$$

Thus,

$$[(a_n)] < [(\overline{c})] < [(b_n)]$$

or equivalently

$$x < c^* < y$$

□

The following proposition should provide the hindsight to make our development of the real numbers from the rational numbers seem natural. We will show that if $x = [(a_n)]$ then $\lim a_n^* = x$. Thus, we have represented a real number x by a sequence (a_n) of rationals which as a sequence of real numbers converges to x. Tracing our way back, for any real number x, we fine that Theorem 3.7 implies that there is a sequence (a_n) of rationals which converges to x in $\mathbb{R}$. Since (a_n) converges in $\mathbb{R}$, it is a Cauchy sequence in $\mathbb{R}$ and, hence, also in the rational field. Also, two sequences converge to the same number if and only if they differ by a null sequence. This suggests defining a real number as an equivalence class of Cauchy sequences of rationals where two sequences in the same equivalence class differ by a null sequence.

3.8 Proposition. *If* $x = [(a_n)]$, *then* $\lim a_n^* = x$.

Proof. Take $\varepsilon > 0$ in $\mathbb{R}$ and let δ be a rational numbers such that $0 < \delta^* < \varepsilon/2$. Since $(a_n) \in \mathfrak{J}$, there exists a positive integer n_0 such that

$$|a_n - a_m| < \delta \text{ whenever } m, n \geq n_0.$$

Thus, for each $n \geq n_0$, we have

$$2\delta - (a_n - a_m) > \delta \text{ and } 2\delta - (a_m - a_n) > \delta \text{ whenever } m \geq n_0.$$

This implies that, for each $n \geq n_0$,

$$[(\overline{a_n}) - (a_m)] < [(\overline{2\delta})] \text{ and } [(a_m) - (\overline{a_n})] < [(\overline{2\delta})],$$

that is,

$$a_n^* - x < 2\delta^* < \varepsilon \text{ and } x - a_n^* < 2\delta^* < \varepsilon.$$

Therefore,

$$|a_n^* - x| < \varepsilon \text{ whenever } n \geq n_0$$

and, hence,

$$\lim a_n^* = x.$$

□

PROBLEMS [2.3]

1. Defining $(a_n)R(b_n)$ if $(a_n - b_n) \in \mathcal{N}$, show that this is an equivalence relation.

2. Verify that the real number system, $\mathbb{R}$, is a commutative ring with unit.

3. If the field $\mathbf{F}$ is isomorphic to the subset $\mathbf{S}'$ of the field $\mathbf{F}'$, show that $\mathbf{S}'$ is a subfield of $\mathbf{F}'$.

4. Show that between any two distinct real numbers there is an infinite number of rational numbers.

5. If A is a ring with unit, a subset I of A is called an **ideal** if

$$(1)\ a, b \in I \Rightarrow a - b \in I$$
$$(2)\ a \in A \text{ and } b \in I \Rightarrow ab \in I.$$

   (a) If we define a relation in A by aRb if $a - b \in I$, show that this is an equivalence relation.

   (b) Show that $\mathcal{N}$ is an ideal in $\mathfrak{J}$.

6. If x is any real number, show that there is a real number greater than x.

7. If x is a real number, show that there is an integer n such that

$$n \leq x < n + 1.$$

2.4 Completeness of The Real Number System

Our objective in the development of the real number system is to obtain a complete ordered field by extending the incomplete ordered field of rationals. In this section we will show that we have gained our objective.

4.1 Theorem *The real field $\mathbb{R}$ is complete.*

Proof. Let (x_n) be a Cauchy sequence of real numbers. For each positive integer n take a rational number a_n such that $x_n < a_n^* < x_n + 1/n$. To show that $(a_n) \in \mathcal{G}$, let ε be any positive rational number. There exists a positive integer $n_0 > 3/\varepsilon$ such that

$$|x_n - x_m| < \varepsilon^*/3 \text{ whenever } m, n \geq n_0.$$

Then

$$|a_n^* - a_m^*| \leq |a_n^* - x_n| + |x_n - x_m| + |x_m - a_m^*| < \varepsilon^* \quad \text{whenever } m, n \geq n_0,$$

and, therefore,

$$|a_n - a_m| < \varepsilon \text{ whenever } m, n \geq n_0.$$

Letting $x = [(a_n)]$, we can show that $\lim x_n = x$. Take $\delta > 0$ in $\mathbb{R}$. Since $\lim a_n^* = x$, there exists a positive integer $m_0 > 2/\delta$ such that

$$|a_n^* - x| < \delta/2 \text{ whenever } n \geq m_0.$$

Then

$$|x_n - x| \leq |x_n - a_n^*| + |a_n^* - x| < \delta \text{ whenever } n \geq m_0$$

and, therefore, $\lim x_n = x$. □

We now prove some other properties of the real number system which are equivalent to completeness and which in some instances are more convenient to use. We call a sequence (x_n) of real numbers **nondecreasing** if $x_n \leq x_{n+1}$ for all n.

4.2 Theorem *A bounded nondecreasing sequence of real numbers converges.*

Proof. Let (x_n) be a bounded nondecreasing sequence of real numbers. We will show that (x_n) is a Cauchy sequence. Suppose (x_n) is not a Cauchy sequence. Then there is a positive number ε such that for every positive integer n there exists positive integers $r, s \geq n$ such that

$$|x_r - x_s| \geq \varepsilon.$$

Since (x_n) is nondecreasing this implies that for every positive integer n there exists a positive integer $r > n$ such that

$$x_r - x_n \geq \varepsilon.$$

Thus, there exists a positive integer $n_1 > 1$ such that

$$x_{n_1} \geq x_1 + \varepsilon.$$

Moreover, there exists a positive integer $n_2 > n_1$ such that

$$x_{n_1} \geq x_{n_1} + \varepsilon \geq x_1 + 2\varepsilon.$$

Continuing in this manner, we obtain a subsequence (x_{n_k}) such that $x_{n_k} \geq x_1 + k\varepsilon$. This contradicts the fact that (x_n) is bounded. Hence, (x_n) is a Cauchy sequence of real numbers and therefore converges. □

If we define a sequence (x_n) of real numbers to be **nonincreasing** if $x_n \geq x_{n+1}$ for all n, then we easily obtain the result: *a bounded nonincreasing sequence of real numbers converges.*

A subset S of $\mathbb{R}$ is said to be **bounded above** if there exists an element $b \in \mathbb{R}$ such that $x \leq b$ for all $x \in S$; in such a case b is called an **upper bound** of S. An element $c \in \mathbb{R}$ is said to be the **least upper bound** of subset S of $\mathbb{R}$ if c is an upper bound of S and no element smaller than c is an upper bound of S. The least upper bound of S is denoted by $\sup S$ (supremum S) or lub S. Thus, $c = \sup S$ if and only if it satisfies:

(1) $x \leq c$ for all $x \in S$

(2) for each $\varepsilon > 0$ there exists an $x \in S$ such that $x > c - \varepsilon$.

The greatest lower bound of a set is defined in a similar way and is denoted by $\inf S$ (infinum S) or glb S.

4.3 Completeness Axiom *If S is a nonempty set of real numbers which is bounded above, then S has a least upper bound.*

Proof. Let A and B be the sets of real numbers which are not and are, respectively, upper bounds of S. Since S is nonempty and is bounded above, A and B are nonempty. Also, any element of A is less than any element of B. Take $a_1 \in A$ and $b_1 \in B$. Construct a nondecreasing sequence (a_n) and a nonincreasing sequence (b_n) as follows: if $\frac{1}{2}(a_n + b_n) \in A$, let $a_{n+1} = \frac{1}{2}(a_n + b_n)$ and $b_{n+1} = b_n$; if $\frac{1}{2}(a_n + b_n) \in B$, let $a_{n+1} = a_n$ and $b_{n+1} = \frac{1}{2}(a_n + b_n)$. Then (a_n) and (b_n) are bounded nondecreasing and nonincresing, respectively, and therefore they converge. Let $a = \lim a_n$ and $b = \lim b_n$. Since $b_n - a_n = \left(\frac{1}{2}\right)^{n-1}(b_1 - a_1)$, $b = a$. Let $c = a = b$.

We now show that $c = \sup S$. For any $x \in S$, $x \leq b_n$ for all n. Therefore, $x \leq b = c$ and c is an upper bound of S. To show that c is the least upper bound of S take $y < c$. Then, for some n, $y < a_n$. Since a_n is not an upper bound of S, y is not an upper bound of S. Thus, $c = \sup S$. □

PROBLEMS [2.4]

1. Prove that a bounded nonincreasing sequence of real numbers converges.

2. If $\lim b_n = b$ and $a \leq b_n$ for all n, show that $a \leq b$.

3. If S is a nonempty set of real numbers which is bounded below, show that S has a greatest lower bound.

4. Show that $\mathbb{R}$ has the Dedekind property: If $\mathbb{R}$ is partitioned into two sets L and U such that each element of L is less than each element of U, then L has a greatest element or U has a least element.

5. A point $c \in \mathbb{R}$ is said to be a **limit point** of the sequence (x_n) of real numbers if for each $\varepsilon > 0$ and positive integer m there is a positive integer $n \geq m$ such that $|x_n - c| < \varepsilon$. Show that a bounded sequence (x_n) of real numbers has at least one limit point.

6. Show that a bounded sequence of real numbers has a convergent subsequence.

7. If $\lim a_n = a, \lim b_n = b$, and $a_n \leq b_n$ for all n, show that $a \leq b$.

8. If (x_n) is a bounded sequence of real numbers, let

$$y_k = \inf\{x_n : n \geq k\} \text{ and } z_k = \sup\{x_n : n \geq k\}.$$

(a) Show that (y_k) and (z_k) are, respectively, nondecreasing and nonincreasing sequences and

$$\lim y_k = \sup\{y_k\} \text{ and } \lim z_k = \inf\{z_k\}.$$

(b) Define the **lower limit** and **upper limit** of (x_n) as follows:

$$\underline{\lim} x_n = \lim_{k\to\infty} \inf\{x_n : n \geq k\} = \sup_k \inf\{x_n : n \geq k\}$$

$$\overline{\lim} x_n = \lim_{k\to\infty} \sup\{x_n : n \geq k\} = \inf_k \sup\{x_n : n \geq k\}.$$

Show that

$$\underline{\lim} x_n \leq \overline{\lim} x_n.$$

(c) Show that $\underline{\lim} x_n$ and $\overline{\lim} x_n$ are limit points of (x_n).

(d) Show that $\lim x_n = x \Longleftrightarrow \underline{\lim} x_n = x = \overline{\lim} x_n$.

2.5 The Topology on The Set of Real Numbers

The simplest sets of real numbers are the intervals. We defined the open interval (a, b) to be the set $\{x : a < x < b\}$. We always take $a < b$ but we consider also the infinite intervals $(a, \infty) = \{x : a < x\}$ and $(-\infty, b) = \{x : x < b\}$. Sometimes we write $(-\infty, \infty)$ for the set of all real numbers. We define the closed interval $[a, b]$ to be the set $\{x : a \leq x \leq b\}$. For closed intervals we take a and b finite. The half-open interval $(a, b]$ is defined to be $\{x : a < x \leq b\}$, and $[a, b) = \{x : a \leq x < b\}$. A generalization of the notion of an open interval is given by that of an open set.

5.1 Definition *A set O of real numbers is called* **open** *if for each $x \in O$ there is a $\delta > 0$ such that each y with $|x - y| < \delta$ belong to O*

Another way of phrasing this definition is to say that a set O is open if for every x in O there is an open sets, and both the empty set $\emptyset$ and the set $\mathbb{R}$ of real numbers are open. We establish some properties of open set.

5.2 Propersition *The intersection $O_1 \cap O_2$ of two open sets O_1 and O_2 is open.*

Proof. Let $x \in O_1 \cap O_2$. Since $x \in O_1$ and O_1 is open, there is a $\delta_1 > 0$ such that all y with $|x - y| < \delta_1$ belong to O_1. Similarly, there is a $\delta_2 > 0$ such that all y with $|x - y| < \delta_2$ belong to O_2. Take δ to be the smaller of δ_1 and δ_2. Then $\delta > 0$, and if $|x - y| < \delta$, then y belongs to both O_1 and O_2, i.e., to $O_1 \cap O_2$.

5.3 Corollary *The intersection of any finite collection of open sets is open.*

5.4 Proposition *The union of any collection $\mathcal{C}$ of open sets is open.*

Proof. Let U be the union of the collection $\mathcal{C}$, and $x \in U$. Then there is an $O \in \mathcal{C}$ with $x \in O$. Since O is open, there is an $\varepsilon > 0$ such that all y with $|x - y| < \varepsilon$ belong to O and hence to U, since $O \subset U$. Thus U is open.

It follows from Proposition 5.2 that the intersection of any *finite* collection of open sets is open. It is not true, however, that the intersection of any collection of open sets is open. Take, for example, O_n to be the open interval $(-\frac{1}{n}, \frac{1}{n})$. Then $\bigcap_{n=1}^{\infty} O_n = \{0\}$, ard $\{0\}$ is not an open set.

Every union of open intervals is an open set by Proposition 5.4. A strong form of the converse of this is also true:

5.5 Proposition *Every open set of real numbers is the union of a countable collection of disjoint open intervals.*

Proof. Since O is open, for each $x \in O$ there is a $y > x$ such that $(x, y) \subset O$. Let $b = \sup\{y : (x, y) \subset O\}$. Let $a = \inf\{z : (z, x) \subset O\}$. Then $a < x < b$, and $I_x = (a, b)$ is an open interval containing x. Now $I_x \subset O$, for if $w \in I_x$, say $x < w < b$, we have by the definition of b a number $y > w$ such that $(x, y) \subset O$, and so $w \in O$. Moreover, $b \in O$, for if $b \in O$, then for some $\varepsilon > 0$ we have $(b - \varepsilon, b + \varepsilon) \subset O$, whence $(x, b + \varepsilon) \subset O$, contradicting the definition of b. Similarly, $a \notin O$. Consider the collection of open intervals $\{I_x\}$, $x \in O$. Since each x in O is contained in I_x, and each I_x is contained in O, we have $O = \bigcup I_x$. Let (a, b) and (c, d) be two intervals in this collection with a point in common. Then we must have $c < b$ and $a < d$. Since c does not belong to O, it does not belong to (a, b) and we have $c \leq a$. Since a does not belong to O and hence not to (c, d), we have $a \leq c$. Thus $a = c$. Similarly, $b = d$, and $(a, b) = (c, d)$. Thus two *different* intervals in the collection $\{I_x\}$ must be disjoint. Thus O is the union of the disjoint collection $\{I_x\}$ of open intervals, and it remains only to show that this collection is countable. But each open interval contains a rational number by the corollary to the Axiom of Archimedes. Since we have a collection of disjoint open intervals, each open interval contains a different rational number, and the collection can be put in one-to-one correspondence with a subset of the rationals. Thus it is a countable collection.

5.6 Proposition(Lindelöf) *Let $\mathcal{C}$ be a collection of open sets of real numbers. Then there is a countable subcollection $\{O_i\}$ of $\mathcal{C}$ such that*

$$\bigcup_{O \in \mathcal{C}} O = \bigcup_{i=1}^{\infty} O_i.$$

Proof. Let $U = \bigcup\{O : O \in \mathcal{C}\}$, and $x \in U$. Then there is an $O \in \mathcal{C}$ with $x \in O$. Since O is open, there is an open interval I_x such that $x \in I_x \subset O$. It follows from Corollay 5.3 that we can find an open interval J_x with rational endpoints such that $x \in J_x \subset I_x$. Since the collection of all open intervals with rational endpoints is countable, the collection $\{J_x\}$, $x \in U$, is countable, and $U = \bigcup_{x \in U} J_x$. For each interval in $\{J_x\}$ choose a set O in $\mathcal{C}$

which contains it. This gives a countable subcollection $\{O_i\}_{i=1}^{\infty}$ of $\mathcal{C}$, and $U = \bigcup_{i=1}^{\infty} O_i$. □

We shall also study the notion of a closed set which generalizes the notion of a closed interval. We begin by defining a point of closure:

5.7 Definition *A real number x is called a **point of closure** of a set E if for every $\delta > 0$ there is a y in E such that $|x - y| < \delta$.*

This is equivalent to saying that x is a point of closure of E if every open interval containing x also contains a point of E. Each point of E is trivially a point of closure of E. We denote the set of points of closure of E by $\overline{E}$. Thus $E \subset \overline{E}$.

5.8 Proposition *If $A \subset B$, then $\overline{A} \subset \overline{B}$. Also $(\overline{A \cup B}) = \overline{A} \cup \overline{B}$.*

Proof. The first part follows immediately from the definition of points of closure. Since $A \subset A \cup B$, we have $\overline{A} \subset (\overline{A \cup B})$. Similarly, $\overline{B} \subset (\overline{A \cup B})$. Thus $\overline{A} \cup \overline{B} \subset (\overline{A \cup B})$. Suppose that $x \in \overline{A} \cup \overline{B}$, Then there is a $\delta_1 > 0$ such that there is no $y \in A$ with $|x - y| < \delta_1$, and there is a $\delta_2 > 0$ such that there is no $y \in B$ with $|x - y| < \delta_2$. Thus if $\delta = \min(\delta_1, \delta_2)$, there is no $y \in A \cup B$ with $|x - y| < \delta$. Consequently, $x \notin (\overline{A \cup B})$, and we have $(\overline{A \cup B}) \subset \overline{A} \cup \overline{B}$.

5.9 Definition *A set F is called **closed** if $F = \overline{F}$.*

Since we always have $F \subset \overline{F}$, a set F is closed if $\overline{F} \subset F$, that is, if F contains all of its points of closure. The empty set $\emptyset$ and the set $\mathbb{R}$ of all real numbers are closed. The closed intervals $[a, b]$ and $[a, \infty)$ are closed. It is customary to use the letter F to denote closed sets (French, *fermé*).

5.10 Proposition *For any set E the set $\overline{E}$ is closed; that is, $\overline{\overline{E}} = \overline{E}$.*

Proof. Let x be a point of closure of $\overline{E}$. Then, given $\delta > 0$, there is a point $y \in E$ with $|x - y| < \delta/2$. Since $y \in E$, there is a $z \in E$ with $|y - z| < \delta/2$. Thus $|x - z| < \delta$, and we see that x is a point of closure of E.

5.11 Proposition *The union $F_1 \cup F_2$ of two closed sets F_1 and F_2 is closed.*

Proof. By proposition 5.8 we have

$$(\overline{F_1 \cup F_2}) = \overline{F_1} \cup \overline{F_2} = F_1 \cup F_2.$$

5.12 Proposition *The intersection of any collection $\mathcal{C}$ of closed sets is closed.*

Proof. Let x be a point of closure of $\bigcap\{F : F \in \mathcal{C}\}$. Then, given $\delta > 0$, there is a $y \in \bigcap\{F : F \in \mathcal{C}\}$ such that $|x - y| < \delta$. Since such a y belong to each $F \in \mathcal{C}$, we see that x is a point of closure of each $F \in \mathcal{C}$. Since each F is closed, we have $x \in F$ for each F in $\mathcal{C}$. Hence $x \in \bigcap\{F : F \in \mathcal{C}\}$.

5.13 Proposition *The complement of an open set is closed and the complement of a closed set is open.*

Proof. Let O be open. If $x \in O$, there is a $\delta > 0$ such that if $|x - y| < \delta$, then $y \in O$. Hence x cannot be a point of closure of $\overline{O}$, since there is no $y \in \overline{O}$ with $|x - y| < \delta$. Thus $\overline{O}$ contains all of its points of closure and is threrefore closed.

On the other hand, let F be closed and $x \in \overline{F}$. Then, since x is not a point of closure of F, there is a $\delta > 0$ such that there is no $y \in F$ with $|x - y| < \delta$. Hence, if $|x - y| < \delta$, then $y \in \overline{F}$. Thus $\overline{F}$ is open.

We say that a collection $\mathcal{C}$ of sets **covers** a set F if $F \subset \bigcup\{O : O \in \mathcal{C}\}$. In this case the collection $\mathcal{C}$ is called a **covering** of F. If each $O \in \mathcal{C}$ is open, we call $\mathcal{C}$ an **open** covering of F. If $\mathcal{C}$ contains only a finite number of sets, we call $\mathcal{C}$ a **finite** covering. This terminology is inconsistent: In 'open covering' the adjective 'open' refers to the sets in the covering; in 'finite covering' the adjective 'finite' refers to the collection and does not imply that the sets in the collection are finite sets. Thus the term 'open covering' is an abuse of language and should properly be 'covering by open sets'. Unfortunately, the former terminology is well established in mathematics. With this terminology we state the following theoem:

5.14 Theorem (Heine-Borel) *Let F be closed and bounded set of real numbers. Then each open covering of F has a finite subcovering. That is, if $\mathcal{C}$ is a collection of open sets such that $F \subset \bigcup\{O : O \in \mathcal{C}\}$, then there is a finite collection $\{O_1, \cdots, O_n\}$ of sets in $\mathcal{C}$ such that*

$$F \subset \bigcup_{i=1}^{n} O_i$$

Proof. Let us first consider the case that F is the closed interval $[a, b]$, where $-\infty < a < b < \infty$. Let E be the set of numbers $x \leq b$ with the property that the interval $[a, x]$ can be covered by a finite number of the sets of $\mathcal{C}$. The set E is bounded by b and so has a least upper bound c. Since $c \in [a, b]$, there is an $O \in \mathcal{C}$ which contains c. Since O is open, there is an $\varepsilon > 0$ such that the interval $(c - \varepsilon, c + \varepsilon)$ is contained in O. Now $c - \varepsilon$ is not an upper bound for E, and so there must be an $x \in E$ with $x > c - \varepsilon$. Since $x \in E$, there is a finite collection $\{O_1, \cdots, O_k\}$ of sets in $\mathcal{C}$ which covers $[a, x]$. Consequently, the finite collection $\{O_1, \cdots, O_k, O\}$ covers $[a, c + \varepsilon)$. Thus each point of $[c, c + \varepsilon)$ would be in E if it were less than or equal to b. Since no point of $[c, c + \varepsilon)$ except c can belong to E, we must have $c = b$ and $b \in E$. Thus $[a, b]$ can be covered by a finite number of sets from $\mathcal{C}$, proving our special case.

Now let F be any closed and bounded set and $\mathcal{C}$ an open covering of F. Since F is bounded it is contained in some closed bounded interval $[a, b]$. Let $\mathcal{C}^*$ be the collection obtained by adding $\overline{F}$ to $\mathcal{C}$; that is, $\mathcal{C}^* = \mathcal{C} \cup \{\overline{F}\}$. Since F is closed, $\overline{F}$ is open, and $\mathcal{C}^*$ is a collection of open sets. By hypothesis $F \subset \bigcup\{O : O \in \mathcal{C}\}$, and so $\mathbb{R} = \overline{F} \cup F \subset \overline{F} \cup \bigcup\{O : O \in \mathcal{C}\} = \bigcup\{O : O \in \mathcal{C}^*\}$. Thus

$\mathcal{C}^*$ is an open covering of $\mathbb{R}$ and therefore of $[a, b]$. By our previous case there is a finite subcollection of $\mathcal{C}^*$ which covers $[a, b]$ and hence F. If this finite subcollection does not contain $\overline{F}$, it is a subcollection of $\mathcal{C}$ and the conclusion of our theorem holds. If the subcollection contains $\overline{F}$, denote it by $\{O_1, \cdots, O_n, \overline{F}\}$. Then $F \subset \overline{F} \cup O_1 \cup \cdots \cup O_n$. Since no point of F is contained in $\overline{F}$, we have $F \subset O_1 \cup \cdots \cup O_n$, and the collection $\{O_1, \cdots, O_n\}$ is a finite subcollection of $\mathcal{C}$ which covers F.

5.15 Proposition *Let $\mathcal{C}$ be a collection of closed sets (of real numbers) with the property that every finite subcollection of $\mathcal{C}$ has a nonempty intersection, and suppose that one of the sets in $\mathcal{C}$ is bounded. Then*

$$\bigcap_{F \in \mathcal{C}} F \neq \emptyset$$

It is often convenient to extend the system of real numbers by the addition of two elements $+\infty$ and $-\infty$. This enlarged set

$$\mathbb{R}^* = \mathbb{R} \cup \{\infty\} \cup \{-\infty\}$$

is called the set of **extended real numbers**. We extend the definition of $<$ to the extended real numbers by postulating $-\infty < x < \infty$, for each real number x. We also define

$$x + \infty = \infty = \infty + x, \quad x + (-\infty) = -\infty = -(\infty) + x$$

for every $x \in \mathbb{R}$, and

$$x\infty = \infty = \infty x, \quad x(-\infty) = -\infty = -(\infty)x$$

for every positive real number x. Furthermore, we set

$$-(\infty) = -\infty, \qquad -(-\infty) = \infty$$

$$\infty + \infty = \infty, \qquad (-\infty) + (-\infty) = -\infty$$

$$\infty\infty = \infty = (-\infty)(-\infty), \qquad \infty(-\infty) = -\infty = (-\infty)\infty$$

$$\infty 0 = 0\infty = (-\infty)0 = 0(-\infty) = 0$$

The operation $\infty - \infty$ is left undefined, but we shall adopt the arbitrary *convention* the $0 \cdot \infty = 0$.

One use of extended real numbers is in the expression "$\sup S$". If S is a nonempty set of real numbers with an upper bound, we define $\sup S$ to be the last upper bound of S. If S has no upper bound, we write $\sup S = \infty$. Then $\sup S$ is defined for all nonempty sets S, and if we define $\sup \emptyset$ to be $-\infty$, then in all eases $\sup E$ is the smallest extended real number which is greater than or equal to each element of E. Similar conventions are adopted with respect to $\inf S$.

Thus $\mathbb{R}^*$ is linearly ordered by $<$ and one of its usefulness is the following proposion which is obvious.

5.16 Proposition *Every subset S of $\mathbb{R}^*$ has a supremum* $\sup(S)$ *and an infimum* $\inf(S)$ *in $\mathbb{R}^*$.*

A function whose values are in the set of extended real unmbers is called an **extended real-valued function**.

Now let us define a function

$$d : \mathbb{R}^2 \to \mathbb{R}$$

from the Cartesian square $\mathbb{R}^2 = \mathbb{R} \times \mathbb{R}$ into $\mathbb{R}$ itself by taking

$$d(x, y) = |y - x|$$

for every member (x, y) of $\mathbb{R}^2$, where $|y - x|$ denotes the absolute value of the real number $y - x$. The following proposition is an immediate consequence,

5.17 Proposition *The function $d : \mathbb{R}^2 \to \mathbb{R}$ has the following properties:*

(*a*) *For every real number x, we have $d(x, x) = 0$.*
(*b*) *For any two real numbers x and y, we have $d(x, y) > 0$.*
(*c*) *For any two real number x and y, we have $d(x, y) = d(y, x)$.*
(*d*) *For any three real numbers x, y, and z, we have*

$$d(x, y) + d(y, z) \geq d(x, z).$$

PROBLEMS [2.5]

1. Is the set of rational numbers open or closed ?

2. What are the sets of real numbers that are both open and closed ?

3. Find two sets A and B such that $A \cap B = \emptyset$ and $\overline{A} \cap \overline{B} \neq \emptyset$.

4. Show that x is a point of closure of E if and only if there is a sequence $< y_n >$ with $y_n \in E$ and $x = \lim y_n$.

5. A number x is called an **accumulation point** of a set E if it is a point of closure of $E \sim \{x\}$. Show that the set E' of accumulation points of E is a closed set.

6. Show that $\overline{E} = E \cup E'$.

7. A set is called **isolated** if $E \cap E' = \emptyset$. Show that every isolated set of real numbers is countable.

8. A set D is called **dense** in $\mathbb{R}$ if $\overline{D} = \mathbb{R}$. Show that the set of rational numbers is dense in $\mathbb{R}$. Prove Propositions 15 and 16 using Proposition 5, 7 and 14.

9. A point x is called an **interior point** of a set A if there is a $\delta > 0$ such that the interval $(x - \delta, x + \delta)$ is contained in A. The set of interior points of A is denoted by A^o. Show that

   (a) A is open if and only if $A = A^o$.

   (b) $A^o = \sim (\overline{\overline{A}})$.

10. Derive Proposition 16 from the Heine-Borel Theorem using DeMorgan's laws.

11. Let $< F_n >$ be a sequence of nonempty closed sets of real numbers with $F_{n+1} \subset F_n$. Show that if one of the sets F_n is bounded, then $\bigcup_{i=1}^{\infty} F_i \neq \emptyset$. Give an example to show that this conclusion may be false if we do not require one of the sets to be bounded.

12. The **Cantor ternary set** C consists of all those real numbers in $[0,1]$ which have ternary expansion $< a_n >$ for which a_n is never 1(If x has two ternary expansions, we put x in the Cantor set if one of the expansions has no term equal to 1). Show that C is a closed set, and that C is obtained by first removing the middle third $(\frac{1}{3}, \frac{2}{3})$ from $[0,1]$, then removing the middle thirds $(\frac{1}{9}, \frac{2}{9})$ and $(\frac{7}{9}, \frac{8}{9})$ of the remaining intervals, and so on.

13. Show that the Cantor set can be put into a one-to-one correspondence with the interval $[0,1]$.

14. Show that the set of accumulation points of the Cantor set is the Cantor set itself.

2.6 Sequences of Real Numbers

Let X be a set. A sequence of elements of X is a "list" of elements from the set X. In this chapter we will be concerned with sequences of real numbers. We first make the notion of sequence precise.

6.1 Definition *Let X be a set. A sequence of elements of X is a function from the set of positive integers into X.*

In particular, a *real sequence*(or sequence of real numbers) is a function

$$\phi : \mathbb{N} \to \mathbb{R}$$

from the set $\mathbb{N}$ of all natural numbers into the real line $\mathbb{R}$. For every natural number $n \in \mathbb{N}$, the real number

$$x_n = \phi(n) \in \mathbb{R}$$

is called the nth *term* of the sequence ϕ. The sequence ϕ is often denoted by

$$\phi = \{x_1, x_2, \cdots, x_n, \cdots\} = \{x_n\}_{n=1}^{\infty}$$

6.2 Definition *A sequence $\{x_n\}$ of real numbers is said to* ***converge*** *to the real number s provided that*

$$\text{for each } \varepsilon > 0 \text{ there exists a number } N \text{ such that} \quad n > N \text{ implies } |x_n - s| < \varepsilon.$$

If $\{x_n\}$ converges to s we will write $\lim_{n\to\infty} x_n = s$, $\lim x_n = s$, or $x_n \to s$. The number s is the *limit* of the sequence $\{x_n\}$. If, in addition, $\{x_n\}$ is monotone, that is, $x_n \leq x_{n+1}$, we write $x_n \uparrow s$.

Let us return to the sequence $\{n/(n+1)\}_{n=1}^{\infty}$, and prove that $\{n/(n+1)\}_{n=1}^{\infty}$ has limit 1. If $\varepsilon > 0$, we are required to show that for some positive integer N

$$\left|\frac{n}{n+1} - 1\right| < \varepsilon$$

if $n \geq N$. Now

$$\left|\frac{n}{n+1} - 1\right| = \frac{1}{n+1}$$

and hence if we choose a positive integer N such that $1/(N+1) < \varepsilon$, then if $n \geq N$,

$$\left|\frac{n}{n+1} - 1\right| = \frac{1}{n+1} \leq \frac{1}{N+1} < \varepsilon$$

and therefore $\{n/(n+1)\}_{n=1}^{\infty}$ has limit 1.

6.3 Theorem *The limit of a sequence is unique. That is, if a sequence $\{x_n\}_{n=1}^{\infty}$ has limit L and limit L', then $L = L'$.*

Proof. We wish to show that the values $\{x_n(x)\}$ cannot be getting arbitrarily close to different values for large n. To prove this, consider $\varepsilon > 0$. By the definition of limit there must exitst N_1 so that

$$n > N_1 \quad \text{implies} \quad |x_n - s| < \frac{\varepsilon}{2}$$

and there must exist N_2 so that

$$n > N_2 \quad \text{implies} \quad |x_n - t| < \frac{\varepsilon}{2}$$

For $n > \max\{N_1, N_2\}$, the Triangle Inequality 3.7 shows that

$$|s-t| = |(s-s_n)+(s_n-t)| < |s-s_n|+|s_n-t| < \frac{\varepsilon}{2}+\frac{\varepsilon}{2} = \varepsilon.$$

This shows that $|s-t| < \varepsilon$ for all $\varepsilon > 0$. It follows that $|s-t| = 0$ and hence $s = t$. $\square$

A sequence that does not converge to some real number is said to *diverge*.

6.4 Definition *Let $\{a_n\}$ be a sequence. We say that $\{a_n\}$ diverges to infinity(minus infinity) and write $\lim_{n\to\infty} a_n = \infty(\lim_{n\to\infty} a_n = -\infty)$ if for every real number M, there exists a positive integer N such that if $n \geq N$, then $a_n > M(a_n < -M)$.*

6.5 Definition *We say that a sequence $\{a_n\}$ is bounded above(below) if there exists a number M such that $a_n \leq M(a_n \geq M$ for every positive integer n. We say that $\{a_n\}$ is bounded if $\{a_n\}$ is bounded both above and below.*

It is easy to see that a sequence $\{a_n\}$ is bounded if and only if there exists a positive number M such that $|a_n| \leq M$ for every positive interger n.

We next show that a convergent sequence is bounded.

6.6 Theorem *If $\{a_n\}$ is a convergent sequence, then $\{a_n\}$ is bounded.*

Proof. Suppose $\lim_{n\to\infty} a_n = L$. If $\varepsilon = 1$, there exists a positive integer N such that if $n \geq N$, then

$$|a_n - L| < 1$$

Thus, if $n \geq N$, we have

$$|a_n| \leq |a_n - L| + |L| < 1 + |L|$$

If $n < N$, then

$$|a_n| \le \max\{|a_1|, |a_2|, \cdots, |a_{N-1}|\}$$

Thus for any positive integer n,

$$|a_n| \le M$$

if we let

$$M = \max\{|a_1|, |a_2|, \cdots, |a_{N-1}|, 1 + |L|\}$$ □

It follows from Theorem 6.6, that the sequence $\{n\}_{n=1}^{\infty}$ diverges since the set of positive integers is not bounded.

The sequence $\{(-1)^n\}_{n=1}^{\infty}$ is bounded and divergent and thus the converse of Theorem 6.6 does not hold. We will show later (Theorem 6.14) that a bounded *monotone* sequence is convergent.

6.7 Theorem *Let $\{a_n\}$ and $\{b_n\}$ be sequences such that $\{a_n\}$ is bounded and $\lim_{n\to\infty} b_n = 0$. Then $\lim_{n\to\infty} a_n b_n = 0$.*

Proof. There exists a positive number M such that $|a_n| \le M$ for every positive integer n. Let $\varepsilon > 0$. There exists a positive integer N such that if $n \ge N$, then

$$|b_n| < \varepsilon/M$$

If $n \ge N$, then

$$|a_n b_n| < M\frac{\varepsilon}{M} = \varepsilon$$ □

In this brief section we collect some miscellaneous theorems which are often useful in proving limits.

6.8 Theorem *Let $\{a_n\}$ be a sequence such that $0 \le a_n$ for every positive integer n and $L = \lim_{n\to\infty} a_n$. Then $0 \le L$.*

Proof. We argue by contradiction. Suppose $L < 0$. There exists a positive interger N such that if $n \geq N$, then

$$|a_n - L| < \frac{-L}{2}$$

Thus

$$a_N - L < \frac{-L}{2}$$

and therefore

$$a_N < \frac{L}{2} < 0$$

which is a contradiction. □

6.9 Theorem *Let $\{a_n\}$ and $\{b_n\}$ be sequences such that $\lim_{n\to\infty} a_n = L$ and $\lim_{n\to\infty} b_n = M$. If $a_n \leq b_n$ for every positive integer n, then $L \leq M$.*

Proof. Since $0 \leq b_n - a_n$ for all $n \in \mathbf{P}$, we conclude that $0 \leq M - L$. □

A special case of Theorem 6.9 is often used with either $\{a_n\}$ or $\{b_n\}$ taken to be a constant sequence. For example, if $a_n \leq M$ for all $n \in \mathbf{P}$ and $\lim_{n\to\infty} a_n = L$, then $L \leq M$.

6.10 Theorem(Squeeze Theorem) *Let $\{a_n\}$, $\{b_n\}$, and $\{c_n\}$ be sequences such that*

$$a_n \leq b_n \leq c_n$$

for every positive integer n. If

$$\lim_{n\to\infty} a_n = L = \lim_{n\to\infty} c_n$$

then

$$\lim_{n\to\infty} b_n = L$$

Proof. Let $\varepsilon > 0$. There exists a positive integer N_1 such that if $n \geq N_1$, then

$$L - \varepsilon < a_n < L + \varepsilon$$

There exists a positive integer N_2 such that if $n \geq N_2$, then

$$L - \varepsilon < c_n < L + \varepsilon$$

Thus if $n \geq \max\{N_1, N_2\}$, we have

$$L - \varepsilon < a_n \leq b_n \leq c_n < L + \varepsilon \qquad \square$$

Examples Since $0 < 1/2^n < 1/n$ and $\lim_{n\to\infty} 1/n = 0$, by Theorem 6.10, $\lim_{n\to\infty} 1/2^n = 0$.

6.11 Theorem *Let $\{a_n\}$ and $\{b_n\}$ be sequences such that $\lim_{n\to\infty} a_n = \infty = \lim_{n\to\infty} b_n$. Then $\lim_{n\to\infty}(a_n + b_n) = \infty$.*

Proof. Let M be a real number. There exists a positive integer N_1 such that if $n \geq N_1$, then

$$\frac{M}{2} < a_n$$

There exists a positive integer N_2 such that if $n \geq N_2$, then

$$\frac{M}{2} < b_n$$

If $n \geq \max\{N_1, N_2\}$, then

$$M = \frac{M}{2} + \frac{M}{2} < a_n + b_n \qquad \square$$

6.12 Definition *Let $\{a_n\}$ be a sequence. We say that $\{a_n\}$ is* ***increasing*** *(* ***decreasing*** *) if $a_n \leq a_{n+1} (a_n \geq a_{n+1})$ for every positive integer n. We say that the sequence $\{a_n\}$ is* **monotone** *if either $\{a_n\}$ is decreasing or $\{a_n\}$ is increasing.*

6.13 Theorem *A monotone sequence $\{a_n\}$ is convergent if and only if $\{a_n\}$ is bounded.*

Proof. Suppose $\{a_n\}$ is increasing(The proof is similar if $\{a_n\}$ is decreasing).

If $\{a_n\}$ is convergent, then $\{a_n\}$ is bounded by Theorem 6.6

Suppose $\{a_n\}$ is bounded. By the least-upper-bound axiom, the least upper bound L of the set

$$X = \{a_n | n \in \mathbf{P}\}$$

exists. Let $\varepsilon > 0$. Since $L - \varepsilon$ is not an upper bound of the set X, there exists a positive integer N such that $L - \varepsilon < a_N$. Since $\{a_n\}$ is increasing, if $n \geq N$ we have $a_N \leq a_n$. Thus if $n \geq N$, we have

$$L - \varepsilon < a_N \leq a_n \leq L \leq L + \varepsilon$$

Therefore $\lim_{n\to\infty} a_n = L$. □

6.14 Theorem *Every monotone sequence in $\mathbb{R}^n$ has a limit in $\mathbb{R}^n$.*

Proof. Let (x_n) be nondecreasing and let $x = \sup\{x_n : n \in N\}$. Then $\lim_{n\to\infty} x_n = x$. □

Before proving the second main results of this section, namely that every bounded sequence has some convergent subsequence, we need the nested-interval property of the set of real numbers. The proof below is again hased on the completeness of the real number system in the form of the least-upper-bound property.

6.15 Theorem(Nested-interval property) *Suppose that $\{I_n\}$ be sequence of closed bounded intervals such that $I_n \supset I_{n+1}$ for each n. Then $\cap_{n=1}^{\infty} I_n \neq \emptyset$.*

Proof. $\{a_n\}$ is a sequence which is monotone increasing and bounded above(b_1 is an upper bound). Therefore $\lim_{n\to\infty} a_n = A$ exists and $A = \sup_n a_n \leq b_k$ for each natural number k. Hence $a_k \leq A \leq b_k$ for every $k \in \mathbb{N}$; that is, A is contained in each interval I_k. Now suppose $B \in I_n$ for every natural number n. Then $a_n \leq B \leq b_n$ for all n, and so $0 \leq B - a_n \leq b_n - a_n$ for each n. Thus by the old squeeze play. $\lim_{n\to\infty}(B - a_n) = 0$. It follows that $B = \lim_{n\to\infty} a_n = A$, and so A is the only real number common to all the intervals I_n. □

6.16 Theorem(Bolzano-Weierstrass Theorem) *Every bounded real sequence has a convergent subsequence.*

Proof. Let $\{a_n\}$ be a bounded sequence. Then there is a closed interval $[c, d]$ such that $a_n \in [c, d]$ for every positive integer n.

Consider the two closed intervals

$$\left[c, \frac{c+d}{2}\right], \left[\frac{c+d}{2}, d\right]$$

One of these intervals must contain a_n for infinitely many positive integers n. We denote this interval by $[c_1, d_1]$.

We repeat this process with the interval $[c_1, d_1]$. One of the intervals

$$\left[c_1, \frac{c_1+d_1}{2}\right], \left[\frac{c_1+d_1}{2}, d_1\right]$$

must contain a_n for infinitely many positive integers n. We denote this interval by $[c_2, d_2]$. Continuing this process, we obtain a sequence

$$[c_1, d_1], [c_2, d_2], \cdots$$

of closed intervals such that

$$[c_1, d_1] \supset [c_2, d_2] \supset [c_3, d_3] \supset \cdots$$
$$d_k - c_k = \frac{d-c}{2^k}, \quad k = 1, 2, \cdots \tag{A}$$

and each interval $[c_k, d_k]$ contains a_n for infinitely many positive integer n.

Choose a positive integer n_1 such that $a_{n_1} \in [c_1, d_1]$. Since $[c_2, d_2]$ contains a_n for infinitely many positive integers n, there exists a positive integer $n_2 > n_1$ such that $a_{n_2} \in [c_2, d_2]$. Continuing this process we obtain elements $a_{n_1}, a_{n_2}, \cdots$ satisfying

$$a_{n_k} \in [c_k, d_k], \quad k = 1, 2, \cdots \tag{B}$$

and

$$n_1 < n_2 < n_3 < \cdots$$

Therefore, $\{a_{n_k}\}_{k=1}^{\infty}$ is a subsequence of $\{a_n\}$. We will show that $\{a_{n_k}\}$ converges.

The sequence $\{c_k\}$ is monotone and bounded so by Theorem 6.13, $\{c_k\}$ is convergent. Let $L = \lim_{k\to\infty} c_k$. Similarly, the sequence $\{d_k\}$ converges to some number M. Using equation (A), we have

$$M - L = \lim_{k\to\infty} d_k - \lim_{k\to\infty} c_k = \lim_{k\to\infty}(d_k - c_k) = \lim_{k\to\infty} \frac{d-c}{2^k} = 0$$

Therefore $L = M$. Because of above (B), we have

$$c_k \le a_{n_k} \le d_k$$

By the squeeze theorem(Theorem 6.10), $\{a_{n_k}\}$ converges to $L = M$. □

Thus if we proved that a convergent sequence is bounded (Theorem 6.6) and we noted that the converse of this theorem is false. The sequence $\{(-1)^n\}$ is bounded and divergent. Although $\{(-1)^n\}$ is divergents, it has a convergent subsequence, $\{(-1)^{2n}\}$. The Bolzano-Weierstrass theorem states that *every* bounded sequence has a convergent subsequence. This theorem states a fundamental property of the real numbers, and, in fact, it can be shown to be equivalent to the least-upper-bound axiom.

We will use the Bolzano-Weierstrass theorem to derive a general condition for convergence of a sequence and to define a kind of limit applicable to any sequence.

6.17 Definition *A sequence $\{x_n\}$ of real numbers is called a* ***Cauchy sequence*** *if*

for each $\varepsilon > 0$ there exists a number N such that
$m, n > N$ implies $|x_n - x_m| < \varepsilon$.

Compare this definition with Definition 2.3.

Now we want to have a convergent test fortrary sequence which does not assume that the value of the limit known

6.18 Lemma *Convergent sequences are Cauchy sequences.*

Proof. Since the proof of this lemma is similar as proposition 2.3, we will left to the reader. □

6.19 Theorem *Every Cauchy sequence is bounded.*

Proof. Since the proof of this theorem is similar as proposition 2.5, we will left to the reader □

6.20 Theorem *Let $\{a_n\}$ be a real sequence. Then $\{a_n\}$ is convergent if and only if $\{a_n\}$ is a Cauchy sequence.*

Proof. Theorem 6.18 states that if $\{a_n\}$ is convergent, then $\{a_n\}$ is a Cauchy sequence and thus we must prove the converse.

Let $\{a_n\}$ be a Cauchy sequence. We have shown that $\{a_n\}$ is bounded, so that we may apply the Bolzano-Weierstrass theorem.

By Theorem 6.16, $\{a_n\}$ has a convergent subsequence $\{a_{n_k}\}$. Suppose

$$\lim_{k\to\infty} a_{n_k} = L$$

Let $\varepsilon > 0$. There exists a positive interger N such that if $m, n \geq N$, then

$$|a_m - a_n| < \frac{\varepsilon}{2}$$

There exists a positive interger N such that if $k \geq N'$, then

$$|a_{n_k} - L| < \frac{\varepsilon}{2}$$

Choose a positive integer K such that $K \geq N'$ and $n_k \geq N$. Now if $n \geq N$, then

$$|a_n - L| \leq |a_n - a_{n_k}| + |a_{n_k} - L| < \frac{\varepsilon}{2} + \frac{\varepsilon}{2} = \varepsilon$$ □

If $\{x_n\}$ is a sequence, we define its limit superior by

$$\overline{\lim} x_n = \inf_n \sup_{k\geq n} x_k.$$

The symbols $\overline{\lim}$ and lim sup are both used for the limit superior. A real number l is the limit superior of the sequence $\{x_n\}$ if and only if

(i) given $\varepsilon > 0$, $^{\exists}n$ such that $x_k < l + \varepsilon$ for all $k \geq n$, and
(ii) given $\varepsilon > 0$, and given n, $^{\exists}k \geq n$ such that $x_k > l - \varepsilon$.

The extended real number ∞ is the limit superior of $\{x_n\}$ if and only if, given Δ and n, there is a $k \geq n$ such that $x_k > \Delta$. The extended real number $-\infty$ is the limit superior of $\{x_n\}$ if and only if $-\infty = \lim x_n$.

We define the limit inferior(lim inf) by

$$\underline{\lim} x_n = \sup_n \inf_{k \geq n} x_k.$$

The sequence $\{x_n\}$ converges to an extended real number l if and only if $l = \underline{\lim} x_n = \overline{\lim} x_n$.

6.21 Example For a sequence $x_n = (1 + 1/n) cos n\pi/2$, find the limit superior and limit inferior.

Since $\{x_n\} = \{0, -3/2, 0, 5/4, 0, -7/6, 0, 9/8, \cdots\}$
if $k = 1, 2, 3, 4, \cdots$ $\sup A_k = 5/4$
if $k = 5, 6, 7, 8, \cdots$ $\sup A_k = 9/8$
. . . .
if $k = 4n - 3, 4n - 2, 4n - 1, 4n$, $\sup A_k = 4n + 1/4n$.
Hence $\inf\{\sup A_k | k \in \mathbb{N}\} = \inf\{4n + 1/4n | n \in \mathbb{N}\} = 1$
That is, $\limsup = \overline{\lim}\, x_n = 1$
Similarly $\liminf = \underline{\lim}\, x_n = -1$

6.22 Proposition *For every sequence $\phi : \mathbb{N} \to \mathbb{R}$, we have*

$$\inf x_n \leq \liminf x_n \leq \limsup x_n \leq \sup x_n.$$

Proof. Since $A_k \subset \{x_n | n \in \mathbb{N}\}$ for all $k \in N$,

$$\sup A_k \leq \sup\{x_n | n \in N\} \text{ for all } k \in N.$$

It follows that $\inf\{\sup A_k | k \in \mathbb{N}\} \leq \sup\{x_n | n \in \mathbb{N}\}$
That is,

$$\overline{\lim} x_n \leq \sup\{x_n | n \in N\}$$

Similary, $\inf\{x_n | n \in \mathbb{N}\} \leq \lim x_n$, and since for all $j, k \in \mathbb{N}$

$$\inf A_j \leq \sup A_k,$$

it follows that

$$\inf A_j \leq \inf\{\sup A_k | k \in \mathbb{N}\} \text{ for } j \in \mathbb{N}.$$

Hence $\sup\{\inf A_j | j \in \mathbb{N}\} \leq \inf\{\sup A_k | k \in \mathbb{N}\}$
That is,

$$\underline{\lim} x_n \leq \overline{\lim} x_n.$$

□

6.23 Proposition $\overline{\lim}(-x_n) = -\underline{\lim} x_n$ *and* $\underline{\lim}(-x_n) = -\overline{\lim} x_n$

Proof. By the difinition,

$$\sup\{-x_n | n \in \mathbb{N}\} = -\inf\{x_n | n \in \mathbb{N}\}$$

and

$$\inf\{-x_n | n \in \mathbb{N}\} = -\sup\{x_n | n \in \mathbb{N}\}$$

It follows that

$$\begin{aligned} \overline{\lim}(-x_n) &= \inf\{\sup[-x_n | n \geq k]\} \\ &= \inf\{-\inf[x_n | n \geq k]\} \\ &= -\sup\{\inf[x_n | n \geq k]\} \\ &= -\underline{\lim} x_n \end{aligned}$$

Similarly, the second statement can be shown easily. □

By the definition, if (x_n) and (y_n) are two sequences, we have

$$\begin{aligned} \underline{\lim} x_n + \underline{\lim} y_n &\leq \underline{\lim}(x_n + y_n) \leq \overline{\lim} x_n + \underline{\lim} y_n \\ &\leq \overline{\lim}(x_n + y_n) \leq \overline{\lim} x_n + \overline{\lim} y_n. \end{aligned}$$

By a **cluster point** of $\{x_n\}$, we mean a real number r such that, for every positive real number ε and every natural k, there exists a natural number $n > k$ satisfying.

$$|x_n - r| < \varepsilon$$

If $\{x_n\}$ is convergent, then its limit r is obviously the only cluster point. On the other hand, the divergent sequence defined by $x_n = (-1)^n$ for every $n \in \mathbb{N}$ has two cluster points, namely, the real numbers 1 and -1.

6.24 Theorem $\limsup x_n$ *is the largest cluster point of* $\{x_n\}$ *and* $\liminf\{x_n\}$ *is the least cluster point of* x_n.

Proof. (i) If $\lim x_n = \infty$, then $\sup A_k = \infty$ for all $k \in \mathbb{N}$ and for all $N_0 \in \mathbb{N}$ there is n such that $x_n \geq N_0$. Hence ∞ is the cluster point of $< x_n >$ and it is maximum.

(ii) If $\lim x_n$ is a real number, then $a = \underline{\lim} x_n = \inf\{\sup[A_k] \mid k \in \mathbb{N}\}$ is a real number and $a = \inf(A)$, here $A = \{\sup[A_k] \mid k \in \mathbb{N}\}$.

Now given $\varepsilon > 0$, $a - \varepsilon < a$. It follows that $a - \varepsilon$ is the lower bound of A. Hence $a - \varepsilon < \sup A_k (k \in \mathbb{N})$

Therefore, for all k there is n such that $[n \geq k \Rightarrow a - \varepsilon < x_n]$ That is, there exist infinitely many n such that $a - \varepsilon < x_n$. Now since $a < a + \varepsilon$ for arbitrary $\varepsilon > 0$, $a + \varepsilon$ is not a lower bound of A. And so, there is k_1 such that $a \leq \sup A_{k_1} < a + \varepsilon$. It means that there is finitely x_n such that $a + \varepsilon \leq x_n$. Therefore, there is no cluster point of $< x_n >$ which is larger than a and there is infinitely many x_n such that $a - \varepsilon < x_n < a + \varepsilon$. Consequently, a is a largest cluster point of $< x_n >$.

Similarly, second statement can be verified easily. □

Obviously the sequences$(\sup_{n \geq k} x_n)_{k=1}^{\infty}$ and $(\inf_{n \geq k} x_n)_{k=1}^{\infty}$ are monotone sequences, so that $\overline{\lim_{n\to\infty}} x_n$ and $\underline{\lim_{n\to\infty}} x_n$ are just their respective limits.

6.25 Example Consider the sequence $a_n = (1 - 1/n)\sin(n\pi/2)$ and let C be the set of all cluster points of a_n. The reader should verify that $C = \{-1, 0, 1\}$ and so $\limsup a_n = -1$ and the sequence $\{a_n\}$ necessarily diverges. But the sequence $\{C_{2k}\}$ converges to 0 and $\{C_{4k+1}\}$ $(k = 0, 1, 2, \cdots)$ converges to 1 and $\{C_{4k-1}\}$ converges to -1 $(k = 1, 2, \cdots)$.

PROBLEMS [2.6]

1. Prove that if $\{a_{n_k}\}$ is a subsequence of a bounded sequence $\{a_n\}$, then
$$\limsup_{k\to\infty} a_{n_k} \le \limsup_{k\to\infty} a_n$$

2. Let $\{a_n\}$ be a bounded sequence. Prove that
$$\limsup_{k\to\infty} a_n = -\liminf_{k\to\infty}(-a_n)$$

3. Let $\{a_n\}$ and $\{b_n\}$ be sequences such that $\{a_n\}$ is convergent and $\{b_n\}$ is bounded. Prove that
$$\limsup_{n\to\infty}(a_n + b_n) = \limsup_{n\to\infty} a_n + \limsup_{n\to\infty} b_n$$
and
$$\liminf_{n\to\infty}(a_n + b_n) = \liminf_{n\to\infty} a_n + \liminf_{n\to\infty} b_n$$

4. Let $\{a_n\}$ be a bounded sequence. Suppose that for every bounded sequence $\{b_n\}$ we have
$$\limsup_{n\to\infty}(a_n + b_n) = \limsup_{n\to\infty} a_n + \limsup_{n\to\infty} b_n$$
Prove that $\{a_n\}$ is convergent.

5. Let $\{a_n\}$ and $\{b_n\}$ be bounded nonnegative sequences. Prove that
$$\begin{aligned}\liminf_{n\to\infty} a_n \liminf_{n\to\infty} b_n &\le \liminf_{n\to\infty} a_n b_n \le \limsup_{n\to\infty} a_n b_n \\ &\le \limsup_{n\to\infty} a_n \limsup_{n\to\infty} b_n\end{aligned}$$

6. Show (by giving examples) that the first and third inequalities of Exercise 5 may fail if the nonnegative hypothesis is omitted.

7. Let $\{a_n\}$ and $\{b_n\}$ be sequences such that $\lim_{n\to\infty} a_n < 0$ and $\{b_n\}$ is a bounded sequence. Suppose that

$$\limsup_{n\to\infty} a_n b_n = \limsup_{n\to\infty} a_n \limsup_{n\to\infty} b_n$$

Prove that $\{b_n\}$ is convergent.

8. Let $\{a_n\}$ be a sequence of positive numbers such that $\lim_{n\to\infty} a_n = L$. Prove that

$$\lim_{n\to\infty} (a_1, a_2, \cdots, a_n)^{1/n} = L$$

9. (Squeeze Theorem) Let $\{a_n\}$, $\{b_n\}$, and $\{c_n\}$ be bounded sequences such that

$$a_n \le b_n \le c_n$$

for every positive integer n and

$$\limsup_{n\to\infty} c_n = \liminf_{n\to\infty} a_n$$

10. Give an example of a sequence $\{a\}$ such that the sequence

$$\left\{\frac{a_1 + a_2 + \cdots + a_n}{n}\right\}$$

converges but $\{a_n\}$ diverges.

2.7 Sequences of Functions

In this section we shall investigate convergence of sequences and series of functions. We begin with the definition of continuity of a function at a point in its domain.

7.1 Definition *f is said to be continuous at x_0 if $\lim_{x\to x_0} f(x) = f(x_0)$; equivalently, given any $\varepsilon > 0$ there is a $\delta > 0$ such that $|f(x) - f(x_0)| < \varepsilon$ whenever $|x - x_0| < \delta$.*

We first point out that in order for a function f to be continuous at x_0, f must be defined on some δ neighborhood of x_0. This is necessary in order for $\lim_{x \to x_0} f(x)$ to exist.

7.2 Proposition *Let f be a real-valued function defined and continuous on a closed and bounded set F. Then f is bounded on F and assumes its maximum and minimum on F; that is, there are points x_1 and x_2 in F such that $f(x_1) \le f(x) \le f(x_2)$ for all x in F.*

Proof. We shall first show that f is bounded on F. Since f is continuous on F, for each $x \in F$ there is an open interval I_x containing x such that $|f(y) - f(x)| < 1$ for $y \in I_x \cap F$. Thus for $y \in I_x \cap F$, we have $|f(y)| \le |f(x)| + 1$ and so f is bounded in I_x. The collection $\{I_x : x \in F\}$ is a collection of open intervals which covers F, and by the Heine-Borel Theorem there is a finite subcollection $\{I_{x_1}, \cdots, I_{x_n}\}$ which covers F. Let $M = 1 + \max[|f(x_1)|, \cdots, |f(x_n)|]$. Then each y in F belongs to some interval I_{x_k} in the finite subcollection, and hence $|f(y)| < 1 + |f(x_k)| \le M$. This shows that f is bounded (by M) on F.

To see that f assumes its maximum on F, let $m = \sup_{x \in F} f(x)$. Since f is bounded, m is finite, and our goal is to show that there is an $x_1 \in F$ such that $f(x_1) = m$. Suppose not. Then $f(x) < m$ for each $x \in F$, and by continuity there is an open interval I_x containing x such that $f(y) < \frac{1}{2}(f(x) + m)$ for all $y \in I_x \cap F$. Again using the Heine-Borel Theorem, we can find a finite number of these intervals $\{I_{x_1}, \cdots, I_{x_n}\}$ which cover F. Set $a = \max[f(x_1), \cdots, f(x_n)]$. Then each $y \in F$ belongs to one such interval I_{x_k} and $f(y) < \frac{1}{2}[f(x_k) + m] \le \frac{1}{2}(a + m)$. Thus $\frac{1}{2}(a + m)$ is bound for f on F. But this is impossible, since $\frac{1}{2}(a + m) < m$. Consequently, there must be an x_1 such that $f(x_1) = m$. Similarly, there is an x_2 at which f assumes its minimum. □

7.3 Proposition *Let f be a real-valued function defined on $(-\infty, \infty)$. Then f is continuous if and only if for each open set O of real numbers $f^{-1}[O]$ is an open set.*

Proof. Suppose $f^{-1}[O]$ is open for each open set O, and let x be an arbitrary real number. Then, given $\varepsilon > 0$, the interval $I = (f(x) - \varepsilon, f(x) + \varepsilon)$ is an open set, and so its inverse image $f^{-1}[I]$ must be open. Since $x \in f^{-1}[I]$, there must be some $\delta > 0$ such that $(x-\delta, x+\delta) \subset f^{-1}[I]$. But this implies that if $|y-x| < \delta$, then $f(y) \in (f(x) - \varepsilon, f(x) + \varepsilon)$; that is, $|f(x) - f(y)| < \varepsilon$. Hence f is continuous at x. Since x was arbitrary, f is continuous.

Suppose now that f is continuous and that O is an open set. Let x be a point of $f^{-1}[O]$. Then $f(x) \in O$, and there is an $\varepsilon > 0$ such that $(f(x) - \varepsilon, f(x) + \varepsilon) \subset O$. Since f is continuous at x, there is a $\delta > 0$ such that $|f(x) - f(y)| < \varepsilon$ for $|x - y| < \delta$. Thus, for every $y \in (x - \delta, x + \delta)$,we have $f(y) \in (f(x) - \varepsilon, f(x) + \varepsilon) \subset O$. Hence $(x - \delta, x + \delta) \subset f^{-1}[O]$, and so $f^{-1}[O]$ is open. □

7.4 Definition *A real-valued function f defined on a set E is said to be* ***uniformly continuous*** *(on E) if, given $\varepsilon > 0$, there is a $\delta > 0$ such that for all x and y in E with $|x - y| < \delta$ we have $|f(x) - f(y)| < \varepsilon$.*

7.5 Proposition *If a real-valued function f is defined and continuous on a closed and bounded set F of real numbers, it is uniformly continuous on F.*

Proof. Given $\varepsilon > 0$ and x in F, there is a $\delta_x > 0$ such that $|x - y| < \delta_x$ implies $|f(x) - f(y)| < \frac{1}{2}\varepsilon$. Let O_x be the interval $(x - \frac{1}{2}\delta_x, x + \frac{1}{2}\delta_x)$. Then $\{O_x : x \in F\}$ is an open covering of F. By the Heine-Borel Theorem there is a finite subcollection $\{O_{x_1}, \cdots, O_{x_n}\}$ which covers F. Let $\delta = \frac{1}{2}\min(\delta_{x_1}, \cdots, \delta_{x_n})$. Then δ is positive. Given two points y and z in F such that $|y - z| < \delta$, the point y must belong to some O_{x_i}, and hence there is an i such that $|y - x_i| < \frac{1}{2}\delta_{x_i}$. Consequently,

$$|z - x_i| \leq |z - y| + |y - x_i| < \frac{1}{2}\delta_{x_i} + \delta \leq \delta_{x_i}.$$

Hence

$$|f(y) - f(x_i)| < \frac{\varepsilon}{2}$$

and

$$|f(z) - f(x_i)| < \frac{\varepsilon}{2},$$

whence

$$|f(z) - f(y)| < \varepsilon,$$

showing that f is uniformly continuous on F. □

We shall be concerned with sequences of real-valued functions defined on some set X. We shall denote a sequence as $\{f_n\}_{n=1}^{\infty}$ (or sometimes just $\{f_n\}$). Thus for each positive integer n we have a real-valued function f_n defined on a set X. The set X will not, in general, be a subset of the real numbers.

7.6 Definition *A sequence $\{f_n\}$ of functions defined on a set E is said to converge **pointwise** on E to a function f if for every x in E we have $f(x) = \lim f_n(x)$; that is, if, given $x \in E$ and $\varepsilon > 0$, there is an N such that for all $n \geq N$ we have $|f(x) - f_n(x)| < \varepsilon$.*

7.7 Definition *A sequence $\{f_n\}$ of functions defined on a set E is said to converge **uniformly** on E if, given $\varepsilon > 0$, there is an N such that for all $x \in E$ and all $n \geq N$ we have $|f_n(x) - f(x)| < \varepsilon$.*

The crucial part of Definition 7.7 is that $|f_n(x) - f(x)| < \varepsilon$ for every x in X(if $n \geq N$). That is, N depends on ε, but not on x.

It is clear that if a sequence of functions converges uniformly it also converges pointwise(verify).

We now prove that contiunity is preserved under uniform convergence.

7.8 Theorem *Let $\{f_n\}$ be a sequence of functions which converges uniformly to a function f on a subset E of $\mathbb{R}$. If each f_n is continuous at a point a in E, then f is continuous at a.*

Proof. Let $\varepsilon > 0$. There exists a positive integer N such that

$$|f_N(x) - f(x)| < \frac{\varepsilon}{3}$$

for every $x \in E$. There exists $\delta > 0$ such that if $|x - a| < \delta$, then

$$|f_N(x) - f_N(a)| < \frac{\varepsilon}{3}$$

Now if $|x - a| < \delta$, then

$$\begin{aligned} |f(x) - f(a)| &\leq |f(x) - f_N(x)| \leq |f_N(x) - f_N(a)| \\ &\leq |f_N(a) - f(a)| < \frac{\varepsilon}{3} + \frac{\varepsilon}{3} + \frac{\varepsilon}{3} = \varepsilon \end{aligned}$$ □

7.9 Corollary *If $\{f_n\}$ is a sequence of continuous functions which converges uniformly to a function, then f is continuous.*

Example. Let

$$f_n(x) = x^n \quad 0 \leq x \leq 1$$

Then $\{f_n\}$ converges pointwise to f on [0,1], where

$$f(x) = \begin{cases} 0 & 0 \leq x \leq 1 \\ 1 & x = 1 \end{cases}$$

SOLUTION. Let $x_0 \in (0,1)$ be arbitrary and let $\varepsilon > 0$ be given. Choose the natural number $N \geq (\ln \varepsilon)/(\ln x_0)$; then $x_0^N \leq \varepsilon$. Now if $n > N$ then

$$|f_n(x_0) - f(x_0)| = |x_0^n - 0| = x_0^n < x_0^N < \varepsilon$$

It is clear that our choice of N in the above example depends not only on the given $\varepsilon > 0$ but also on the point x_0. If the point x_0 is closer to 1 then the N will have to be chosen appropriately larger. The question arises: Is there an n such that $|f_n(x) - f(x)| < \varepsilon$ for every $x \in (0,1)$? To show that the answer to this question is no, let $\varepsilon = \frac{1}{2}$ and choose x_n such that $1 > x_n \geq 1/\sqrt[n]{2}$. Then $|f_n(x) - f(x)| = x_n^n \geq \frac{1}{2} = \varepsilon$. Thus there is no natural number n for which $|f_n(x) - f(x)| < \frac{1}{2}$ for every $x \in (0,1)$ (See Fig. 2.1)

The geometric interpretation of uniform convergence is given in Fig. 2.2 Picture the limit function $f(x)$ and an ε band around the graph of $f(x)$; that is, $(f(x) - \varepsilon, f(x) + \varepsilon)$ for every x. If

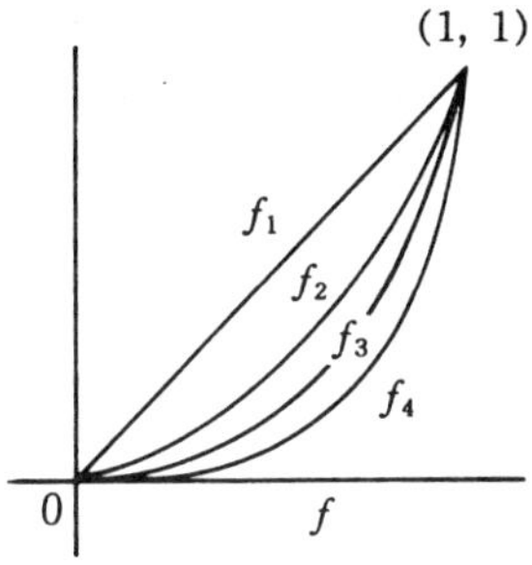

Figure 2.1

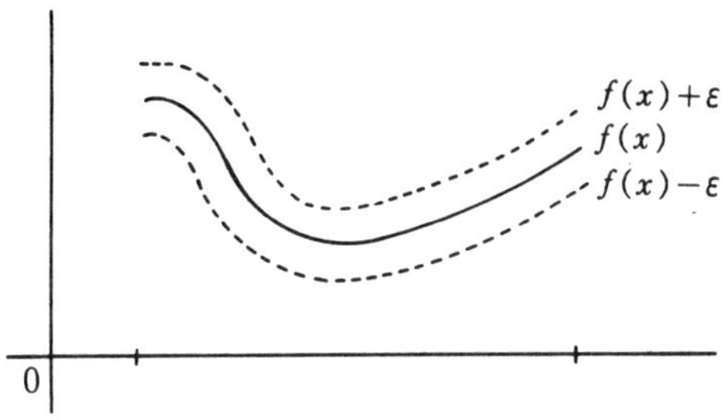

Figure 2.2

$\{f_n(x)\}$ converges uniformly to $f(x)$ then there must be a natural number N such that $f_n(x)$ lies in the ε band for every $n > N$.

Uniform convergence is stronger than pointwise convergence. If $\{f_n(x)\}$ converges uniformly to $f_n(x)$ on I then $\{f_n(x)\}$ converges pointwise to $f_n(x)$ on I (see Exercise 3). If $\{f_n(x)\}$ converges pointwise to $f(x)$ on I then $\{f_n(x)\}$ may or may not converge uniformly on I, but if it does then necessarily $f_n(x)$ is the uniform limit. To test a specific sequence of functions for uniform convergence first find the pointwise limit (if it exists) and then check to see whether the convergence is uniform.

It follows from Corollary 7.9 that the sequence $\{f_n\}$ of Example does not converge uniformly to f. Thus a sequence may converge pointwise, but fail to converge uniformly.

The conclusion of Corollary 7.9 may also be written

$$\lim_{x\to a}\lim_{n\to\infty} f_n(x) = \lim_{n\to\infty}\lim_{x\to a} f_n(x)$$

For $\lim_{n\to\infty} f_n(x) = f(x)$ and since f is continuous, $\lim_{x\to a} f(x) = f(a)$. Also $\lim_{x\to a} f_n(x) = f_n(a)$ since each function f_n is contiuous, and $\lim_{n\to\infty} f_n(x) = f(a)$. Therefore, Corollary 7.9 provides another example of a result which deals with interchanging limits.

Example. We let $C[a,b]$ denote the set of continuous real-valued functions on $[a,b]$. We define a metric d on $C[a,b]$ by the formula

$$d(f,g) = \text{lub}\{|f(x) - g(x)| : x \in [a,b]\}$$

Since the function $f - g$ is continuous on $[a, b]$, it is bounded, and therefore the least upper bound of the above set exists. The rest of the proof that d is metric on $C[a, b]$ is straightforward (Exercise). When we refer to $C[a, b]$, unless otherwise specified, we will assume that the metric on $C[a, b]$ is that defined above. This metric is often called the **metric of uniform convergence** because of the next theorem.

7.10 Theorem *Let $\{f_n\}$ be a sequence of functions in $C[a, b]$. Then $\lim_{n\to\infty} f_n = f$ in the metric space $C[a, b]$ if and only if f_n converges uniformly to f on $[a, b]$.*

Proof. Suppose $\{f_n\}$ converges to f in the metric sapce $C[a, b]$. Let $\varepsilon > 0$. There exists a positive integer N such that if $n \geq N$, then $d(f_n, f) < \varepsilon$. This means $d(f_n, f) = \text{lub}\{|f_n(x) - f(x)| : x \in [a, b]\} < \varepsilon$, or equivalently, $|f_n(x) - f(x)| < \varepsilon$ for every x in $[a, b]$. By Definition 7.6 this says that f_n converges uniformly to f on $[a, b]$.

Conversely, suppose $\{f_n\}$ is a sequence in $C[a, b]$ which converges uniformly to a function f on $[a, b]$. By Corollary 7.8 we have that $f \in C[a, b]$. We must show that $\lim_{n\to\infty} f_n = f$ in the metric of $C[a, b]$.

Let $\varepsilon > 0$. Since f_n converges uniformly to f on $[a, b]$, there exists a positive integer N such that if $n \geq N$, then

$$|f_n(x) - f(x)| < \varepsilon/2 \text{ for every } x \in [a, b]$$

Thus if $n \geq N$, $\varepsilon/2$ is an upper bound for the set

$$\{|f_n(x) - f(x)| : x \in [a, b]\}$$

and so $d(f_n, f) \leq \varepsilon/2 < \varepsilon$. Therefore, $\lim_{n\to\infty} f_n = f$. □

PROBLEMS [2.7]

1. Let $\{f_n\}$ be a sequence of continuous functions defined on a set E. Prove that if $\{f_n\}$ converges uniformly to f on E then f is continuous on E.

2. Let f be that function defined by setting

$$f(x) = \begin{cases} x & \text{if } x \text{ irrational} \\ p \sin \frac{1}{q} & \text{if } x = \frac{p}{q} \text{ in lowest terms} \end{cases}$$

At what points is f continuos?

3. (a) Show that if f and g are continuous functions, then the functions $f + g$ and fg are continuous.
   (b) Show that if f and g are continuous, then so is $f \circ g$.
   (c) Let $f \vee g$ be the function defined by $(f \vee g)(x) = f(x) \vee g(x)$, and define $f \wedge g$ similarly. Show that if f and g are continuous so are $f \vee g$ and $f \wedge g$.
   (d) If f is continuous, then so is $|f|$.

4. A real-valued function f is said to **monotone increasing** if $f(x) \leq f(y)$ whenever $x \leq y$. and **strictly** monotone increasing if $f(x) < f(y)$ whenever $x < y$. It is called monotone (or strictly monotone) if f or $-f$ is monotone (or strictly monotone) increasing. Let f be a continuous function on the interval $[a, b]$. Then there is a continuous function g such that $g(f(x)) = x$ for all $x \in [a, b]$ if and only if f is strictly monotone. In this case we also have $f(g(y)) = y$ for each y between $f(a)$ and $f(b)$. A function f which has a continuous inverse is called a **homeomorphism** (between its domain and its range).

5. A continuous function φ on $[a, b]$ is called **polygonal** (or piecewise linear) if there is a subdivision $a = x_0 < x_1 < \cdots < x_n = b$ such that φ is linear on each interval $[x_i, x_{i+1}]$. Let f be an arbitrary continuous function on $[a, b]$ and ε a

positive number. Show that there is a polygonal function φ on $[a, b]$ with $|f(x) - \varphi(x)| < \varepsilon$ for all $x \in [a, b]$.

6. Let F be a closed set of real numbers and f a real-valued function which is defined and continuous on F. Show that there is a function g defined and continuous on $(-\infty, \infty)$ such that $f(x) = g(x)$ for each $x \in F$.

   [Hint: Take g to be linear in each of the intervals of which $\overline{F}$ is composed.]

7. Let f be a real-valued function with domain E. Prove that f is continuous if and only if for each open set O there is an open set U such that $f^{-1}[O] = E \cap U$.

Chapter 3

LEBESGUE MEASURE

The concept of the measure mE of a set E is a natural generaliztion of such concepts as

1) The length $l(\Delta)$ of a line segment Δ ;
2) The area $A(F)$ of a plane figure F;
3) The volume $V(G)$ of a space figure G;
4) The increment $\varphi(b) - \varphi(a)$ of a nondecreasing function $\varphi(t)$ over a half-open interval $[a, b)$;
5) The integral of a nonnegative function over a set on the line or over a region in the plane or in space.

Although the notion of measure first arose in the theory of functions of a real variable, it was subsequently used extensively in functional analysis. probability theory, the theory of dynamical systems, and other branches of mathematics.

The reader will easily confirm that the consideration in Sec.2 are of a general nature and carry over to the case of the more abstract theory without essential changes.

3.1 Algebra and Sigma algebra

1.1 Definition *Let X be a set. A collection $\mathcal{A}$ of subsets of X is said to be* ***algebra*** *(or* ***Boolean algebra*** *) of sets in X. If*

(*a*) $A \in \mathcal{A} \Rightarrow A^c \in \mathcal{A}$ *that is* $S - A \in \mathcal{A}$.
(*b*) $A, B \in \mathcal{A} \Rightarrow A \cup B \in \mathcal{A}$

Note that if the set A belongs to $\mathcal{A}$, then X, since it is the union of A and A^c, must also belong to $\mathcal{A}$.

By DeMorgan's law it follows that if $\mathcal{A}$ is an algebra of the sets in $\wp(X)$, then

(c) $A, B \in \mathcal{A} \Rightarrow A \cap B \in \mathcal{A}$

If X is a set and if $\mathcal{A}$ is a family of subsets of X that is closed under complementation, then X belongs to $\mathcal{A}$ if and only if $\emptyset$ belongs to $\mathcal{A}$. Thus in the definitions of algebras and σ-algebras given above we can replace condition (a) with the requirement that $\emptyset$ be a member of $\mathcal{A}$. Furthemore, if $\mathcal{A}$ is a family of subsets of X that is non-empty, closed complementation, and closed the formation of finite or countable unions, then $\mathcal{A}$ must contain X: Thus in our definitions of algebras and σ-algebras we can replace condition (a) with the requirement that $\mathcal{A}$ be non-empty.

Thus our equivalent definition can be obtained as follows.

Let X be an arbitrary set. A collection $\mathcal{A}$ of subsets of X is an **algebra** on X if

(a) $X \in \mathcal{A}$ (or $\emptyset \in \mathcal{A}$)

(b) for each set A that belongs to $\mathcal{A}$ the set A^c belongs to $\mathcal{A}$

(c) for each finite sequence $A_1, \cdots, A_n$ of sets that belong to $\mathcal{A}$ the set $\cup_{i=1}^n A_i$ belongs to $\mathcal{A}$, and

(d) for each finite sequence $A_1, \cdots, A_n$ of sets that belong to $\mathcal{A}$ the set belongs to $\cap_{i=1}^n A_i$ belongs to $\mathcal{A}$.

1.2 Definition *Let X be a set. A collection $\mathcal{A}$ of subsets of X is said to be σ-**algebra**(or **Boolean σ-algebra**)*

(i) $A \in \mathcal{A}$ implies $A^c \in \mathcal{A}$

(ii) $A_n \in \mathcal{A}$ for $n = 1, 2, 3, \cdots$ imples $\cup_{n=1}^\infty A_n \in \mathcal{A}$

As in the case of algebra, σ-algebra can be defined as follow.

Again let X be an arbitrary set. A collection $\mathcal{A}$ of subsets of X is a *σ-algebra* on X if

(a) $X \in \mathcal{A}$ (or $\emptyset \in \mathcal{A}$)

(b) for each set A that belongs to $\mathcal{A}$ the set A^c belongs to $\mathcal{A}$

(c) for each infinite sequence $\{A_i\}$ of sets that belong to $\mathcal{A}$ the set $\cup_{i=1}^{\infty} A_i$ belongs to $\mathcal{A}$, and

(d) for each infinite sequence $\{A_i\}$ of sets that belong to $\mathcal{A}$ the set $\cap_{i=1}^{\infty} A_i$ belongs to $\mathcal{A}$.

Thus a σ-algebra on X is a family of subsets of X that contains X and is closed under complementation, under the formation of countable unions, and the formation of countable intersections.

Each σ-algebra on X is an algebra on X since, for example, the union of the finite sequence $A_1, A_2, \cdots, A_n$ is the same as the union the infinite sequence $A_1, A_2, \cdots, A_n, A_n, A_n, \cdots$.

We turn to some examples.

1. Let X be a set, and let $\mathcal{A} = \wp(X)$ be the collection of all subsets of X. then $\mathcal{A}$ is σ-algebra on X. That is, $\wp(X)$ is the largest σ-algebra on X.

2. Let X be a set, and let $\mathcal{A} = \{\emptyset, X\}$. Then $\mathcal{A}$ is a σ-algebra on X. ($\mathcal{A}$ is the smallest σ-algebra on X)

Note that any σ-algebra in X always lies between the σ-algebra $\{\emptyset, X\}$ and $\wp(X)$.

3. Let X be an infinite set, and let $\mathcal{A}$ be the collection of all finite subsets of X. Then $\mathcal{A}$ does not contain X and is not closed under complementation, and so is not an algebra(or a σ-algebra) on X.

4. Let X be an infinite set, and let $\mathcal{A}$ be the collection of all subsets A of X such that either A or A^c is finite. Then $\mathcal{A}$ is an algebra on X (check this), but is not closed under the formation of countable unions, and so is not a σ-algebra.

5. Let X be an uncountable set, and let $\mathcal{A}$ be the collection of all countable (i.e., finite or countably infinte) subsets of X. Then $\mathcal{A}$ does not contain X and is not closed under complementation, and so is not an algebra.

6. Let X be a set, and let $\mathcal{A}$ be the collection of all subsets A of X such that either A or A^c is countable. Then $\mathcal{A}$ is a σ-algebra.

7. Let $\mathscr{A}$ be the collection of all subsets of $\mathbb{R}$ that are unions of finitely many intervals of the form $(a, b]$, $(a, +\infty)$, or $(-\infty, b]$.

It is easy to check that each set that belongs to $\mathscr{A}$ is the union of a finite disjoint cllection of intervals of the types listed above, and then to check that $\mathscr{A}$ is an algebra on $\mathbb{R}$ (the empty set belongs to $\mathscr{A}$, since it is the union of the empty, and hece finite, collection of intervals). The algebra $\mathscr{A}$ is not a σ-algebra; for example, the bounded open subintervals of $\mathbb{R}$ are unions of sequences of sets in $\mathscr{A}$, but do not themselves belong to $\mathscr{A}$.

Next we consider ways of constructing σ-algebra.

1.3 Proposition *Let X be a set. Then the intersection of an arbitrary non-empty collection of σ-algebra on X is σ-algebra on X.*

Proof. Let $\mathscr{J}$ be a non-empty collection of σ-algebras on X, and let $\mathscr{A}$ be the intersection of the σ-algebras that belong to $\mathscr{J}$. It is enough to check that $\mathscr{A}$ contains X, is closed under complementation, and is closed under the formation of countable unions. The set X belongs to $\mathscr{A}$, since it belongs to each σ-algebra that belongs to $\mathscr{J}$. Now suppose that $A \in \mathscr{A}$. Each σ-algebra that belongs to $\mathscr{J}$ contains A and so contains A^c, thus A^c belongs to the intersection $\mathscr{A}$ of these σ-algebras. Finally, suppose that $\{A_i\}$ is a sequence of sets that belong to $\mathscr{A}$, and hence to each σ-algebra in $\mathscr{J}$. Then $\cup A_i$ belongs to each σ-algebra in $\mathscr{J}$, and so to $\mathscr{A}$.

The reader should note that the union of a family of σ-algebras can fail to be a σ-algebra (see Exercise 5).

Proposition 1.3 implies the following result, which is a basic tool for the construction of σ-algebras.

1.4 Corollary *Let X be a set, and let $\mathcal{F}$ be a family of subsets of X. Then there is a smallest σ-algebra on X that includes $\mathcal{F}$.*

Of course, to say that $\mathscr{A}$ is the smallest σ-algebra on X that includes $\mathcal{F}$ is to say that $\mathscr{A}$ is a σ-algebra on X that includes $\mathcal{F}$, and that every σ-algebra on X that includes $\mathcal{F}$ also includes $\mathscr{A}$. This smallest σ-algebra on X that includes $\mathcal{F}$ is clearly unique; it is called the **σ-algebra generated by $\mathcal{F}$** and is often denoted by $\sigma(\mathcal{F})$.

Proof. Let $\mathscr{C}$ be the collection of all σ-algebras on X that include $\mathcal{F}$. Then $\mathscr{C}$ is non-empty, since it contains the σ-algebra that consists of all subsets of X. The intersection of the σ-algebras that belong to $\mathscr{C}$ is, according to Proposition 1.3, a σ-algebra; it includes $\mathcal{F}$ and is included in every σ-algebra on X that includes $\mathcal{F}$.

1.5 Example Let $F = \{A \in \mathbb{R} | A : \text{countable set}\}$ and $\mathscr{A} = \{B \in \mathbb{R} | B : \text{countable or } B^c : \text{countable}\}$. Then $\mathscr{A}$ is the smallest σ-algebra containing F.

We now use the preceding corollary to define an important family of σ-algebras.

< **Borel Sets** >

Although the intersection of any collection of closed sets is closed and the union of any finite collection of closed sets is closed, the union of a countable collection of closed sets need not be closed. For example, the set of rational numbers is the union of a countable collection of closed sets each of which contains exactly one number.Thus if we are interested in σ-algebras of sets which contain all of the closed sets we must consider more general types of sets than the open and closed sets. This leads us to the following definition:

Let $\mathscr{A}$ be the collection of all open sets of real line $\mathbb{R}$. Then by above Corollary, there is a smallest σ-algebra $\mathscr{B}$ of subsets of which contains $\mathscr{A}$. The elements of $\mathscr{B}$ are called **Borel sets**.

1.6 Definition *The Borel σ-algebra on $\mathbb{R}^d$ is the σ-algebra on $\mathbb{R}^d$ generated by the collection of open subsets of $\mathbb{R}^d$, and is denoted by $\mathscr{B}(\mathbb{R}^d)$. The Borel subsets of $\mathbb{R}^d$ are those that belong to $\mathscr{B}(\mathbb{R}^d)$. In case $d = 1$, one generally writes $\mathscr{B}(\mathbb{R})$ in place of $\mathscr{B}(\mathbb{R}^1)$.*

Borel sets are sometimes defined as the members of the smallest σ-algebra containing all closed sets.

1.7 Proposition *The σ-algebra $\mathscr{B}(\mathbb{R})$ of Borel sets of $\mathbb{R}$ is generated by each of the following collections of sets:*

(*a*) *the collection of all closed subsets of $\mathbb{R}$;*

(*b*) *the collection of all subintervals of $\mathbb{R}$ of the form $(-\infty, b]$;*

(*c*) *the collection of all subintervals of $\mathbb{R}$ of the form $(a, b]$.*

Proof. Let $\mathscr{B}_1$, $\mathscr{B}_2$, and $\mathscr{B}_3$ be the σ-algebras generated by the collections of sets in parts (a), (b), and (c) of the proposition. We shall show that $\mathscr{B}(\mathbb{R}) \supset \mathscr{B}_1 \supset \mathscr{B}_2 \supset \mathscr{B}_3$, and then that $\mathscr{B}_3 \supset \mathscr{B}(\mathbb{R})$; this will establish the proposition. Since $\mathscr{B}(\mathbb{R})$ includes the family of open subsets of $\mathbb{R}$ and is closed under complementation, it includes the family of closed subsets of $\mathbb{R}$: thus it includes the σ-algebra generated by the closed subsets of $\mathbb{R}$, namely $\mathscr{B}_1$. The sets of the form $(-\infty, b]$ are closed and so belong to $\mathscr{B}_1$, consequently $\mathscr{B}_2 \subset \mathscr{B}_1$. Since

$$(a, b] = (-\infty, b] \cap (-\infty, a]^c,$$

each set of the form $(a, b]$ belongs to $\mathscr{B}_2$; thus $\mathscr{B}_3 \subset \mathscr{B}_2$. Finally, note that each open subinterval of $\mathbb{R}$ is the union of a sequence of sets of the form $(a, b]$, and that each open subset of $\mathbb{R}$ is the union of a sequence of open intervals. Thus each open subset of $\mathbb{R}$ belongs to $\mathscr{B}_3$, and so $\mathscr{B}(\mathbb{R}) \subset \mathscr{B}_3$. □

1.8 Proposition *The σ-algebra $\mathscr{B}(\mathbb{R}^n)$ of Borel subsets of $\mathbb{R}^n$ is generated by each of the following collections of sets:*

(*a*) *the collection of all closed subsets of $\mathbb{R}^n$;*

(*b*) *the collection of all closed half-spaces in $\mathbb{R}^n$ that have the form $\{(x_1, \cdots, x_n) : x_i \le b\}$ for some index i and some b in $\mathbb{R}$;*

(*c*) *the collection of all rectangles in $\mathbb{R}^n$ that have the form $\{(x_1, \cdots, x_n) : a_i < x_i \le b_i \text{ for } i = 1, \cdots n\}$.*

Proof. This proposition can be proved with essentially the argument that was used for Proposition 1.7 and so most of the proof is omitted. To see that the σ-algebra generated by the rectangles of part (c) is included in the σ-algebra generated by the half-spaces of part (b), note that each strip that has the form

$$\{(x_1, \cdots, x_n) : a < x_i \leq b\}$$

for some i is the difference of two of the half-spaces in part (b), and that each of the rectangles in part (c) is the intersection of n such strips. □

1.9 $\mathfrak{F}_\sigma$ and $\mathfrak{G}_\delta$ Sets. *Although the intersection of any collection (countable or uncountable) of sets in $\mathbb{R}$ is closed and the union of a finite collection of closed sets is closed, the union of a countable (infinite) collection of closed sets need not be closed. Similarly, the intersection of a countable(infinite) collection of open sets need not be open (cf.I-5). Thus we are motivated to define two new classes of sets,* ***$\mathfrak{F}_\sigma$-sets and $\mathfrak{G}_\delta$-set*** *, different form those of open sets and closed sets, and consequently many others.*

1.10 Definition *A set which is a countable (finite or infinite) union of closed sets os called an* ***$\mathfrak{F}_\sigma$-set.***

Note : The class of all $\mathfrak{F}_\sigma$-sets is denoted by $\mathfrak{F}_\sigma$. In $\mathfrak{F}_\sigma$, $\mathfrak{F}$ stands for **ferme** (closed) and σ for **summe** (sum).

1.11 Example Each of the follwing is an $\mathfrak{F}_\sigma$-set:

(i) A closed set.
(ii) A countable set.
(iii) An open interval (a, b) since

$$(a, b) = \bigcup_{n=1}^{\infty} \left[a + \frac{1}{n}, b - \frac{1}{n}\right],$$

and hence an open set.

1.12 Definition *A set which is a countable intersection of open sets is a $\mathfrak{G}_\delta$-set.*

Note : The class of all $\mathfrak{G}_\delta$-sets is denoted by $\mathfrak{G}_\delta$. In $\mathfrak{G}_\delta$, $\mathfrak{G}$ stands for **gebiet** (region) and δ for **durchschnitt** (intersection).
Remark : The complement of an $\mathfrak{F}_\sigma$-set is a $\mathfrak{G}_\delta$-set, and conversely.

1.13 Example Each of the following is a $\mathfrak{G}_\delta$-set.
(i) An open set, and, in particular, an open interval.
(ii) A closed interval $[a, b]$ since

$$[a, b] = \bigcap_{n=1}^{\infty} \left(a - \frac{1}{n}, b + \frac{1}{n}\right).$$

(iii) A closed set.
(iv) A countable intersection of $\mathfrak{G}_\delta$-set.

Remark : Each of the classes $\mathfrak{F}_\sigma$ and $\mathfrak{G}_\delta$ of sets is wider than the classes of open and closed sets.

The $\mathfrak{F}_\sigma$ and $\mathfrak{G}_\delta$ are relatively simple types of Borel sets. We could also consider sets of type $\mathfrak{F}_{\sigma\delta}$, which are the intersection of countable collections of sets each of which is an $\mathfrak{F}_\sigma$. Similarly, we can construct the classes $\mathfrak{G}_{\delta\sigma}$, $\mathfrak{F}_{\sigma\delta\sigma}$, etc. Thus the classes in the two sequences

$$\mathfrak{F}_\sigma, \mathfrak{F}_{\sigma\delta}, \mathfrak{F}_{\sigma\delta\sigma}, \cdots, \mathfrak{G}_\delta, \mathfrak{G}_{\delta\sigma}, \mathfrak{G}_{\delta\sigma\delta}, \cdots$$

are all classes of Borel sets.

Thus we can formulate these properties.

1.14 Proposition
1) *Each closed subset of* $\mathbb{R}^d$ *is a* $\mathfrak{G}_\delta$, *and each open subset of* $\mathbb{R}^d$ *is an* $\mathfrak{F}_\sigma$.
2) *Every closed subset of* $\mathbb{R}^n$ *is an* $\mathfrak{F}_\sigma$ *set.*
3) *The complement of* $\mathfrak{F}_\sigma$ *set is a* $\mathfrak{G}_\delta$ *set, and vice versa.*
4) *A countable union of sets of type* $\mathfrak{F}_\sigma$ *is again an* $\mathfrak{F}_\sigma$ *set.*
5) *Sets of type* $\mathfrak{G}_\delta$ *set or* $\mathfrak{F}_\sigma$ *are Borel sets and*
$\mathfrak{G}_\delta, \mathfrak{G}_{\delta\sigma}, \mathfrak{G}_{\delta\sigma\delta}, \cdots, \mathfrak{F}_\sigma, \mathfrak{F}_{\sigma\delta}, \mathfrak{F}_{\sigma\delta\sigma}, \cdots$
are Borel sets.

Proof. 1) Suppose that $\mathfrak{F}$ is a closed subset of $\mathbb{R}$. We need to construct a sequence $\{U_n\}$ of open subsets of $\mathbb{R}$ such that $\mathfrak{F} = \cap_n U_n$. For this define U_n by

$$U_n = \{x \in \mathbb{R}^d : ||x - y|| < 1/n \text{ for some } y \text{ in } \mathfrak{F}\}.$$

(Note that U_n is empty if $\mathfrak{F}$ is empty) It is clear that each U_n is open and that $\mathfrak{F} \subset \cap_n U_n$. The reverse inclusion follows from the fact that $\mathfrak{F}$ is closed (note that each point in $\cap_n U_n$ is the limit of a sequence of points in $\mathfrak{F}$). Hence each closed subsets of $\mathbb{R}^d$ is a $\mathfrak{G}_\delta$.

If U is open, then U^c is closed, and so is a $\mathfrak{G}_\delta$. Thus there is a sequence $\{U_n\}$ of open sets such that $U^c = \cap_n U_n$. The sets U_n^c are then closed, and $U = \cup_n U_n^c$; hence U is an $\mathfrak{F}_\sigma$.

2) $\sim$ 5): The proofs are left to the reader. □

For an arbitrary family $\mathfrak{G}$ of sets let $\mathfrak{G}_\sigma$ be the collection of all unions of sequences of sets in $\mathfrak{G}$, and let $\mathfrak{G}_\delta$ be the collection all intersections of sequences of sets in $\mathfrak{G}$. We can iterate the operations represented by σ and δ, obtaining from the class $\mathfrak{G}$ the classes $\mathfrak{G}_\delta, \mathfrak{G}_{\delta\sigma}, \mathfrak{G}_{\delta\sigma\delta}, \cdots$ and from the class $\mathfrak{F}$ the classes $\mathfrak{F}_\sigma, \mathfrak{F}_{\sigma\delta}, \mathfrak{F}_{\sigma\delta\sigma}, \cdots$ (Note that $\mathfrak{G} = \mathfrak{G}_\sigma$ and $\mathfrak{F} = \mathfrak{F}_\delta$. Note also that $\mathfrak{G}_{\delta\delta} = \mathfrak{G}_\delta$, that $\mathfrak{F}_{\sigma\sigma} = \mathfrak{F}_\sigma$, and so on.) It now follows (see Proposition 1.14) that all the inclusions in dicated in the following diagram are valid.

$$\begin{array}{ccccccccc}
\mathfrak{G} & \subset & \mathfrak{G}_\delta & \subset & \mathfrak{G}_{\delta\sigma} & \subset & \mathfrak{G}_{\delta\sigma\delta} & \subset & \cdots \\
 & \times & & \times & & \times & & & \\
\mathfrak{F} & \subset & \mathfrak{F}_\sigma & \subset & \mathfrak{F}_{\sigma\delta} & \subset & \mathfrak{F}_{\sigma\delta\sigma} & \subset & \cdots
\end{array}$$

It turns out that no two of these classes of sets are equal, and that there are Borel sets that belong to none of them.

1.15 Theorem *Let $f : \mathbb{R} \to \mathbb{R}$ be a continuous function and $\beta = \{E \subset \mathbb{R} | f^{-1}(E) : \text{Borel set}\}$, then β is a σ-algebra. Furthermore, β contains Borel σ-algebra.*

Proof. It E_i belong to β for each i, then $f^{-1}(E_i)$ is a Borel set for each i. Since $f^{-1}[\cup E_i] = \cup f^{-1}(E_i)$, $f^{-1}[\cup E_i]$ is a Borel set.

It follows that $\cup E_i$ belongs to β. (1)

And since $f^{-1}(\sim E) = \sim f^{-1}(E)$, if E belongs to β then $f^{-1}(E)$ is Borel set and $\sim f^{-1}(E)$ is a Borel set. Hence $\sim E$ belongs to β. (2)

Thus, by (1),(2), β is a σ-algebra. And for each open set O, $f^{-1}(O)$ is open. It follows that $f^{-1}(O)$ is a Borel set and O belons to β. Thus β is a σ-algebra containing all open sets. Also, since Borel σ-algebra is the smallest σ-algebra containg all open sets, β contains Borel σ-algebra. □

PROBLEMS [3.1]

1. Find the σ-algebra on $\mathbb{R}$ that is generated by the collection of all one-point subsets of $\mathbb{R}$.

2. Show that $\mathcal{B}(\mathbb{R})$ is generated by the collection of intervals $(-\infty, t]$ for which the end-point t is a rational number.

3. Show that $\mathcal{B}(\mathbb{R})$ is generated by the collection of all compact subsets of $\mathbb{R}$.

4. Show that if $\mathcal{A}$ is an algebra of sets, and if $\cup_n A_n$ belongs to $\mathcal{A}$ whenever $\{A_n\}$ is a sequence of disjoint sets in $\mathcal{A}$, then $\mathcal{A}$ is a σ-algebra.

5. Show by example that the union of a collection of σ-algebra on a set X can fail to be a σ-algebra on X.

   [Hint: There are examples in which X is a small finite set.]

6. Find an infinite collection of subsets of $\mathbb{R}$ that contains $\mathbb{R}$, is closed under the formation of countable unions, and is closed under the formation of countable intersections, but is not a σ-algebra.

7. Let $\mathcal{J}$ be a collection of subsets of the set X. Show that for each A in $\sigma(\mathcal{J})$ there is a countable subfamily $\mathcal{J}_0$ of $\mathcal{J}$ such that $A \in \sigma(\mathcal{J}_0)$.

[Hint: Let $\mathcal{A}$ be the union of the σ-algebras $\sigma(\mathcal{J})$, where $\mathcal{J}$ ranges over the countable subfamilies of $\mathfrak{I}$, and show that $\mathcal{A}$ is a σ-algebra that satisfies $\mathfrak{I} \subset \mathcal{A} \subset \sigma(\mathfrak{I})$]

8. Find all σ-algebras on $\mathbb{N}$.

9. (a) Show that $\mathbb{Q}$ is an F_σ, but not a G_δ, in $\mathbb{R}$.
   [Hint: Use the Baire category theorem]
   (b) Find a subsets of $\mathbb{R}$ that is neither an F_σ nor a G_δ.

3.2 Lebesgue Outer Measure

The **length** of an interval I, written $l(I)$, is defined to be the difference of the endpoints of the interval I. i.e., for an interval I with a anb b as its endpoints the length $l(I)$ is $b - a$, where $a < b$. Thus, length is an example of a **set function**, i.e., a function which associates an extended real number to each set in some collection of sets. In the case of length, the domain is the collection of all intervals. The set function l clearly satisfies the following properties.

2.1 Proposition *Let l be the length of interval, then*

(1) $l(I) \geq 0$, *for all interval I.*

(2) *For two interval I, J, if $I \subset J$ then $l(I) \leq l(J)$.*

(3) *If x is any fixed real number, then*

$$l(I) = l(I + x).$$

(4) *For all interval I, $l(I) = l(-I)$ where $-I = \{x | -x \in I\}$.*

(5) *If $\{I_i\}$ is a countable collection of intervals*

$$l\left(\bigcup_{i=1}^{\infty} I_i\right) \leq \sum_{i=1}^{\infty} l(I_i).$$

in particular, $\{I_i\}$ are mutually disjoint intervals,

$$l\left(\bigcup_{i=1}^{\infty} I_i\right) = \sum_{i=1}^{\infty} l(I_i).$$

Proof. It is clear by definition, so left to the reader. □

We would now like to extend the notion of length to more complicated and arbitrary sets than intervals. We should like to extend the notion of length to more complicated sets than intervals.

Let O be an open set in $\mathbb{R}$. Then O can be written as a countable union of mutually disjoint open intervals $\{I_i\}$, unique except so far as order is concerned; i.e.,

$$O = \bigcup_{i=1}^{\infty} I_i$$

The **length of the open set** O is defined by

$$l(O) = \sum_{i=1}^{\infty} l(I_i).$$

The length $l(O)$ is well defined since the sum on the right does not depend on the order of the terms used in the summing process. Thus, the length of an open set is the sum of the lengths of the intervals (of course open and mutually disjoint) comprising O.

So far, we have extended the concept of length to open and closed sets. And since the classes of these sets are too restricted, we would like to extend the concept of length to a wider classs of sets in $\mathbb{R}$, if possible, to the class of all sets in $\mathbb{R}$. In this regard, we imagine a function m which assigns to each set E in $\mathbb{R}$, a nonnegative extended real number, written $m(E)$, called the *measure* of E (an extension of the notion of length function), satisfying the following properties:

1. $m(E)$ is defined for all sets $E \in \wp(\mathbb{R})$.
2. $m(E) \geq 0$ for all sets E.

3. $m(I) = l(I)$, for an interval I.

4. m is **countable additivity**; i.e., If $\{E_i\}$ is a sequence of disjoint sets, then

$$m\left(\bigcup_{i=1}^{\infty} E_i\right) = \sum_{i=1}^{\infty} m(E_i)$$

5. m is **monotonicity**; i.e. $A \subset B \Rightarrow m(A) \leq m(B)$

6. m is **translation invariant**; that is, $m(E+y) = m(E)$, where y is any fixed number.

Unfortunately, it is impossible to construct a set function having all four properties (1) to (4). In fact, one cannot construct such a measure satisfying the properties (1) to (4). As a result one of these four properties must be sacrificed or weakend at least. Following Henri Lebesgue(1875-1941), who made many contributions to measure theory and integration, it is most useful to retain the last three properties, i.e., (2) to (4), and to weaken the condition, given in (1) so that $m(E)$ need not be defined for all sets E in $\mathbb{R}$. Still, of course, we shall be interested in defining $m(E)$ for as many sets as possible.

All the sets considered in this chapter are contained in $\mathbb{R}$, unless stated otherwise. We shall be concerned particularly with intervals I of the form (a, b) unless otherwise specified.

Let us consider the family $\mathfrak{F}$ of all countable collections of open intervals. For any arbitrary $\mathcal{J} \in \mathfrak{F}$, the sum $\sum_{I\in\mathcal{J}} l(I)$ is a nonnegative extended real number. Further, this sum depends only on $\mathcal{J}$ and not on the order used in the summing process.

Now, let E be an arbitrary set. Consider the subfamily $\mathcal{C}$ of $\mathfrak{F}$ consisting of countable collections $\mathcal{J}$ of open intervals $\{I_i\}$ such that $E \subset \cup_i I_i$;i.e.

$$\mathcal{C} = \{\mathcal{J} : \mathcal{J} \in \mathcal{F} \text{ and } \mathcal{J} \text{ cover } E\}.$$

The subfamily $\mathcal{C}$ is obviously nonempty. Thus we obtain a well defined number $m^*(E)$ in the set of all nonnegative extended real numbers given by

$$m^*(E) = \inf\left\{\sum_{I\in\mathcal{J}} l(I) : \mathcal{J} \in \mathcal{C}\right\}$$

For a more precise understanding, see Definition 2.1.

2.2 Definition *The* ***Lebesgue outer measure*** *or briefly the* ***outer measure*** $m^*(E)$ *of an arbitrary set* E *is given by*

$$m^*(E) = \inf \sum_i l(I_i),$$

where the infimum is taken over all cuntable collections $\{I_i\}$ *of open intervals such that* $E \subset \cup_i I_i$.

Remark : The outer measure m^* is a set function which is defined from the power set $\wp(\mathbb{R})$ into the set of all non-negaive extended real numbers.

2.3 Theorem

(*a*) $m^*(A) \geq 0$, *for all sets* A.
(*b*) $m^*(\emptyset) = 0$.
(*c*) *If* A *and* B *are two sets with* $A \subset B$, *then* $m^*(A) \leq m^*(B)$. (*This property is known as* ***monotonicity***.)
(*d*) $m^*(A) = 0$, *for every singleton set* A.
(*e*) *The function* m^* *is* ***translation invariant***, *i.e.*, $m^*(A + x) = m^*(A)$, *for every set* A *and for every* $x \in \mathbb{R}$.
(*f*) *Let* $\{E_n\}$ *be a countable collection of sets. Then* $m^*(\bigcup_n E_n) \leq \sum_n m^*(E_n)$. (***countable subadditivity***.)

Proof. The proofs of (a) and (b) are obvious.

(c) Let $\{I_n\}$ be a countable collection of disjoint open intervals such that $B \subset \cup_n I_n$. Then $A \subset \cup_n I_n$ and therefore

$$m^*(A) \leq \sum_{n=1}^{\infty} l(I_n).$$

This inequality is true for any coverings $\{I_n\}$ of B. Hence the result follows.

(d) Let $A = \{x\}$ be an arbitrary singleton set. Since

$$I_n = \left(x - \frac{1}{n}, x + \frac{1}{n}\right)$$

is an open covering of A for each $n \in \mathbb{N}$, and $l(I_n) = \frac{2}{n}$, the result follows in view of (a).

(e) Given any interval I with end points a and b, the set $I + x$ defined by

$$I + x = \{y + x : y \in I\}$$

is clearly an interval with endpoints $a + x$ and $b + x$. Moreover,

$$l(I + x) = l(I).$$

Now, let $\varepsilon > 0$ be given. Then there is a countable collection $\{I_n\}$ of open intervals such that $A \subset \cup_n(I_n)$ and satisfies

$$\sum_{i=1}^{\infty} l(I_n) < m^*(A) + \varepsilon.$$

Clearly $A + x \subset \cup(I_n + x)$. Therefore

$$m^*(A + x) \leq \sum_{n=1}^{\infty} l(I_n + x) = \sum_{i=1}^{\infty} l(I_n) < m^*(A) + \varepsilon$$

Since $\varepsilon > 0$ is arbitrary, we have $m^*(A+x) \leq m^*(A)$. The reverse inequality follows by considering $A = (A + x) - x$ and using the above.

(f) If $m^*(E_n) = \infty$ for some $n \in \mathbb{N}$, the inequality holds trivially. Let us assume that $m^*(E_n) < \infty$, for each $n \in \mathbb{N}$. Then, for each n, and for a given $\varepsilon > 0$, $\exists$ a countable collection $\{I_{n,i}\}_i$ of open intervals such that $E_n \subset \cup I_{n,i}$ satisfying

$$\sum_i l(I_{n,i}) < m^*(E_n) + 2^{-n}\varepsilon$$

Then

$$\bigcup_n E \subset \bigcup_n \bigcup_i I_{n,i}.$$

However, the collection $\{I_{n,i}\}_{n,i}$ forms a countable collection of open intervals, as the countable union of countable sets is countable and covers $\cup E_n$. Therefore

$$\begin{aligned} m^*\left(\bigcup_n E_n\right) &\leq \sum_{n,i} l(I_{n,i}) \\ &< \sum_n (m^*(E_n) + 2^{-n}\varepsilon) \\ &= \sum_n m^*(E_n) + \varepsilon. \end{aligned}$$

But $\varepsilon > 0$ being arbitrary, the result follows. □

2.4 Corollary *If E is a countable set, then $m^*(E) = 0$.*

Proof. Since the set E is countable, we may express it as

$$E = \{a_1, a_2, \cdots, a_n, \cdots\}.$$

Let $\varepsilon > 0$ be given. Then enclosing each a_i in an open interval I_i with $l(I_i) = 2^{-i}\varepsilon$, $(i = 1, 2, \cdots)$ we get

$$m^*(E) \leq \sum_{i=1}^{\infty} l(I_i) = \sum_{i=1}^{\infty} 2^{-i}\varepsilon = \varepsilon.$$

Letting $\varepsilon \to 0$, the result follows. □

Note : Each of the sets $\mathbb{N}$, $\mathbb{I}$, $\mathbb{Q}$ and A (the set of all algebraic numbers) has outer measure zero since each one of these is countable.

The converse of corollary 2.4, i.e., "a set with outer measure zero is countable" is not always ture.

Example. The Cantor set $\mathbf{C}$ is uncountable with outer measure zero.

Let $\mathbf{C}_n$ denote the union of the intervals left at the nth stage while constructing the Cantor set $\mathbf{C}$. One may note that $\mathbf{C}_n$ consists of 2^n closed intervals, each of length 3^{-n}. Therefore

$$m^*(\mathbf{C}_n) \leq 2^n \cdot 3^{-n},$$

by Theorems 2.3(f) and 2.8. But any point of $\mathbf{C}$ must be in one of the intervals comprising the union $\mathbf{C}_n$, for each $n \in \mathbb{N}$, and as such $\mathbf{C} \subset \mathbf{C}_n$, for all $n \in \mathbb{N}$. Hence

$$m^*(\mathbf{C}) \leq \left(\frac{2}{3}\right)^n.$$

This being true for each $n \in \mathbb{N}$, letting $n \to \infty$ gives $m^*(\mathbf{C}) = 0$.

Problem 1 If $m^*(A) = 0$, then $m^*(A \cup B) = m^*(B)$. In particular, if $B \subset A$, then $m^*(B) = 0$.

Solution. By Theorem 2.5, we have

$$m^*(A \cup B) \leq m^*(A) + m^*(B) = m^*(B).$$

But $B \subset A \cup B$ verifies (cf. Theorem 2.3(c))

$$m^*(B) \leq m^*(A \cup B).$$

Hence the result follows.

As a negation of Corollary 4, we have the following.

2.5 Corollary *Any set with the outer measure different from zero is uncountable.*

2.6 Corollary *The set* $[0, 1]$ *is uncountable.*

2.7 Corollary *Every interval is uncountable.*

To answer affirmatively the question whether m^* is a generalization of the length function defined for the intervals, we prove the following theorem.

2.8 Theorem *The outer measure of an interval is its length.*

Proof. *Case* 1: Suppose I is a closed finite interval, (say) [a,b]. Since, for each $\varepsilon > 0$, the open interval $\left(a - \frac{\varepsilon}{2}, b + \frac{\varepsilon}{2}\right)$ contains $[a, b]$, we have

$$m^*(I) \leq l\left(\left(a - \frac{\varepsilon}{2}, b + \frac{\varepsilon}{2}\right)\right) = b - a + \varepsilon.$$

This being true for each $\varepsilon > 0$, we must have

$$m^*(I) \leq b - a - l(I)$$

To complete the proof of the result, we need to show that

$$(1) \qquad m^*(I) \geq b - a$$

Let $\varepsilon > 0$ be given. Then there exists a countable collection $\{I_n\}$ of open intervals convering $[a, b]$ such that

$$(2) \qquad m^*(I) > \sum_n l(I_n) - \varepsilon$$

By the Heine-Borel Theorem (cf.II-5.14), any collection of open intervals covering $[a, b]$ contains a finite subcollection which also covers $[a, b]$, and since the sum of the lengths of the finite subcollection is not greater than the sum of the lengths of the original collection, it suffices to establish the inequality (2) for finite collection $\{I_n\}$ which cover $[a, b]$.

Since $a \in [a, b]$ implies that $a \in \cup_n I_n$, there must be one of the intervals I_n which contains a. Let it be (a_1, b_1) Then $a_1 < a < b_1$. If $b_1 \leq b$ then $b_1 \in [a, b]$, and since $b_1 \notin (a_1, b_1)$ there must be an interval (a_2, b_2) in the finite collection $\{I_n\}$ such that $b_1 \in (a_2, b_3)$; that is $a_2 < b_1 < b_2$. Continuing in this manner, we get intervals $(a_1, b_1), (a_2, b_2), \cdots$ from the collection $\{I_n\}$ such that

$$a_i < b_{i-1} < b_i, \quad i = 1, 2, \cdots$$

where $b_0 = a$. Since $\{I_n\}$ is a finite collection, this process must terminate with some interval (a_k, b_k) in the collection which is possible only when $b \in (a_k, b_k)$. Thus

$$\begin{aligned}
\sum_n l(I_n) &\geq \sum_{i=1}^{k} l((a_i, b_i)) \\
&= (b_k - a_k) + (b_{k-1} - a_{k-1}) + \cdots + (b_1 - a_1) \\
&= b_k - (a_k - b_{k-1}) - \cdots - (a_2 - b_1) - a_1 \\
&> b_k - a_1 \\
&> b - a,
\end{aligned}$$

since $a_i - b_{i-1} < 0$, $b_k > b$ and $a_1 < a$. This, in view of (2), verifies that

$$m^*(I) > b - a - \varepsilon.$$

Hence $m^*(I) \geq b - a$.

Case 2: Suppose I is any finite interval. Then given an $\varepsilon > 0$, there exists a closed finite interval $J \subset I$ such that

$$l(J) > l(I) - \varepsilon.$$

Therefore.

$$\begin{aligned}
& l(I) - \varepsilon < l(J) = m^*(J) \leq m^*(I) \leq m^*(\overline{I}) = l(\overline{I}) = l(I) \\
\Rightarrow \quad & l(I) - \varepsilon < m^*(I) \leq l(I).
\end{aligned}$$

This is true for each $\varepsilon > 0$. Hence $m^*(I) = l(I)$.

Case 3: Suppose I is an infinite interval. Then given any real number $K > 0$, there exists a closed finite interval $J \subset I$ such that $l(J) = K$. Thus $m^*(I) \geq m^*(J) = l(J) = K$, that is $m^*(I) \geq K$ for any arbitrary real number $K > 0$. Hence $m^*(I) = \infty = l(I)$. □

We now turn to a result which gives useful relations between the outer measure of an arbitrary set and the outer measure of open sets and $\mathfrak{G}_\delta$-sets which contain the set.

2.9 Theorem *Let E be any set. Then:*

(a) Given $\varepsilon > 0$, $\exists$ an open set $O \supset E$ such that

$$m^*(O) < m^*(E) + \varepsilon$$

and hence $m^*(E) = \inf m^*(O)$, *where the infimum is taken over all open sets* $O \supset E$.

(b) $\exists$ *a* $\mathfrak{G}_\delta$*-set* $G \supset E$ *such that*

$$m^*(E) = m^*(G).$$

Proof. (a) There exists a countable collection $\{I_n\}$ of open intervals such that $E \subset \cup_n I_n$ and

$$\sum_{n=1}^{\infty} l(I_n) < m^*(E) + \varepsilon.$$

Set $O = \cup_{n=1}^{\infty} I_n$. Clearly O is an open set and

$$\begin{aligned} m^*(O) &= m^*\left(\bigcup_{n=1}^{\infty} I_n\right) \\ &\leq \sum_{n=1}^{\infty} m^*(I_n) = \sum_{i=1}^{\infty} l(I_n) < m^*(E) + \varepsilon. \end{aligned}$$

(b) Choose $\varepsilon = \frac{1}{n}$, $n \in \mathbb{N}$ in (a). Then, for each $n \in \mathbb{N}$, $\exists$ an open set $O_n \supset E$ such that

$$m^*(O_n) < m^*(E) + \frac{1}{n}.$$

Define $G = \cap_{n=1}^{\infty} O_n$. Clearly, G is a $\mathfrak{G}_\delta$-set and $G \supset E$. Moreover, we observe that

$$m^*(E) \leq m^*(G) \leq m^*(O_n) < m^*(E) + \frac{1}{n}, \qquad n \in \mathbb{N}.$$

Hence letting $n \to \infty$ verifies $m^*(G) = m^*(E)$. □

Remark. The significance of Theorem 2.9(b) is that any arbitrary set E can be included in a set which is of relatively simple type, namely a $\mathfrak{G}_\delta$-set, with the same outer measure.

2.10 Example The Cantor set $\mathbf{C}$ is uncountable with outer measure zero and Non-Borel measurable set.

PROBLEMS [3.2]

1. Let A be the set of rational numbers between 0 and 1, and $\{I_n\}$ be a finite collection of open intervals covering A. Prove that $\sum l(I_n) \geq 1$.

2. Prove that the outer measure of the generalized Cantor set $\mathbf{C}(\alpha)$, $0 < \alpha < 1$, is $I - \alpha$.

3. Prove the if $m^*(A) = 0$. Then $m^*(A \cup B) = m^*(B)$. In particular, if $B \subset A$, then $m^*(B) = 0$

4. Let l be the length of interval. Find $l\left(\cup_{n=1}^{\infty}\left[\frac{1}{n+1}, \frac{1}{n}\right]\right)$.

5. Let $E, F \subset \mathbb{R}$ such that $E \subset F$. Show that if $m^*(E) < \infty$, then $m^*(F \sim E) \leq m^*(F) - m^*(E)$.

3.3 Measurable Sets and Lebesgue Measure

An outer measure on X is a *monotone and countably subaddilive* finction from $\wp(X)$ to $[0, +\infty]$ whose value at $\emptyset$ is 0.

Note that a measure can fail to be an outer measure; in fact, a measure on X is an outer measure if and only if its domain is $\wp(X)$. That is, outer measuer has the advantage that it is defined for all sets, it is not countably additive. If we suitably reduce the family of sets on which it is defined, it becomes countably additive. Thus an outer measure generally fails to be countably additive, and so fails to be a measure. If we insist on countable additive, we have to restrict the domain of the function m^* to some subset $\mathfrak{M}$ of the set $\wp(\mathbb{R})$ of all subsets $\mathbb{R}$. The members of $\mathfrak{M}$ are called the **(Lebesgue) measurable subset** of $\mathbb{R}$.

Following Carathéodory, we define the measurable subsets of $\mathbb{R}$ as follows.

3.1 Definition *A set E is said to be **Lebesgue measurable** or briefly **measurable** if for each set A, we have*

$$m^*(A) = m^*(A \cap E) + m^*(A \cap E^c).$$

Since $A = (A \cap E) \cup (A \cap E^c)$ and m^* is subadditive, we always have

$$m^*(A) \leq m^*(A \cap E) + m^*(A \cap E^c)$$

Thus, in order to establish that E is measurable, we need only to show, for any set A, that

$$m^*(A) \geq m^*(A \cap E) + m^*(A \cap E^c)$$

Hence we have the following lemma.

3.2 Lemma *A subset E of $\mathbb{R}$ is measurable iff, for every subset A of $\mathbb{R}$, we always have*

$$m^*(A) \geq m^*(A \cap E) + m^*(A \cap E^c)$$

3.3 Definition *The set function $m : \mathfrak{M} \to \mathbb{R}^*$, obtained by restricting the set functions m^* to the subset $\mathfrak{M}$ of the domain of definition $\wp(\mathbb{R})$ of m^*; that is, $m = m^*|_{\mathfrak{M}}$, is called* ***Lebesgue measure*** *for the sets in $\mathfrak{M}$. And $(X, \mathfrak{M})$ is called* ***Lebesgue measurable space****, also $(X, \mathfrak{M}, m)$ is called* ***Lebesgue measure space****. The set in $\mathfrak{M}$ is called the* ***Lebesgue measurable set****.*

For each $E \in \mathfrak{M}$, $m(E) = m^*(E)$. The extended real number $m(E)$ is called the **Lebesgue measure** or simply **measure** of the set E. Thus m is the set function obtained by restricting the set function m^* to the family $\mathfrak{M}$ of measurable sets.

We shall now show, in the following sections, that the measure m satisfies all the properties as desired in 2.1; namely, m is countably additive (Theorem 3.20); m is invariant under translation, m is a generalization of the "length" and many others.

< Properties of Measurable Sets >

3.4 Theorem

(*a*) *If E is a measurable set, then so is E^c.*

(*b*) *The sets $\emptyset$ and $\mathbb{R}$ are measurable sets.*

Proof. The proof is evident from Definetion 3.1. □

3.5 Theorem *If E has the outer measure zero, then E is a measurable set. Furthermore, every subset of E is measurable.*

Proof. Let A be any set. Then

$$m^*(A) \geq m^*(A \cap E) + m^*(A \cap E^c)$$

since

$$A \cap E \subset E \Rightarrow m^*(A \cap E) \leq m^*(E) = 0$$

and

$$A \cap E^c \subset A \Rightarrow m^*(A \cap E^c) \leq m^*(A).$$

Hence E is measurable. The other part is clear. □

3.6 Corollary *Every countable set is measurable and its measure is zero.*

Proof. It follows from Corollary 2.4 and Theorem 3.5 □

3.7 Corollary *The Cantor set* **C** *and all its subsets are measurable, and each of them has measure zero.*

Proof. It follows from Example in sec. 3.2 and Theorem 3.5 □

3.8 Theorem *If E_1 and E_2 are measurable sets, then so is $E_1 \cup E_2$.*

Proof. Using the fact that E_1 and E_2 are measurable sets, for any set A, we have

$$\begin{aligned} m^*(A) &= m^*(A \cap E_1) + m^*(A \cap E_1^c) \\ &= m^*(A \cap E_1) + m^*([A \cap E_1^c] \cap E_2) + m^*([A \cap E_1^c] \cap E_2^c) \\ &= m^*(A \cap E_1) + m^*([A \cap E_2] \cap E_1^c) + m^*(A \cap E_1^c \cap E_2^c) \\ &= m^*(A \cap E_1) + m^*(A \cap E_2 \cap E_1^c) + m^*(A \cap [E_1 \cup E_2]^c) \\ &\geq m^*(A \cap [E_1 \cup E_2]) + m^*(A \cap [E_1 \cup E_2]^c), \end{aligned}$$

since $A \cap (E_1 \cup E_2) = [A \cap E_1] \cup [A \cap E_2 \cap E_1^c]$. Hence $E_1 \cup E_2$ is a measurable set. □

3.9 Proposition *If two subsets A and B of $\mathbb{R}$ are measurable, so is $A \cap B$.*

Proof. Let X denote any subset of $\mathbb{R}$. Since B is measurable, we have

$$m^*(X \cap A) = m^*(X \cap A \cap B) + m^*[(X \cap A) \sim B].$$

Next, since $X \sim (A \cap B) = (X \sim A) \cup [(X \cap A) \sim B]$, it follows from 2.2 that

$$m^*[X \sim (A \cap B)] \leq m^*(X \sim A) + m^*[(X \cap A) \sim B].$$

Consequently, we obtain

$$\begin{aligned} & m^*[X \cap (A \cap B)] + m^*[X \sim (A \cap B)] \\ \leq \;& m^*(X \cap A \cap B) + m^*(X \sim A) + m^*[(X \cap A) \sim B] \\ = \;& m^*(X \cap A) + m^*(X \sim A) \\ = \;& m^*(X) \end{aligned}$$

because A is measurable. According to 3.2, this proves that $A \cap B$ is measurable. □

As an immediate consequence of Theorem 3.8 and 3.9, we have the following result.

3.10 Corollary *The family $\mathfrak{M}$ of all measurable sets (subsets of $\mathbb{R}$) is an algebra of sets in $\wp(\mathbb{R})$. In particylar, if $\{E_1, E_2, \cdots, E_n\}$ is any finite collection of measurable sets, then so are $\cup_{i=1}^n E_i$ and $\cap_{i=1}^n E_i$.*

3.11 Theorem *Let $E_1, E_2, \cdots, E_n$ be a finite sequence of disjoint measurable sets. Then, for any set A,*

$$m^*\left(A \cap \left[\bigcup_{i=1}^n E_i\right]\right) = \sum_{i=1}^n m^*(A \cap E_i).$$

Proof. We prove it by induction on n. For $n = 1$, the result is trivially true. Assuming it to be true for $(n-1)$ sets, we would have

$$m^*\left(A \cap \left[\bigcup_{i=1}^{n-1} E_i\right]\right) = \sum_{i=1}^{n-1} m^*(A \cap E_i).$$

Adding $m^*(A \cap E_n)$ on both the sides, we get

$$m^*\left(A \cap \left[\bigcup_{i=1}^{n-1} E_i\right]\right) + m^*(A \cap E_n) = \sum_{i=1}^{n} m^*(A \cap E_i)$$

$$\Rightarrow \quad m^*\left(A \cap \left[\bigcup_{i=1}^{n} E_i\right] \cap E_n^c\right) + \left(A \cap \left[\bigcup_{i=1}^{n} E_i\right] \cap E_n\right) = \sum_{i=1}^{n} m^*(A \cap E_i),$$

since the sets E_i $(i = 1, 2, \cdots, n)$ are disjoint. But the measurability of the set E_n, by taking $A \cap [\cup_{i=1}^{n} E_i]$ as a test set, implies

$$m^*\left(A \cap \left[\bigcup_{i=1}^{n} E_i\right]\right) = m^*\left(A \cap \left[\bigcup_{i=1}^{n} E_i\right] \cap E_n\right) + m^*\left(A \cap \left[\bigcup_{i=1}^{n} E_i\right] \cap E_n^c\right)$$

Hence

$$m^*\left(A \cap \left[\bigcup_{i=1}^{n} E_i\right]\right) = \sum_{i=1}^{n} m^*(A \cap E_i).$$

□

3.12 Corollary *If $E_1, E_2, \cdots, E_n$ is a finite sequence of disjoint measurable sets, then*

$$m^*\left(\bigcup_{i=1}^{n} E_i\right) = \sum_{i=1}^{n} m^*(E_i)$$

Proof. Taking $A = \mathbb{R}$ in Theorem 3.11 the result follows in view of Corollary 3.10 □

The result in Corollary 12 verifies that the measure function m on $\mathfrak{M}$ is finitely additive. However, we would extend it to a more general situation to the case of countably additive. But before we do so we prove that the family $\mathfrak{M}$ is closed under countable union.

3.13 Theorem *A countable union of measurable sets is a measurable set.*

The proof of Theorem 3.13 requires a result on algebra of sets which we state and prove below in the form of a lemma.

3.14 Lemma *Let $\mathcal{A}$ be an algebra of subsets of a set S. If $\{A_i\}$ is a sequence of sets in $\mathcal{A}$, then there exists s sequence $\{B_i\}$ of mutually disjoint sets in $\mathcal{A}$ such that*

$$\bigcup_{i=1}^{\infty} B_i = \bigcup_{i=1}^{\infty} A_i.$$

Proof. If the sequence $\{A_i\}$ is finite, the result follows trivially. Let us assume $\{A_i\}$ to be an infinite sequence. Set $B_1 = A_1$ and for each $n \in \mathbb{N}$, define

$$\begin{aligned} B_n &= A_n - \left[\bigcup_{i=1}^{n-1} A_i\right] \\ &= A_n \cap A_1^c \cap A_2^c \cap \cdots \cap A_{n-1}^c \end{aligned}$$

Now, we note the following facts:

(i) $B_n \in \mathcal{A}$, for each $n \in \mathbb{N}$, since $\mathcal{A}$ is closed under the complementation and finite intersection of sets in $\mathcal{A}$

(ii) $B_n \subset A_n$, for each $n \in \mathbb{N}$.

(iii) The sets B_n are mutually disjoint. Consider B_n and B_m to be two such sets and suppose $m < n$. Then, because $B_m \subset A_m$, we have

$$\begin{aligned} B_m \cap B_n &\subset A_m \cap B_n \\ &= A_m \cap [A_n \cap A_1^c \cap \cdots \cap A_m^c \cap \cdots \cap A_{n-1}^c] \\ &= [A_m \cap A_m^c] \cap \cdots \\ &= \emptyset \cap \cdots \\ &= \emptyset \end{aligned}$$

(iv) $\cup_{i=1}^{\infty} B_i = \cup_{i=1}^{\infty} A_i$.

Since $B_i \subset A_i$, for each $i \in \mathbb{N}$, we have

$$\cup_{i=1}^{\infty} B_i \subset \cup_{i=1}^{\infty} A_i$$

To prove the reverse inclusion, let $x \in \cup_{i=1}^{\infty} A_i$. Clearly, x must be in at least one of the sets A_i's. Let n be the least value of i such that $x \in A_i$. Then $x \in B_n$, and so $x \in \cup_{n=1}^{\infty} B_n$. Hence

$$\cup_{i=1}^{\infty} A_i \subset \cup_{i=1}^{\infty} B_i$$

This completes the proof of the lemma. □

Proof of Theorem 3.13 : Let $\{E_i\}$ be a sequence of measurable sets and let $E = \cup_{i=1}^{\infty} E_i$. To prove E to be a measurable set, by lemma 3.14, we may assume, without any loss of generality, that the sets E_i are mutually disjoint.

For each $n \in \mathbb{N}$, define $F_n = \cup_{i=1}^{n} E_i$. Since $\mathfrak{M}$ is an algebra of sets and $E_1, E_2, \cdots, E_n$ are in $\mathfrak{M}$, the sets F_n are measurable. Therefore, for any set A, we have

$$\begin{aligned} m^*(A) &= m^*(A \cap F_n) + m^*(A \cap F_n^c) \\ &\geq m^*(A \cap F_n) + m^*(A \cap E^c), \end{aligned}$$

since

$$F_n^c = \left[\bigcup_{i=n+1}^{\infty} E_i \right] \cup E^c \supset E^c.$$

But, by Theorem 3.11, we observe that

$$m^*(A \cap F_n) = \sum_{i=1}^{n} m^*(A \cap E_i)$$

Threrfore,

$$m^*(A) \geq \sum_{i=1}^{n} m^*(A \cap E_i) + m^*(A \cap E^c).$$

This inequality holds for every $n \in \mathbb{N}$ and since the left-hand side is independent of n, letting $n \to \infty$, we obtain

$$\begin{aligned} m^*(A) &\geq \sum_{i=1}^{\infty} m^*(A \cap E_i) + m^*(A \cap E^c) \\ &\geq m^*(A \cap E) + m^*(A \cap E^c), \end{aligned}$$

in view of the countable subadditivity of m^*, cf. Theorem 2.3. Hence E is a measurable set. □

As an outcome of Theorem 3.13 in view of Corollary 3.10, we have the following result.

3.15 Theorem *The collection $\mathfrak{M}$ of measurable sets is a σ-algebra; that is, the complement of a measurable set is measurable and the union (and intersection) of a countable collection of measurable sets is measurable.*

3.16 Lemma *The interval (a, ∞) is measurable.*

Proof. Let A be any set, $A_1 = A \cap (a, \infty)$, $A_2 = A \cap (-\infty, a]$. Then we must show $m^*A_1 + m^*A_2 \leq m^*A$. If $m^*A = \infty$, then there is nothing to prove. If $m^*A < \infty$, then, given $\varepsilon > 0$ there is a countable collection $\{I_n\}$ of open intervals which cover A and for which

$$\sum l(I) \leq m^*A + \varepsilon.$$

Let $I'_n = I_n \cap (a, \infty)$ and $I''_n = I_n \cap (-\infty, a]$. Then I'_n and I''_n are intervals (or empty) and

$$l(I_n) = l(I'_n) + l(I''_n) = m^*I'_n + m^*I''_n.$$

Since $A_1 \subset \cup I'_n$, we have

$$m^*A_1 \leq m^*(\cup I'_n) \leq \sum m^*I'_n,$$

and, since $A_2 \subset \cup I''_n$, we have

$$m^*A_2 \leq m^*(\cup I''_n) \leq \sum m^*I''_n.$$

Thus

$$\begin{aligned} m^*A_1 + m^*A_2 &\leq \sum (m^*I'_n + m^*I''_n) \\ &\leq \sum l(I_n) \leq m^*A + \varepsilon. \end{aligned}$$

But ε was an arbitrary positive number, and so we must have $m^*A_1 + m^*A_2 \leq m^*A$. □

3.17 Theorem *Every Borel set in $\mathbb{R}$ is measurable; that is, $\mathscr{B} \subset \mathfrak{M}$.*

Proof. We prove the theorem in several setps by using the fact that $\mathfrak{M}$ is a σ-algebra.

Step 1: The interval (a, ∞) is measurable by Lemma 3.16

Step 2: The interval $(-\infty, a]$ is measurable since $(-\infty, a] = (a, \infty)^c$

Step 3: The interval $(-\infty, b)$ is measurable since it can be expressed as a countable union of the interval of the form as in Step 2; that is,

$$(-\infty, b) = \bigcup_{n=1}^{\infty} \left(-\infty, b - \frac{1}{n}\right].$$

Step 4: Since any open interval (a, b) can be expressed as

$$(a, b) = (-\infty, b) \cap (a, \infty),$$

it is measurable.

Step 5: Every open set is measurable. It is so because it can be expressed as a countable union of open intervals (disjoint).

Hence, in view of Step 5, the σ-algebra $\mathfrak{M}$ contains all the open sets in $\mathbb{R}$. Since $\mathscr{B}$ is the smallest σ-algebra containing all the open sets, we conclude that $\mathscr{B} \subset \mathfrak{M}$. This completes the proof of the theorem. □

Remark : The Borel sets do not exhaust all measurable sets; that is, the inclusion $\mathscr{B} \subset \mathfrak{M}$ is proper. This assertion follows form the fact that the cardinality of $\mathfrak{M}$ is 2^c (Theorem 3.30) whereas the cardinality* of $\mathscr{B}$ is c. However, there exist examples of sets which are measurable sets but not Borel sets.

3.18 Proposition *Every subset E of $\mathbb{R}$ with $m^*(E) = 0$ is measurable*

Proof. Let X denote any subset of $\mathbb{R}$. Since $X \cap E \subset E$, it follows from 2.3(c) that

$$0 \leq m^*(X \cap E) \leq m^*(E) = 0$$

and hence $m^*(X \cap E) = 0$. On the other hand, since $X \supset X \sim E$, we have $m^*(X) \geq m^*(X \sim E) = m^*(X \cap E) + m^*(X \sim E)$

According to 3.2, this proves that E is measurable. □

3.19 Corollary *Each of the sets in $\mathbb{R}$: an open set, a closed set, an $\mathfrak{F}_\sigma$-sets and $\mathfrak{G}_\delta$-set is measurable.*

Some important properties of Lebesgue measure are summarized by the following propositions:

3.20 Theorem *The function $m^* : \wp(\mathbb{R}) \to R^*$ is countably additive on the subset $\mathfrak{M}$ of $\wp(\mathbb{R})$, that is, for every countable family $\mathfrak{F}$ of disjoint measurable subsets of $\mathbb{R}$, we have*

$$m^*\left(\bigcup_{E\in\mathfrak{F}} E\right) = \sum_{E\in\mathfrak{F}} m^*(E).$$

3.21 Theorem *Let E_1 and E_2 be measurable sets with $E_2 \subset E_1$ If $m^*(E_2) < \infty$. Then $m^*(E_1 - E_2) = m^*(E_1) - m^*(E_2)$.*

Proof. Since E_2 is measurable, E_2^c is measurable. And so if E_1 and E_2 are measurable, then by the above theorem $E_1 \cap E_2^c = E_1 \sim E_2$ is measurable. Also since $E_1 = (E_1 \sim E_2) \cup E_2$ and $(E_1 \sim E_2) \cap E_2 = \emptyset$. By theorem 3.12. $m^*(E_1) = m^*(E_1 \sim E_2) + m^*(E_2)$. Since $m^*(E_2) < \infty$, it follows that $m^*(E_1 \sim E_2) = m^*(E_1) - m^*(E_2)$. □

Now we formulate the properties of Lebesgue measure by followings.

Let $(X, \mathfrak{M}, m)$ be a Lebesgue measure space. Then

(1) m is monotonicity; i.e.

$$A \subset B \Rightarrow m(A) \leq m(B)$$

(2) m is countably subadditivity; i.e.

$$m\left(\bigcup E_i\right) \leq \sum m(E_i)$$

(3) $\{E_n\}$ is infinite decreasing sequence

$$\Rightarrow \qquad m\left(\bigcap E_n\right) = \lim_{n\to\infty} m(E_n)$$

(4) $\{E_n\}$ is infinite increasing sequence

$$\Rightarrow \qquad m\left(\bigcup E_n\right) = \lim_{n\to\infty} m(E_n)$$

3.22 Proposition *Let $(X, \mathfrak{M}, m)$ be a measure space. Then m is monotonicity; i.e. If $A \subset B$ for all $A, B \in \mathfrak{M}$, then $m(A) \leq m(B)$.*

Proof. Since $A \subset B$, $m(B) = m(A) + m(B - A) \geq m(A)$ □

3.23 Proposition *Let $\{E_i\}$ be a sequence of measurable sets. Then*

$$m\left(\bigcup E_i\right) \leq \sum mE_i.$$

If the sets E_n are pairwise disjoint, then

$$m\left(\bigcup E_i\right) = \sum mE_i$$

Proof. The inequality is simply a restatement of the subadditivity of m^* given by Proposition 2.3. If $\{E_i\}$ is a finite sequence of disjoint measurable sets, then Theorem 3.11 with $A = \mathbb{R}$ implies that

$$m\left(\bigcup E_i\right) = \sum mE_i$$

and so m is finitely additive. Let $\{E_i\}$ be an infinite sequence of pairwise disjoint measurable sets. Then

$$\bigcup_{i=1}^{\infty} E_i \supset \bigcup_{i=1}^{n} E_i,$$

and so

$$m\left(\bigcup_{i=1}^{\infty} E_i\right) \geq m\left(\bigcup_{i=1}^{n} E_i\right) = \sum_{i=1}^{n} mE_i.$$

Since the left side of this inequality is independent of n, we have

$$m\left(\bigcup_{i=1}^{\infty} E_i\right) \geq \sum_{i=1}^{\infty} mE_i.$$

The reverse inequality follows from countable subadditivity, and we have

$$m\left(\bigcup_{i=1}^{\infty} E_i\right) = \sum_{i=1}^{\infty} mE_i. \qquad \square$$

The following two theorems reveal a basic property of Lebesgue measure that concerns its bahaviour for monotone sequence of sets.

3.24 Theorem *Let $\{E_i\}$ ba an infinite decreasing sequence of measureable sets; that is, a sequence with $E_{i+1} \subset E_i$ for each $i \in \mathbb{N}$. Let $m(E_i) < \infty$ for at least one $i \in \mathbb{N}$. Then*

$$m\left(\bigcap_{i=1}^{\infty} E_i\right) = \lim_{n\to\infty} m(E_n).$$

Proof. Let p be the least integer such that $m(E_p) < \infty$, Then $m(E_i) < \infty$ for all $i \geq p$. Set $E = \bigcap_{i=1}^{\infty} E_i$ and $F_i = E_i - E_{i+1}$. Then the sets F_i are pairwise disjoint, and

$$E_p - E = \bigcup_{i=p}^{\infty} F_i.$$

Therefore,

$$m(E_p - E) = \sum_{i=p}^{\infty} m(F_i) = \sum_{i=p}^{\infty} m(E_i - E_{i+1})$$

But $m(E_p) = m(E) + m(E_p - E)$ and $m(E_i) = m(E_{i+1}) + m(E_i - E_{i+1})$, for all $i \geq p$, since $E \subset E_p$ and $E_{i+1} \subset E_i$. Further, using the fact that $m(E_i) < \infty$, for all $i \geq p$, it follows that

$$m(E_p - E) = m(E_p) - m(E)$$

and

$$m(E_i - E_{i+1}) = m(E_i) - m(E_{i+1}), \forall i \geq p.$$

Hence,

$$\begin{aligned} m(E_p) - m(E) &= \sum_{i=p}^{\infty}(m(E_i) - m(E_{i+1})) \\ &= \lim_{n\to\infty}\sum_{i=p}^{n}(m(E_i) - m(E_{i+1})) \\ &= \lim_{n\to\infty}\{m(E_p) - m(E_n)\} \\ &= m(E_p) - \lim_{n\to\infty} m(E_n). \end{aligned}$$

Since $m(E_p) < \infty$, it gives

$$m(E) = \lim_{n\to\infty} m(E_n).$$

This proves the result. □

Remark : The condition $m(E_i) < \infty$, for at least one $i \in \mathbb{N}$, in Theorem 24 cannot be dropped.

3.25 Example Consider the sets E_n given by $E_n = (n, \infty)$, $n \in \mathbb{N}$. Then $\{E_n\}$ is a decreasing sequence of measurable sets such that $m(E_n) = \infty$ for each $n \in \mathbb{N}$ and $\cap_{n=1}^{\infty} E_n = \emptyset$. Therefore,

$$\lim_{n\to\infty} m(E_n) = \infty, \qquad \text{while } m(\emptyset) = 0.$$

3.26 Theorem *Let $\{E_i\}$ be an infitinite increasing sequence of measurable sets; that is, a sequence with $E_{i+1} \supset E_i$ for each $i \in \mathbb{N}$. Then*

$$m\left(\bigcup_{i=1}^{\infty} E_i\right) = \lim_{n\to\infty} m(E_n).$$

Proof. If $m(E_p) = \infty$ for some $p \in \mathbb{N}$, then the result is trivially true, since

$$m\left(\bigcup_{i=1}^{\infty} E_i\right) \geq m(E_p) = \infty,$$

and $m(E_n) = \infty$, for each $n \geq p$. Let $m(E_i) < \infty$, for each $i \in \mathbb{N}$. Set

$$E = \bigcup_{i=1}^{\infty} E_i, \quad F_i = E_{i+1} - E_i.$$

Then the sets F_i are pairwise disjoint, and

$$\begin{aligned} E - E_1 &= \bigcup_{i=1}^{\infty} F_i \\ \Rightarrow \quad m(E - E_1) &= m(\bigcup_{i=1}^{\infty} F_i) = \sum_{i=1}^{\infty} m(F_i) = \sum_{i=1}^{\infty} m(E_{i+1} - E_i) \\ \Rightarrow m(E) - m(E_1) &= \lim_{n\to\infty} \sum_{i=1}^{n} \{m(E_{i+1}) - m(E_i)\} \\ &= \lim_{n\to\infty} \{m(E_{n+1}) - m(E_1)\} \\ \Rightarrow \quad m(E) &= \lim_{n\to\infty} m(E_n). \end{aligned}$$

□

Note : Theorems 25 and 26 establish that the set function m is **conditionally continuous** from above as well as below.

We are always interested, whenever possible, in measureble sets. However, proving the measurability of a given set is sometimes defficult in practice. It may be desirable in that situation to apply theorems about outer measure. A particularly useful result is the following theorem which, in fact, improve Theorem 3.26

3.27 Theorem *Let* $\{E_i\}$ *be an infinite increasing sequence of sets (not necessarily measurable). Then*

$$m^*\left(\bigcup_{i=1}^{\infty} E_i\right) = \lim_{n\to\infty} m^*(E_n).$$

Proof. For each E_i, there exists a $\mathfrak{G}_\delta$-set G_i (of course measurable) such that $E_i \subset G_i$ and $m^*(E_i) = m(G_i)$. cf. Theorem 2.8 (b). For each $n \in \mathbb{N}$, define

$$F_n = \bigcap_{i=n}^{\infty} G_i.$$

Clearly $\{F_n\}$ is an increasing sequence of measurable sets. Therefore, by Theorem 3.26, it follows that

$$m\left(\bigcup_{n=1}^{\infty} F_n\right) = \lim_{n\to\infty} m(F_n).$$

But $E_i \subset E_{i+1}$ implies that $E_n \subset G_i$, for all $i \geq n$ and as such $E_n \subset F_n \subset G_n$, for each $n \in \mathbb{N}$. Consequently

$$\begin{aligned} & m^*(E_n) \leq m(F_n) \leq m(G_n) = m^*(E_n) \\ \Rightarrow \quad & m^*(E_n) = m(F_n), \qquad \forall n \in \mathbb{N}. \end{aligned}$$

Hence $\lim_{n\to\infty} m^*(E_n)$ exists, and

$$\lim_{n\to\infty} m^*(E_n) = m\left(\bigcup_{n=1}^{\infty} F_n\right).$$

Also, we observe that

$$\bigcup_{n=1}^{\infty} F_n \supset \bigcup_{n=1}^{\infty} E_n,$$

and, therefore,

$$\lim_{n\to\infty} m^*(E_n) \geq m^*\left(\bigcup_{n=1}^{\infty} E_n\right).$$

The reverse inequality is obvious from the fact that

$$E_n \subset \bigcup_{n=1}^{\infty} E_n, \forall n \in \mathbb{N}$$

and this completes the proof. □

In analogy with limit superior and the limit inferior of a sequence of real numbers (cf. I-6), let us define the concepts such as the limit superior and the limit inferior of a sequence sets.

Let $\{E_i\}$ be any sequence of sets. Define

$$\begin{cases} \limsup_{i\to\infty} E_i = \bigcap_{n=1}^{\infty} \bigcup_{i=n}^{\infty} E_i \\ \liminf_{i\to\infty} E_i = \bigcup_{n=1}^{\infty} \bigcap_{i=n}^{\infty} E_i. \end{cases}$$

If $\limsup E_i = \liminf E_i$, the sequence $\{E_i\}$ of sets is said to be **convergent**, and the common limit is written by $\lim E_i$.

3.28 Corollary *If $\{E_n\}$ is a convergent sequence of measurable sets and $E_n \subset F$ for each $n \in \mathbb{N}$, where $m^*(F) < \infty$, then $\lim_{n\to\infty} m(E_n)$ exists and*

$$m(\lim_{n\to\infty} E_n) = \lim_{n\to\infty} m(E_n).$$

Problem. Prove the second part of Theorem 3.28 and show that the additional assumption stated therein is necessary.

< Characterizations of Measurable Sets >

We now give certain characterizations of the measurability of an arbitrary set E in $\mathbb{R}$ in terms of certain relatively simple types of sets each of which is a Borel set.

3.29 Theorem *Let E be a given set. Then the following statements are equivalent:*

(*a*) *E is measurable.*

(*b*) *Given $\varepsilon > 0$, there is an open set $O \supset E$ such that $m^*(O - E) < \varepsilon$.*

(*c*) *There is a $\mathfrak{G}_\delta$-set $G \supset E$ such that $m^*(G - E) = 0$.*

(*d*) *Given $\varepsilon > 0$, there is a closed set $F \subset E$ such that $m^*(E - F) < \varepsilon$.*

(*e*) *There is $\mathfrak{F}_\sigma$-set $F \subset E$ such that $m^*(E - F) = 0$.*

If m^*E is finite, the above statements are equivalent to:

(*f*) *Given $\varepsilon > 0$, there is a finite union B of open intervals such that*

$$m^*(E \Delta B) < \varepsilon.$$

Proof. (a) $\Rightarrow$ (b) : Suppose first that $m(E) < \varepsilon$. By Theorem 2.14(a), there is an open set $O \supset E$ such that

$$m^*(O) < m^*(E) + \varepsilon.$$

This, since both the sets O and E are measurable, verifies (cf. Problem 6).

$$m^*(O - E) = m^*(O) - m^*(E) < \varepsilon.$$

Consider now the case when $m(E) = \infty$. Write the set $\mathbb{R}$ of real numbers as a union of disjoint finite intervals; that is, $\mathbb{R} = \cup_{n=1}^{\infty} I_n$. Then, if $E_n = E \cap I_n$, $m(E_n) < \infty$. We can, thus, find open sets $O_n \supset E_n$ such that

$$m^*(O_n - E_n) < \frac{\varepsilon}{2^n}.$$

Difine $O = \cup_{n=1}^{\infty} O_n$. Clrealy O is an open set such that $O \supset E$ and satisfies

$$O - E = \bigcup_{n=1}^{\infty} O_n - \bigcup_{n=1}^{\infty} E_n \subset \bigcup_{n=1}^{\infty} (O_n - E_n).$$

Hence

$$m^*(O - E) \leq \sum_{n=1}^{\infty} m^*(O_n - E_n) < \varepsilon.$$

(b) $\Rightarrow$ (c): Given $\varepsilon = 1/n$, there is an open set $O_n \supset E$ with $m^*(O_n - E) < 1/n$. Define $G = \cap_{n=1}^{\infty} O_n$. Then G is $\mathfrak{G}_\delta$-set such that $G \supset E$ and

$$m^*(G - E) \leq m^*(O_n - E) < \frac{1}{n}, \qquad \forall n \in \mathbb{N}.$$

This on letting $n \to \infty$ proves (c).

(c) $\Rightarrow$ (a) : Write $E = G - (G - E)$. But the sets G and $G - E$ measurable since G is a Borel set and $G - E$ is of outer measure zero. Hence E is measurable.

(a) $\Rightarrow$ (d) : E^c is measurable and so, in wiew of (c), there is an open set $O \supset E^c$ such that $m^*(O - E^c) < \varepsilon$, But $O - E^c = E - O^c$. Taking $F = O^c$, the assertion (d) follows.

(d) $\Rightarrow$ (e) : Given $\varepsilon = 1/n$, there is a closed set $F_n \subset E$ with $m^*(E - F_n) < 1/n$. Define $F = \cup_{n=1}^{\infty} F_n$. Then F is $\mathfrak{F}_\sigma$-set such that $F \subset E$ and

$$m^*(E - F) \leq m^*(E - F_n) < \frac{1}{n}, \quad \forall n \in \mathbb{N}.$$

Hence the result in (e) follows on letting $n \to \infty$.

(e) $\Rightarrow$ (a) : The proof is similar to (c) $\Rightarrow$ (a).

(f) $\Leftrightarrow$ (a) : Suppose E be measurable, and let $\varepsilon > 0$ be given. Then, by Thoerem 2.8(b), there-exists an open set $O \supset E$ with $m^*(O - E) < \varepsilon/2$. As $m^*(E)$ is finite, so is $m^*(O)$. Further, since the open set O can be expressed as the union of countable (disjoint) open intervals $\{I_i\}$, there exists an $n \in \mathbb{N}$ such that

$$\sum_{i=n+1}^{\infty} l(I_i) < \frac{\varepsilon}{2},$$

in view of Theorem 3.22 and the fact that $m^*(O) < \infty$.

Write $B = \cup_{i=1}^{n} I_i$. Then

$$E \Delta B = (E - B) \cup (B - E) \subset (O - B) \cup (O - E).$$

Hence

$$m^*(E \Delta B) \leq m^* \left(\bigcup_{i=n}^{\infty} I_i \right) + m^*(O - E) < \varepsilon.$$

Conversely, assume that for a given $\varepsilon > 0$, there is a finite union. $B = \cup_{n=1}^{n} I_i$, of open intervals with $m^*(E \Delta B) < \varepsilon$. By Theorem 3.28(a), there is an open set $O \supset E$ such that

$$m^*(O) < m^*(E) + \varepsilon. \tag{1}$$

If we can show that $m^*(O - E)$ is arvitrarily small, then by Theorem 3.29 it follows that E is a measurable set.

Write $S = \cup_{i=1}^{n} (I_i \cap O)$. Then $S \subset B$ and so

$$S \Delta E = (E - S) \cup (S - E) \subset (E - S) \cup (B - E).$$

However,

$$E - S = (E \cap O^c) \cup (E \cap B^c) = E - B$$

since $E \subset O$. Therefore

$$S \Delta E = (E - B) \cup (B - E) = E \Delta B,$$

and as such $m^*(S \Delta E) < \varepsilon$. However, $E \subset S \cup (S \Delta E)$ and so

$$m^*(E) < m^*(S) + \varepsilon. \tag{2}$$

Also $O - E \subset (O - S) \cup (S \Delta E)$ gives

$$m^*(O - E) < m^*(O) - m^*(S) + \varepsilon.$$

Hence, in view of (1) and (2), we get

$$m^*(O - E) < 2\varepsilon.$$

This proves the theorem. □

Towards the close of this section we would like to examine: How big could thd family $\mathfrak{M}$ of measurable sets be?

3.30 Theorem *The cardinality of* $\mathfrak{M}$ *is* 2^c.

Proof. Since $\mathfrak{M} \subset \wp(\mathbb{R})$, it follows that $\#\mathfrak{M} \leq 2^c$. The reverse inequality is available from the fact that the family $\wp(\mathbf{C})$, of all subsets of the Cantor set $\mathbf{C}$, is contained in $\mathfrak{M}$ (cf. Corollary 3.7) and that $\#\wp(\mathbf{C}) = 2^c$, since the set $\mathbf{C}$ is uncountable having cardinality c □

Remark : Though all the sets in $\wp(\mathbb{R})$ may not be measurable [18] still we observe that the cardinalities of each of the families $\mathfrak{M}$ and $\wp(\mathbb{R})$ of subsets of $\mathbb{R}$ is the same. This reveals the fact, inspite of the inclusion $\mathfrak{M} \subset \wp(\mathbb{R})$ being proper, that the family $\mathfrak{M}$ is very rich; so much so that there are as many measurable sets as in $\wp(\mathbb{R})$.

P R O B L E M S [3.3]

1. Show that the condition $mE_1 < \infty$ is necessary in Proposition 22 by giving a decreasing sequence $< E_n >$ of measurable sets with $\emptyset = \cap E_n$ and $mE_n = \infty$ for each n.

2. Let $< E_i >$ be a sequence of disjoint measurable sets and A any set. Then $m^*(A \cap \cup_{i=1}^{\infty} E_i) = \sum_{i=1}^{\infty} m^*(A \cap E_i)$.

3. (a) Show that the Cantor ternary set has measure zero.

(b) Let F be a subset of $[0,1]$ constructed in the same manner as the Cantor ternary set except that each of the intervals removed at the nth step has length $\alpha 3^{-n}$ with $0<\alpha<1$. Then F is a closed set, $\overline{F}$ dense in $[0,1]$ and $mF=1-\alpha$. Such a set F is called a **generalized Cantor set**.

4. Show that if E is a measurable set, then each translate $E+y$ of E is also measurable.

5. Show that if E_1 and E_2 are measurable, then $m(E_1\cup E_2)+m(E_1\cap E_2)=mE_1+mE_2$.

6. Let E_1 and E_2 be measurable sets each having finite measure. Prove that the following statements are equivalent.

   (a) $m(E_1 \Delta E_2)=0$
   (b) $m(E_1\sim E_2)=0=m(E_2\sim E_1)$
   (c) $m(E_1)=m(E_1\cap E_2)=m(E_2)$

3.4 Measurable Functions

The class of measurable functions plays a role of central importance in Lebesgue Theory of integration. Since not all sets are measurable, it is of great importance to know that sets which arise naturally in certain constructions are measurable. If we start with a function f the most important sets which arise from it are those listed in the following proposiotion:

4.1 Proposition *Let f be an extended real-valued function whose domain is measurable. Then the following statements are equivalent:*

i. For each real number α the set $\{x|f(x)>\alpha\}$ is measurable.
ii. For each real number α the set $\{x|f(x)\geq\alpha\}$ is measurable.
iii. For each real number α the set $\{x|f(x)<\alpha\}$ is measurable.
vi. For each real number α the set $\{x|f(x)\leq\alpha\}$ is measurable.

These statements imply

v. For each extended real number α the set $\{x|f(x) = \alpha\}$ is measurable.

Proof. Let the domain of f be D. We have (i) $\Rightarrow$ (iv), since $\{x|f(x) \leq \alpha\} = D \sim \{x|f(x) > \alpha\}$ and the difference of two measurable sets is measurable. Similarly, (iv) $\Rightarrow$ (i) and (ii) $\Leftrightarrow$ (iii).

Now (i) $\Rightarrow$ (ii), since $\{x : f(x) \geq \alpha\} = \bigcap_{n=1}^{\infty}\{x : f(x) > \alpha - 1/n\}$, and the intersection of a sequence of measurable sets is measurable, Similarly, (ii) $\Rightarrow$ (i), since $\{x : f(x) > \alpha\} = \bigcup_{n=1}^{\infty}\{x : f(x) > \alpha + 1/n\}$ and the union of a sequence of measurable sets is measurable. This shows that the first four statements are equivalent. If α is a real number, $\{x|f(x) = \alpha\} = \{x|f(x) \geq \alpha\} \cap \{x|f(x) \leq \alpha\}$ and so (ii) and (iv) $\Rightarrow$ (v) for α real. Since

$$\{x : f(x) = \infty\} = \bigcap_{n=1}^{\infty}\{x : f(x) \geq n\}$$

(ii) $\Rightarrow$ (v) for $\alpha = \infty$. Similarly, (iv) $\Rightarrow$ (v) for $\alpha = -\infty$, and we have (ii) & (iv) $\Rightarrow$ (v). □

4.2 Definition *An extended real-valued function f is said to be (Lebesgue) measurable if its domain is measurable and if it satisfies one of the first four statements of Proposition 4.1.*

Note : Throughout this present chapter "measurable" means "measurable with respect to the Lebesgue measure"

4.3 (Example of non-measurable function) *Let S be a non-measurable set and consider the characteristic function defined by*

$$\chi(x) = \begin{cases} 1 & \text{if } x \in S \\ 0 & \text{if } x \notin S \end{cases}$$

Then the function $\chi(x)$ is non-measurable, because for $a = 1/2$, $\{x \in E : f(x) \geq 1/2\} = S$ is noe-measurable.

Thus if we restrict ourselves to measurable functions, the most important sets connected with them are measurable. It should be noted that a continuous function (with a measurable domain) is measurable, and of course each step function is measurable. If f is a measurable function and E is a measurable subset of the domain of f, then the function obtained by restricting f to E is also measurable. The following proposition tells us that certain operations performed on measurable functions lead again to measurable functions:

4.4 Propositon *Let c be a constant and f and g two measurable real-valued functions defined on the same domain. Then the functions $f + c$, cf, $f + g$, $g - f$, and fg are also measurable.*

Proof. We shall use condition (iii) of Propositon 4.1 Then

$$\{x : f(x) + c < \alpha\} = \{x : f(x) < \alpha - c\}$$

and so $f + c$ is measurable when f is. A similar argument shows cf to be measurable.

If $f(x) + g(x) < \alpha$, then $f(x) < \alpha - g(x)$ and by the corollary to the axiom of Archimedes there is a rational number r such that

$$f(x) < r < \alpha - g(x)$$

Hence

$$\{x : f(x) + g(x) < \alpha\} = \bigcup_r (\{x : f(x) < r\} \cap \{x : g(x) < \alpha - r\}).$$

Since the rationals are countable, this set is measurable and so $f + g$ is measurable. Since $-g = (-1)g$ is measurable when g is, we have $f - g$ measurable.

The function f^2 is measurable, since

$$\{x : f^2(x) > \alpha\} = \{x : f(x) > \sqrt{\alpha}\} \cup \{x : f(x) < -\sqrt{\alpha}\}$$

for $\alpha \geq 0$ and

$$\{x | f^2(x) > \alpha\} = D$$

if $\alpha < 0$, where D is the domain of f. Thus

$$fg = \frac{1}{2}[(f+g)^2 - f^2 - g^2]$$

is measurable. □

We will often want to use properties 4.4 for extended real-valued functions f and g. Unfortunately, $f+g$ is not defined at points where it is of the form $\infty - \infty$. However, fg is always measurable and $f+g$ is measurable if we always take the same value for $f+g$ at points where it is undefined. Also, $f+g$ is measurable no matter what values we take at the points where it is not defined, provied these points are a set of measure zero.

4.5 Corollary *Let f and g be two measurable real-valued functions defined on the same domain. Then the functions f^2, $f \div g$ (g ranishes nowhere on the domain) are also measurable.*

Also constant function is measurable.

4.6 Propositon *For any measurable function $f : E \to \mathbb{R}^*$, the following two statements are true*

(*a*) *for every interval I, the inverse image $f^{-1}(I)$ is measurable.*

(*b*) *For every extended real number x, the inverse image $f^{-1}(x)$ is measurable.*

Proof. One can easily verify this proppproposition by means of

$$\begin{aligned}
f^{-1}[r,\infty] &= \cap f^{-1}(r - I/n, \infty], \\
f^{-1}[-\infty, r) &= E \sim f^{-1}[r, \infty], \\
f^{-1}[-\infty, r] &= \cap f^{-1}(-\infty, r + I/n), \\
f^{-1}(r, \infty] &= E \sim f^{-1}[-\infty, r]. \\
f^{-1}(I) &= f^{-1}(a,b) = \{x \in E | \alpha < x < \beta\} \\
&= \{x \in E : f(x) > a\} \cap \{x \in E | f(x) < b\}
\end{aligned}$$

where $I = (a,b)$, $f(\alpha) = a$, $f(\beta) = b$ □

4.7 Theorem *If $f(x)$ is measurable on a set E_1 and if E_2 is measurable subset of E_1, then the restriction $g = f|_{E_2}$ is a measurable function.*

Proof. Since for each $\alpha \in \mathbb{R}$, $\{x \in E_2 | g(x) > \alpha\} = \{x \in E_1 | f(x) > \alpha\} \cap E_2$ the result is clear. □

4.8 Theorem *If $f(x)$ is measurable on a countable class of disjoint sets $E_1, E_2, \cdots$ then it is measurable on their union $\bigcup_{k=1}^{\infty} E_k$.*

Proof. Let $f(x)$ be measurable on E_i for each i, and let $E_i \cap E_j = \emptyset$ if $i \neq j$ with $E = \cup E_i$

Then since $\{x \in E_i : f(x) > \alpha\} = f^{-1}(\alpha, \infty) \cap E_i$. It follows that

$$\begin{aligned} & [f^{-1}(\alpha,\infty) \cap E_1] \cup [f^{-1}(\alpha,\infty) \cap E_2] \cup \cdots \\ = \; & f^{-1}(\alpha,\infty) \cap [E_1 \cup E_2 \cup \cdots] \\ = \; & f^{-1}(\alpha,\infty) \cap [\cup E_i] \\ = \; & f^{-1}(\alpha,\infty) \cap E = \{x \in E | f(x) > \alpha\}. \end{aligned}$$

Since for each i $\{x \in E_i | f(x) > \alpha\}$ is measurable, the set $\{x \in E | f(x) > \alpha\}$ is measurable. □

4.9 Lemma *for any two measurable functions $f, g : E \to \mathbb{R}^*$, the following sets are measurable:*

$$\begin{aligned} A(f,g) &= \{x \in E | f(x) < g(x)\}, \\ B(f,g) &= \{x \in E | f(x) \leq g(x)\}, \\ C(f,g) &= \{x \in E | f(x) = g(x)\}. \end{aligned}$$

Proof. If $f(x) < g(x)$, then it follows from the Archimedean property of $\mathbb{R}$ that there exists a rational number r satisfying

$$f(x) < r < g(x).$$

Hence we have

$$A(f,g) = \bigcup_{r \in \mathbb{Q}} \{f^{-1}[-\infty, r) \cap g^{-1}(r, \infty]\}$$

Since $\mathbb{Q}$ is countable, it follows form (III.3.10) that $A(f,g)$ is measurable. The measurability of $B(f,g)$ and $C(f,g)$ follows from relations:

$$B(f,g) = E \sim A(f,g)$$
$$C(f,g) = B(f,g) \cap B(g,f).$$

□

The set $\{x \in E | f(x) > \alpha\}$ can be written $E[f(x) > \alpha]$ briefly.

4.10 Theorem *If f is a measurable function defined on E, then the set $E(f = \alpha)$ is measurable for each extended real number α.*

Proof. For a real number x, we have

$$E(f = \alpha) = E(f \geq \alpha) \cap E(f \leq \alpha).$$

Since f is a measurable function, the set $E(f = \alpha)$ is measurable for each real number α. For $\alpha = \infty$

$$E(f = \infty) = \bigcap_{n=1}^{\infty} E(f > n);$$

and for $\alpha = -\infty$,

$$E(f = -\infty) = \bigcap_{n=1}^{\infty} E(f < -n).$$

Since the countable intersection of measurable sets is again a measurable set, the result follows in both the cases,i.e. when $\alpha = \pm\infty$.
□

Remark : The converse of Theorem 10 need not be true.

4.11 Example Let P be a nonmeasurable set in $E = (0,1)$. Consider the function f with domain E and defined by

$$f(x) = \begin{cases} x^2 & \text{if } x \in P \\ -x^2 & \text{if } x \in E - P \end{cases}$$

Then, for each real number α, the set $E(f = \alpha)$ consists of at most two elements and is hence measurable. However, the set $E(f > 0)$ is, in fact, the set P which is nonmeasurable. Consequently, f is not a measurable function.

4.12 Theorem *If $f(x)$ is measurable on E, then $|f(x)|$ is measurable on E.*

Proof. i) For any $k \geq 0$ $\{x \in E : |f(x)| > k\}$

$$
\begin{aligned}
&= \{x \in E : f(x) > k \text{ or } f(x) < -k\} \\
&= \{x \in E | f(x) < -k\} \cup \{x \in E : f(x) > k\}
\end{aligned}
$$

and the union of countable sets is countable. Hece the result is proved.

ii) For any $k < 0$ $\{x \in E : |f(x)| > k\} = E$ and since E is measurable, the result holds clearly. □

Remark : The converse of Theorem 4.12 may not be true.

4.13 Example Let P be a non-measurable subset of $E = (0, 1)$. Define a function $f : E \to \mathbb{R}$ by

$$
f(x) = \begin{cases} 1 & \text{if } x \in P \\ -1 & \text{if } x \notin P. \end{cases}
$$

The function f is not measurable, since $E(f > 0)$ $(= P)$ is a non-measurable set. But $|f|$ is measurable as the set

$$
E(|f| > \alpha) = \begin{cases} E & \text{if } \alpha < 1 \\ \emptyset & \text{if } \alpha \geq 1. \end{cases}
$$

is measurable.

4.14 Theorem *For any two measurable functuons $f, g : E \to \mathbb{R}^*$. the function*

$$
f \vee g, f \wedge g : E \to \mathbb{R}^*
$$

defined by

$$
\begin{aligned}
f^*(x) &= (f \vee g)(x) = \max\{f(x), g(x)\}, \\
f_*(x) &= (f \wedge g)(x) = \min\{f(x), g(x)\}
\end{aligned}
$$

for every $x \in E$ are measurable.

Proof. Let α be any real number. We note that

$$E(f^* > \alpha) = E(f > \alpha) \cup E(g > \alpha)$$
$$E(f_* > \alpha) = E(f > \alpha) \cap E(g > \alpha).$$

Since f and g are measurable functions. the sets $E(f > \alpha)$ and $E(g > \alpha)$ are measurable. Also, the union and intersection of two measurable sets are again measurable sets. This proves that f^* and f_* are measurable functions. □

4.15 Theorem *Let $\{f_n\}$ be a sequence of measurable functions (with the same domain of definition). Then the functions* $\sup\{f_1, \cdots, f_n\}$, $\inf\{f_1, \cdots, f_n\}$, $\sup_n f_n$, $\inf_n f_n$, $\overline{\lim} f_n$ *and* $\underline{\lim} f_n$ *are all measurable.*

Proof. If h is defined by $h(x) = \sup\{f_1(x), \cdots, f_n(x)\}$, then $\{x : h(x) > \alpha\} = \bigcup_{i=1}^{n}\{x : f_i(x) > \alpha\}$. Hence the measurability of the f_i implies that of h. Similarly, if g is defined by $g(x) = \sup f_n(x)$, then $\{x : g(x) > \alpha\} = \bigcup_{n=i}^{\infty}\{x : f_n(x) > \alpha\}$, and so g is measurable. A similar argument establishes the corresponding statements for inf. Call

$$F_1(x) = \sup\{f_1(x), f_2(x), \cdots\}$$
$$F_2(x) = \sup\{f_2(x), f_3(x), \cdots\}$$
$$F_3(x) = \sup\{f_3(x), f_4(x), \cdots\}$$

and so on, It follows that $F_1(x) \geq F_2(x) \geq F_3(x) \geq \cdots$ and Since $\overline{\lim} f_n = \inf_n \sup_{k \geq n} f_k$, we have $\overline{\lim} f_n$ measurable, and similarly

$$\underline{\lim} f_n(x) = \lim F_n(x)$$ □

4.16 Corollary *If $\{f_n\}$ is a sequence of measurable functions converging to f on E, then f is a measurable function.*

Proof. Since $f_n \to f$, we have

$$\overline{\lim} f_n = \underline{\lim} f_n = f.$$

In view of Theorem 4.15, f is a measurable function. □

4.17 Theorem *A continuous function defined on a measurable set is measurable.*

Proof. Let f be a function defined and continuous on E (measurable). Let α be any arbitrary real number. Consider the set

$$A = \{x \in E | f(x) \leq \alpha\}.$$

Let x_0 be a limit point of A. Then, $\exists$ a sequence $\{x_n\}$ of points in A with $x_n \to x_0$. therefore,

$$f(x_n) \leq \alpha, \qquad \forall n.$$

By the continuity of f at x_0, we have

$$f(x_0) \leq \alpha$$

and, as such, $x_0 \in A$. Hence the set A closed and, therefore, measurable. This proves that f is a measurable function. □

Remark : The converse of Theorem 4.17 may not be true; more precisely, a measurable function need not be continuous.

4.18 Example Consider a function $f : \mathbb{R} \to \{0, 1\}$ given by

$$f(x) = \begin{cases} 1 & \text{if } 0 \leq x < 1 \\ 0 & \text{otherwise} . \end{cases}$$

Clearly, the function f is measurable but not continuous. The point at $x = 0$ is a point of discontinuity of f.

In order to give examples of measurable functions, let us first introduce the terminollgy "almost everywhere" which will be frequently used in the sequel.

A property is said to hold **almost everywhere** (abbreviated a.e.) if the set of points where it fails to hold is a set of measure zero.

In other word, Let $p(x)$ denote a statement concerning the points $x \in E$. Then we asy that $p(x)$ *holds almost everywhere* iff there exists a subset D of E with $m^*(D) = 0$ such that $p(x)$ holds for every point x of $E\backslash D$. By (3.27), D is measurable and is of measure 0. Hence $p(x)$ holds almost everywhere in E iff it holds everywhere in E except possibly at a subset D of measure 0.

For example, a function $f : E \to \mathbb{R}^*$ is said to be **continuous a.e.** in E iff there exists a subset D of E of measure 0 such that f is continuous at every point of $E\backslash D$.

For another example, we say that $f = g$ a.e. if f and g have the same domain and $m\{x : f(x) \neq g(x)\} = 0$. Similarly, we say that f_n converges to g a.e. if there is a set E of measure zero such that $f_n(x)$ converges to $g(x)$ for each x not in E. One consequence of equality a.e. is the following:

4.19 Propositon *If f is a measurable function and $f = g$ a.e., then g is measurable.*

Proof. Let E be the set $\{x : f(x) \neq g(x)\}$. By hypothesis $mE = 0$. Now

$$\begin{aligned}\{x|g(x) > \alpha\} &= \{x : f(x) > \alpha\} \cup \{x \in E : g(x) > \alpha\} \\ &\quad \sim \{x \in E : g(x) \leq \alpha\}.\end{aligned}$$

The first set on the right is measurable, since f is a measurable function. The last two sets on the right are measurable since they are subsets of E and $mE = 0$. Thus $\{x : g(x) > \alpha\}$ is measurable for each α, and so g is measurable. □

Now if f is continuous, then f is measurable by Theorem 4.17. Therefore we have the following result.

4.20 Corollary *If a function f defined on E is continuous a.e., then f is measurable on E.*

Proof. It runs on the lines of proof of Theorem 4.17 and hence left to the reader. □

4.21 Example Define function $f, g : \mathbb{R} \to \mathbb{R}$ by

$$f(x) = \cos x \qquad \text{for } x \in \mathbb{R}$$
$$g(x) = \begin{cases} \sin x & \text{for } x \in \mathbb{Q} \\ \cos x & \text{for } x \in \mathbb{R} - \mathbb{Q}. \end{cases}$$

Then $f(x)$ is a continuous function and $m\{x \in \mathbb{R} | f(x) \neq g(x)\} = m(\mathbb{Q}) = 0$. Hence $g(x)$ is a measurable (w.r.t Lebesgue measure).

4.22 Propositon *If φ is a measurable function on a set E (measurable) and f a function defined and continuous on the range of φ, then $f \circ \varphi$ is a measurable function on E.*

Proof. Let α be an arbitrary real number. Then

$$E(f \circ \varphi > \alpha) = \{x \in E : \varphi(x) \in G\},$$

where $G = \{u : f(u) > \alpha\}$. But G is an open set since f is continuous. Hence the set $E(f \circ \varphi > \alpha)$ is measurable. This proves that $f \circ \varphi$ is a measurable function. □

3.5 Properties of Measurable Functions

With the class $\mathfrak{M}$ of measurable sets in mind, we introduce a rich class of function; namely, the class of measurable functions which includes the class of continuous functions as a proper subclass(Theorem 4.17 and Example 4.18).

5.1 Definition *A function $\varphi : [a, b] \to \mathbb{R}$ is said to be a **step function** if $\exists$ a partition*

$$\{a = x_0 < x_1 < x_2 < \cdots < x_n = b\}$$

of the interval $[a, b]$ such that in every subinterval (x_{i-1}, x_i), the function φ is constant, i.e.

$$\varphi(x) = c_i, \quad \forall x \in (x_{i-1}, x_i), i = 1, 2, \cdots, n.$$

Note : A step function is one which is defined in a closed interval assuming only a finite number of values. Each of these values is assumed by the steop function in an open subinterval. We can disregard the values of the function at the partitioning points x_i or assign values to the function thereat arbitrarily, since these are finite in number and as such form a set of measure zero (finite sets or more generally sets of measure zero, if omitted, do not matter in our consideration).

5.2 Example

1. A function $f : [a,b] \to \mathbb{R}$ defined by

$$f(x) = \begin{cases} \alpha & \text{if } a \le x < c \\ \beta & \text{if } c \le x \le b, \end{cases}$$

where α and β are constants, is a step function.

2. The Signum function S defined by

$$S(x) = \begin{cases} 1 & \text{if } x > 0 \\ 0 & \text{if } x = 0 \\ -1 & \text{if } x < 0 \end{cases}$$

is a step function.

3. The greatest integer function f defined, on anly open interval, by $f(x) = [x]$ is a step function.

Since a constant function on a measurable set is measurable, by Theorem 4.7 of sec.4 we have the following corollary.

5.3 Corollary *A step function is a measurable function.*

5.4 Definition *Let E be a given set. If A is a subset of E, then the **characteristic function** χ_A of A is a real-valued function defined on E by*

$$\chi_A(x) = \begin{cases} 1 & \text{if } x \in A \\ 0 & \text{if } x \notin A. \end{cases}$$

Note : The characteristic function χ_A of the set A is also called the **indicator function** of A.

For an example, define a function $f : \mathbb{R} \to \mathbb{R}$ by

$$f(x) = \begin{cases} 1 & \text{if } x \text{ is rational} \\ 0 & \text{if } x \text{ is irrational} \end{cases}$$

Then f is a characteristic function.

A few simple properties of a characteristic function are given in the following theorem, the proof which is straightforward and hence omitted.

5.5 Theorem *Let A and B be subsets of E. Then:*

(*a*) $\chi_\emptyset = 0$ *and* $\chi_E = 1$.

(*b*) $A \subset B \Rightarrow \chi_A \le \chi_B$.

(*c*) $\chi_{A\cup B} = \chi_A + \chi_B - \chi_{A\cap B}$.

(*d*) $\chi_{A\cap B} = \chi_A \cdot \chi_B$.

(*e*) $\chi_{A^c} = 1 - \chi_A$.

(*f*) $\chi_A = \sum\limits_{i=1}^{\infty} \chi_{A_n}$ *where* $A = \bigcup\limits_n A_n$ *and the sequence* $\{A_n\}$ *consists of disjoint subsets of* E.

(*g*) $\chi_{A-B} = \chi_A - \chi_{A\cap B}$.

(*h*) $\chi_{A\times B} = \chi_A \cdot \chi_B$. (*The set B can be considered to be a subset of some other set F.*)

We define the measurability of a function f over a set E, provided the domain set E is measurable. If E is a nonmeasurable set, the function f is nonmeasurable. But it does not mean that if E is a measurable set, the function f is always measurable. In fact, we prove the following theorem.

5.6 Theorem *Let E be a measurable set. Then the set $A \subset B$ and its characteristic function χ_A are simultaneously measurable or nonmeasurable.*

Proof. Let α be any real number. Then

$$E(\chi_A > \alpha) = \begin{cases} \emptyset & \text{if } \alpha \geq 1 \\ A & \text{if } 0 \leq \alpha < 1 \\ E & \text{if } \alpha < 0. \end{cases}$$

If A is a measurable set, then the set $E(\chi_A > \alpha)$ is measurable for each real number α and hence χ_A is a measurable function. On the other hand, if χ_A is a measurable function, the set A turns out to be measurable by choosing an α such that $0 \leq \alpha < 1$. □

Remark : Theorem 5.6 asserts that the characteristic function of nonmeasurable sets are nonmeasurable even though the domain set is measurable.

A very useful concept in the theory of measurable functions is that of a simple function. A function is said to be a simple function if it assumes only a finite number of values, each of them being finite. We put more precisely as follows.

5.7 Definition *A real valued function f is called a* ***simple function*** *if it is measurable and assumes only a finite number of values such that*

$$f(x) = a_i, x \in E_i, 1 \leq i \leq n.$$

Note that every simple functions are bounded.

5.8 Example

1. Each step function is a simple function.

2. Each characteristic function of a measurable set is a simple function.

3. Any linear combination of the form $\sum_{i=1}^{n} a_i\chi_{E_i}$ represents a simple function defined on the set $E = \bigcup_{i=1}^{n} E_i$ (The sets E_i's are measurable but not necessarily disjoint).

Note : The representation of the function φ described in Example 8 above is not unique. On the other hand. if φ is a simple function defined on a set E and the set $\{a_1, a_2, \cdots, a_m\}$ its range, then

$$\varphi = \sum_{i=1}^{m} a_i\chi_{A_i},$$

where $A_i = \{x \in E : \varphi(x) = a_i\}$, $i = 1, 2, \cdots, m$. This representation of φ is called the **canonical representation** and it is characterized by the fact that the A_i's are disjoint and the a_i's distinct and nonzero.

5.9 Propositon *Every simple function on E is a linear combination of characteristic function of measurable subsets of E, and vice versa.*

It is clear that by the definition of simple function, the sum, product, and difference of two simple functions are simple.

5.10 Corollary *For any real number c and any two simple functions $f, g : E \to \mathbb{R}$, the four functions $f+g$, $f-g$, cf, and fg are simple.*

Proof. Let the range of f be $f(D) = \{\alpha_1, \alpha_2, \cdots, \alpha_m\}$ and let the range of g be $g(D) = \{\beta_1, \beta_2, \cdots, \beta_n\}$ $(n \leq m)$. Consider the set of non-zero elements of the set $\{\alpha_i + \beta_i | 1 \geq i, j \leq n\}$ and denoted by $\{r_1, r_2, \cdots, r_n\}$. If we put $D_i = \{x \in D : f(x) + g(x) = r_i\}$ $(1 < i < k)$. Then

$$f + g = \sum_{i=1}^{k} r_i\chi_{D_i}.$$

Thus $f+g$ is a simple function. Similarly we can prove the remaining part as follows.

$$cf = c\sum \alpha_i \chi_{E_i} = \sum c\alpha_i \chi_{E_i}.$$

$$fg = \left(\sum_{i=1}^{m} \alpha_i \chi_{A_i}\right)\left(\sum_{j=1}^{n} \beta_j \chi_{B_j}\right) = \sum (\alpha_i \beta_j) \chi_{A_i \cap B_j}.$$

□

The following proposition tells us that a measurable function is "almost" a continuous function. The proof is left to the reader

Concluding this section, we discuss Littlewood's three principles, roughly expressible in the following terms: Every (measurable) set is nearly a finite union of intervals; every [measurable] function is nearly continuous; every convergent sequence of [measurable] function is nearly uniformly convergent. Various forms of the first principle are given by Problems 5.11 $\sim$ 5.14. And second is given by proposition 5.18, 5.19. A third is given by proposition 5.20. A slightly stronger form is given by Egoroff's theorem (Problem 20), but you will generally find the weak form adequate.

If we restrict the measure space to the Lebesgue measure space, then the following two propositions tell us that the Lebesgue measurable function can be considered a continuous function approximately. The proofs are left to the readers (Ref.[9]).

5.11 Theorem *For any measurable function $f : [a,b] \to \mathbb{R}$ defined on a bounded closed interval $[a,b]$ and for any positive real number ε, there exist a measurable subset D of $[a,b]$ with $\mu(D) < \varepsilon$, a step function $g : [a,b] \to \mathbb{R}$, and a continuous function $h : [a,b] \to \mathbb{R}$ such that*

$$|g(x) - f(x)| < \varepsilon, \quad |h(x) - f(x)| < \varepsilon$$

hold for every $x \in E \backslash D$. If $c \le f(x) \le d$ holds for every $x \in [a,b]$, the function f and g can be so chosen that

$$c \le g(x) \le d, \quad c \le h(x) \le d$$

are satisfied for every $x \in [a,b]$.

5.12 Propositon *Let f be a measurable function defined on an interval $[a,b]$, and assume that f takes the values $\pm\infty$ only on a set of measure zero. Then given $\varepsilon > 0$, we can find a steop function g and a continuous function h such that*

$$|f-g| < \varepsilon \text{ and } |f-h| < \varepsilon$$

except on a set of measure less than ε; i.e., $m\{x : |f(x)-g(x)| \geq \varepsilon\} < \varepsilon$ and $m\{x : |f(x)-h(x)| \geq \varepsilon\} < \varepsilon$. If in addition $m \leq f \leq M$, then we may choose the functions g and h so that $m \leq g \leq M$ and $m \leq h \leq M$.

5.13 Theorem *For every measurable function $f : E \to \mathbb{R}$, there exists a sequence*

$$\{f_n : E \to \mathbb{R} | n \in \mathbb{N}\}$$

of simple functions $f_n : E \to \mathbb{R}$ which converges (piontwise) to f. In case $f \geq 0$, the sequence can be so chosen that

$$0 \leq f_n \leq f_{n+1}$$

holds for every $n \in \mathbb{N}$. In case f is bounded, the sequence cna be so chosen that the convergence is uniform.

Proof. We shall left this proof to the readers (Ref [10]). □

5.14 Theorem (Lusin Theorem) *Let f be a measurable function defined on E. Then for each $\varepsilon > 0$. $\exists$ a closed set $F \subset E$ with $m(E-F) < \varepsilon$ such that f is continuous on F.*

Proof. We will left to the readers. □

5.15 Theorem *Let f be a measurable function defined on E. Then for each $\varepsilon > 0$, $\exists$ a continuous function g defined on $\mathbb{R}$ such that $m(x \in E : f(x) \neq g(x)) < \varepsilon$. Further, if $|f(x)| \leq M$ on E, then g may be chosen so that $|g(x)| \leq M$ on $\mathbb{R}$.*

Proof. By the Lusin Theorem, for each $\varepsilon > 0$, there exists a closed set $F \subset E$ such that $m(E - F) < \varepsilon$ and f is continuous on F. Using the result: "Let f be a continuous function defined on a closed set F. Then there exists a continuous function g defined on $\mathbb{R}$ such that $g|_F = f$. Moreover, if f is bounded, g can be constructed so that it attains a maximum equal to that of f" studied in real analysis, the required function g exists on $\mathbb{R}$ which is continuous on $\mathbb{R}$ and satisfies $g|_F = f$. Therefore the set

$$A = \{x \in E : f(x) \neq g(x)\} \subset E - F.$$

Clearly, the set A is measurable since f and g both are measurable functions and $m(A) < \varepsilon$.

Finally, the boundedness of g follows directly from its construction. □

5.16 Propositon *Let E be a measurable set of finite measure, and $\{f_n\}$ a sequence of measurable functions defined on E. Let f be a measurable real-valed function such that $f_n(x) \to f(x)$ for each x in E. Then, given $\varepsilon > 0$ and $\delta > 0$, there is a measurable set $A \subset E$ with $mA < \delta$ an integer N such that for all $x \notin A$ and all $n \geq N$,*

$$|f_n(x) - f(x)| < \varepsilon.$$

Proof. Let

$$G_n = \{x \in E : |f_n(x) - f(x)| \geq \varepsilon\},$$

and set

$$E_n = \bigcup_{n=N}^{\infty} G_n = \{x \in E : |f_n(x) - f(x)| \geq \varepsilon \text{ for some } n \geq N\}.$$

We have $E_{N+1} \subset E_N$, and for each $x \in E$ there must be some E_N to which x does not belong, since $f_n(x) \to f(x)$. Thus $\cap E_N = \emptyset$, and so, $\lim mE_N = 0$. Hence given $\delta > 0$, $\exists N$ so that $mE_N < \delta$; that is,

$$m\{x \in E : |f_n(x) - f(x)| \geq \varepsilon \text{ for some } n \geq N\} < \delta.$$

If we write A for this E_N, then $mA < \delta$ and

$$\overline{A} = \{x \in E : |f_n(x) - f(x)| < \varepsilon \text{ for all } n \geq N\}.$$ □

If, as in the hypothesis of the proposition, we have $f_n(x) \to f(x)$ for each x, we say that the sequence $< f_n >$ converges **pointwise** to f on E. If there is a subset B of E with $mB = 0$ such that $f_n \to f$ pointwise on $E \sim B$, we say that $f_n \to f$ a.e. on E. We have the following trivial modification of the last proposition:

5.17 Propositon *Let E be a measurable set of finite measure, and $< f_n >$ a sequence of measurable functions which converge to a real-valued function f a.e. on E. Then, given $\varepsilon > 0$ and $\delta > 0$, there is a set $A \subset E$ with $mA < \delta$, and an N such that for all $x \notin A$ and all $n \geq N$,*

$$|f_n(x) - f(x)| < \varepsilon.$$

Proof. Let $\{x \in E : f_n(x) \to f(x)\} \subset D$ and let $m(D) = 0$. If we put $E' = E - D$, then since $f_n(x) \to f(x)$ for all $x \in E'$, we obtain the subset A of E' and a natural number N by means of applying Proposition 1.29 on $f|_E$ and $f|_{E'}$

Hence for $A = A' \cup D$. $m(A) \leq m(A') + m(D) = m(A') < \delta$. But if $x \in A$ and $n \geq N$, then $|f_n(x) - f(x)| < \varepsilon$. □

Remark : In above Propositon 5.13 and 5.14, If we exclude the condition $m(E) < \infty$, then the above two theorems may not true.

5.18 Example For any natural number n, if we define a function

$$f_n : \mathbb{R} \to \mathbb{R} \text{ by}$$

$$x \mapsto f_n(x) = \begin{cases} 0 & (x < n) \\ 1 & (x \geq n) \end{cases}$$

Then f_n is a measurable function and for each real number x, $f_n(x) \to 0$ holds.

Assume that $f = 0$, it follows that

$$m[\{x \in \mathbb{R} : |f_n(x) - f(x)| \geq 1\}] = m[[n, \infty)] = \infty.$$

Hence Theorem 5.13 and 5.14 does not hold. Here we note that the measure $m(\mathbb{R})$ of the domain of this function is not finite.

5.19 Definition *A sequence $\{f_n\}$ of measurable functions is said to* **converge almost uniformly** *to a measurable function f defined on a measurable set E if for each $\varepsilon > 0$, $\exists$ a measurable set $A \subset E$ with $m(A) < \varepsilon$ such that $\{f_n\}$ converges to f uniformly on $E - A$.*

Note : One should not confuse the term 'almost uniform convergence' with 'uniform convergent almost everywhere'. 'almost uniform convergence' implies 'convergence almost everywhere'. Towards its converse, we prove the following result.

5.20 Theorem (Egoroff's Theorem) *Let E be a measurable set with $m(E) < \infty$ and $\{f_n\}$ a sequence of measurable functions which coverge to f a.e. on E. Then, given $\eta > 0$, there is a set $A \subset E$ with $m(A) < \eta$ such that the sequence $\{f_n\}$ converges to f uniformly on $E - A$.*

Proof. Applying Theorem with $\varepsilon = 1$, $\delta = \eta/2$, we get a measurable set $A_1 \subset E$ with $m(A_1) < \delta$ and a positive integer N_1, such that

$$|f_n(x) - f(x)| < 1,$$

for all $n \geq N_1$ and $x \in E_1 (= E - A_1)$. Again, taking $\varepsilon = 1/2$ and $\delta = \eta/2^2$, we get another measurable set $A_2 \subset E_1$ with $m(A_2) < \delta$ and a positive integer N_2 such that

$$|f_n(x) - f(x) < \frac{1}{2},$$

for all $n \geq N_2$ and $x \in E_2 (= E_1 - A_2)$; and so on. At the nth stage, we get measurable set $A_n \subset E_{n-1}$ with $m(A_n) < \delta$ and a positive integer N_n such that

$$|f_n(x) - f(x)| < \frac{1}{n},$$

for all $n \geq N_n$ and $x \in E_n (= E_{n-1} - A_n)$. Setting

$$A = \bigcup_{n=1}^{\infty} A_n,$$

we note that

$$m(A) \leq \bigcup_{n=1}^{\infty} m(A_n) < \bigcup_{n=1}^{\infty} \frac{\eta}{2^n} = \eta.$$

Also,

$$\begin{aligned} E - A &= E - \bigcup_n A_n \\ &= \bigcap_n (E - A_n) = \bigcap_n (E_{n-1} - A_n) \\ &= \bigcap_n E_n. \end{aligned}$$

Let $x \in E - A$. Then $x \in E_n$ for all n and so

$$|f_n(x) - f(x)| < \frac{1}{n}, \quad \forall n \geq N$$

Choosing n such that $\frac{1}{n} < \varepsilon$, we get

$$|f_n(x) - f(x)| < \varepsilon,$$

for all $x \in E - A$ and $n \geq N_n = N$. □

Remark : Theorem 5.20 is no longer true if E is of infinite measurable. A counter-example is obtained by taking $E = \mathbb{R}$ and $f_n = \chi_{[n,n+1]}$, for each n.

PROBLEMS [3.5]

1. Show that (v) does not imply (iv) in Propositon 4.1 by constructing a function f such that $\{x : f(x) > 0\} = E$, a given nonmeasurable set, and such that f assumes each value at most once.

2. Let D be a dense set of real numbers, that is, a set of real numbers such that every interval contains an element of D. Let f be an extended real-valued function on $\mathbb{R}$ such that $\{x : f(x) > \alpha\}$ is measurable for each $\alpha \in D$. Then f is measurable.

3. Show that the sum and product of two simple functions are simple. Show that

$$\begin{aligned} \chi_{A\cap B} &= \chi_A \cdot \chi_B \\ \chi_{A\cup B} &= \chi_A + \chi_B - \chi_A \cdot \chi_B \\ \chi_{\overline{A}} &= 1 - \chi_A \end{aligned}$$

4. (a) Let D and E be measurable sets and f be a function with domain $D \cup E$. Show that f is measurable if and only if its restrictions to D and E are measurable.

   (b) Let f be a function with measurable domain D. Show that f is measurable iff the function g defined by $g(x) = f(x)$ for $x \in D$ and $g(x) = 0$ for $x \notin D$ is measurable.

5. (a) Let f be an extended real-valued function with measurable domain D, and let $D_1 = \{x : f(x) = \infty\}$, $D_2 = \{x : f(x) = -\infty\}$. Then f is measurable if and only if D_1 and D_2 are measurable and the restriction of f to $D \sim (D_1 \cup D_2)$ is measurable.

   (b) Prove that the product of two measurable extended real-valued functions is measurable.

   (c) If f and g are measurable extended real-valued functions and α a fixed number, then $f + g$ is measurable if we define $f + g$ to be α whenever it is of the form $\infty - \infty$ or $-\infty + \infty$.

   (d) Let f and g be measurable extended real-valued functions which are finite almost everywhere. Then $f+g$ is measurable no matter how it is defined at points where it has form $\infty - \infty$.

6. Prove Propositon 5.11 by establishing the following lemmas:

   (a) Given a measurable function f on $[a, b]$ which takes the values $\pm\infty$ only on a set of measure zero, and given $\varepsilon > 0$, there is an M such that $|f| < M$ except on a set of measure less than $\varepsilon/3$.

(b) Let f be a measurable function on $[a, b]$. Given $\varepsilon > 0$ and M, there is a simple function φ such that $|f(x) - \varphi(x)| < \varepsilon$ except where $|f(x)| \geq M$. If $m \leq f \leq M$, then we may take φ so that $m \leq \varphi \leq M$.

(c) Given a simple function φ on $[a, b]$, there is a step function g on $[a, b]$ such that $g(x) = \varphi(x)$ except on a set of measure less than $\varepsilon/3$. [Hint: Use Propositon 3.27.] If $m \leq \varphi \leq M$, then we can take g so that $m \leq g \leq M$.

(d) Given a step function g on $[a, b]$, there is a continuous function h such that $g(x) = h(x)$ except on a set of measure less than $\varepsilon/3$. If $m \leq g \leq M$, then we may take h so that $m \leq h \leq M$.

7. Let f be measurable and B a Borel set. Then $f^{-1}[B]$ is a measurable set.

   [Hint: The class of sets for which $f^{-1}[E]$ is measurable is a σ-algebra.]

8. Show that if f is a measurable real-valued function and g a continuous function defined on $(-\infty, \infty)$, then $g \circ f$ is measurable.

9. How much of the preceding problem can be carried out if we replace the class $\mathcal{B}$ of Borel sets by an arbitrary σ-algebra $\mathfrak{a}$ of sets?

Chapter 4

LEBESGUE INTEGRAL

The concept of the Riemann integral is familiar from calculus and the theory of Riemann integration though very useful and adequate for solving many mathematical problems, both pure and applid, is not free from defects. It does not meet the needs of a number of important branches of mathematics and physics of comparatively recent development. First of all, the Riemann integral of a function is defined on a closed interval and cannot be defined on an arbitrary set. Investigations in probability theory, partial differential equations, hydromechanics and quantum mechanics often pose problems which require integration over sets. Second and more important is the fact that the Riemann integrability depends upon the continuity of the function. Of course, there are functions which are discontinuous and yet Riemann-integrable, but these functions are continuous almost everywhere.

4.1 The Riemann Integrals

The classical definition of an integral, given first by Cauchy and later developed by Riemann, runs as follows.

Let f be a bounded real-valued function defined on the interval

$[a, b]$ and let

$$P = \{a = x_0 < x_1 < \cdots < x_n = b\}$$

be a partition(or subdivision) of $[a, b]$. For each partition P of $[a, b]$, consider the sums

$$S(P) = \sum_{i=1}^{n}(x_i - x_{i-1})M_i \qquad < \text{Upper Riemann Sum} >$$

and

$$s(P) = \sum_{i=1}^{n}(x_i - x_{i-1})m_i \qquad < \text{Lower Riemann Sum} >$$

where $M_i = \sup\{f(x) : x \in [x_{i-1}, x_i]\}$ and $m_i = \inf\{f(x) : x \in [x_{i-1}, x_i]\}$, for $i = 1, 2, \cdots, n$.

The upper Riemann integral of f over $[a, b]$ is defined by

$$\Re\overline{\int_a^b} f(x)dx = \inf S(P)$$

and the lower Riemann integral of f over $[a, b]$ is defined by

$$\Re\underline{\int_a^b} f(x)dx = \sup s(P),$$

where the supremum and infimum are taken over all possible partitions P of $[a, b]$.

Then $S(P)$ and $s(P)$ depend on the partition P, and for any partition P, $s(P) \leq S(P)$.

Also the upper integral is always at least as large as the lower integral.

1.1 Definition *For any bounded function $f : [a, b] \to \mathbb{R}$ If*

$$\Re\overline{\int_a^b} f(x)dx = \Re\underline{\int_a^b} f(x)dx,$$

*we say that f is **Riemann integrable** on $[a, b]$ and call this common value the **Riemann integral** of f. We shall denote it by*

$$\mathcal{R}\int_a^b f(x)dx$$

to distinguish it from the Lebesgue integral, which we shall consider later.

Remark : The Riemann integral can also be defined as a limit of finite sums(Ref.[22]).

Not every bounded measurable function on $[a, b]$ is Riemann integrable. It may be noted that in order that the function f be Riemann integrable, it is necessary for it to be bounded.

1.2 Example Let f be a function defined on $\mathbb{R}$ by

$$f(x) = \begin{cases} 0 & \text{if } x \text{ is irrational} \\ 1 & \text{if } x \text{ is rational.} \end{cases}$$

Then f is bounded and measurable

Let

$$P = \{a = x_0 < x_1 < x_2 < \cdots < x_n = b\}$$

be a partition(or subdivision) of $[a, b]$.

The upper Riemann integral of f over $[a, b]$

$$\Re \overline{\int_a^b} f(x)dx = \inf \ S(P) = b - a$$

and the lower Riemann integral of f over $[a, b]$ is defined by

$$\Re \underline{\int_a^b} f(x)dx = \sup \ s(P) = 0$$

It follows that f is not Riemann integrable on $[a, b]$.

Remark : The above example tells us that the simple funciton, which is the elementary function, fails to be Riemann integrable. Thus, f fails to be Riemann integrable. In fact, Lebesgue gave the following characterization of the Riemann integrable functions in terms of their discontinuities.

1.3 Theorem *A bounded functions is Riemann integrable if and only if it is continuous almost everywhere.*

Proof. For the proof, refer to [11]. □

4.2 The Lebesgue Integral of Bounded Functions

The concept of the Riemann integral, applies only to functions which are either continuous or else do not have "too many" points of discontinuity. Hence we cannot form the Riemann integral of a general measurable function f. In fact, f may be discontinuous everywhere, or it may even be meaningless to talk about the continuity of f in the case where f is defined on an abstract set. For such functions, Lebesgue introduced the concept of an integral based on the measure theory that generalizes the Riemann integral.

Let us try to look at the definition of the Riemann integral of a bounded function through a slightly different angle; namely, through step functions, as below.

Let ψ be a step function(cf. III-5) defined on the closed interval $[a, b]$. Then

$$\psi(x) = c_i, \quad x_{i-1} < x < x_i \quad (i = 1, 2, \cdots, n)$$

where $\{a = x_0 < x_1 < x_2 < \cdots < x_n = b\}$ is a partition of $[a, b]$. Define the **elementary integral** of ψ over $[a, b]$ as

$$\int_a^b \psi(x)dx = \sum_{i=1}^{n} c_i(x_i - x_{i-1}).$$

The elementary integral of ψ is defined since the representation of ψ is unique and the sum is finite.

It means that the upper Riemann sum $S(P)$ and the lower Riemann sum $s(P)$ are the integrals of a step function ϕ which is determined by the function f and a partition P. That is, if we define step functions ϕ and ψ by

$$\psi(x) = \begin{cases} M_i, & x_{i-1} \leq x < x_i, (i = 1, 2, 3, \cdots, n) \\ M_n, & x = b. \end{cases}$$

$$\phi(x) = \begin{cases} m_i, & x_{i-1} \leq x < x_i, (i = 1, 2, 3, \cdots, n) \\ m_n, & x = b. \end{cases}$$

$$\text{where} \quad \begin{aligned} M_i &= \sup\{f(x) | x_{i-1} \leq x < x_i\} \\ m_i &= \inf\{f(x) | x_{i-1} \leq x < x_i\} \end{aligned}$$

Then $\phi(x) \leq f(x) \leq \psi(x)$ and

$$S(P) = \mathcal{R}\int \psi(x)dx, \quad s(P) = \mathcal{R}\int \phi(x)dx$$

This means that the Riemann integral of bounded functions can be considered as the infimum or supremum of Riemann integral of step functions. Furthermore, in order to obtain a class of integrable functions wider than that of Riemann integrable functions, we replace the class of step functions by larger class of functions known as simple functions (c.f. III.5)

To introduce the Lebesgue integral, let us first consider an arbitrarily given measurable subset E of the real line $\mathbb{R}$ with $m(E) < \infty$.

Let us begin with the set $S(E)$ of all simple functions $\phi : E \to \mathbb{R}$ defined on the set E.

2.1 Definition *Let $\varphi \in S(E)$. If φ vanishes outside a set of finite measure, we define the integral of φ by*

$$\int \varphi(x)dx = \sum_{i=1}^{n} a_i m(A_i)$$

when φ has the canonical representation

$$\varphi = \sum_{i=1}^{n} a_i \chi_{A_i}$$

We call this integral the ***Lebesgue integral of a simple function*** *φ with respect to the measure m on X, and sometimes abbreviate the expression for this integral to $\int \varphi$*

2.2 Definition *If E is a measurable set, we define the integral of φ over E by*

$$\int_E \varphi = \int \varphi \chi_E.$$

The elementary integral of a simple function has been defined through its canonical representation. However, we prove that the elementary integral is independent of the choice of the representation of the simple function.

2.3 Lemma *Let* $\varphi = \sum_{i=1}^{n} a_i \chi_{E_i}$, *where each* E_i *is a measurable set of finite measure and* $E_i \cap E_j = \emptyset$ *for* $i \neq j$. *Then*

$$\int \varphi = \sum_{i=1}^{n} a_i m(E_i).$$

Proof. Clearly, φ is a simple function defined on $\cup_{i=1}^{n} E_i$. Let a be an element in the range of φ. Then the canonical representataion of φ is given by

$$\varphi = \sum_{a} a \chi_{A_a},$$

where a varies over the range of φ and the set A_a is given by

$$A_a = \{x : \varphi(x) = a\} = \bigcup_{a_i = a} E_i.$$

Oberve, in view of the additivity of m, that

$$am(A_a) = \sum_{a_i = a} a_i m(E_i).$$

Hence

$$\begin{aligned} \int \varphi &= \sum_{a} am(A_a) \\ &= \sum_{a} \sum_{a_i = a} a_i m(E_i) \\ &= \sum_{i=1}^{m} a_i m(E_i) \end{aligned}$$

□

Note : In view of Lemma 2.3, we can write the elementary integral for a simple function even when its representation is not canonical.

2.4 Theorem *Let φ and ψ be simple functions which vanish outside a set of finite measure. Then:*

(a) $\int a\varphi + b\psi = a\int\varphi + b\int\psi$, *for all reals a and b*

(b) *If $\varphi \geq \psi$ a.e., then $\int\varphi \geq \int\psi$.*

Proof. (a) Let $\{A_i\}$ and $\{B_i\}$ be the sets occurring in the canonical representations of φ and ψ. Let A_0 and B_0 be the sets where φ and ψ, respectively, vanish. Then

$$\begin{aligned} a\varphi + b\psi &= a\sum_{i=1}^{m} \alpha_i \chi_{A_i} + b\sum_{j=1}^{n} \beta_j \chi_{B_j} \\ &= \sum_{i=0}^{m}\sum_{j=0}^{n} (a\alpha_i + b\beta_j)\chi_{A_i \cap B_j}. \end{aligned}$$

Since the collection of sets obtained by taking all the intersection $A_i \cap B_j (i = 0, 1, 2, \cdots, m; j = 0, 1, 2, \cdots, n)$ forms a finite disjoint collection of measurable sets, by Lemma 2.3, we get

$$\begin{aligned} \int a\varphi + b\psi &= \sum_{i=0}^{m}\sum_{j=0}^{n} (a\alpha_i + b\beta_j) m(A_i \cap B_j) \\ &= \sum_{i=0}^{m}\sum_{j=0}^{n} a\alpha_i m(A_i \cap B_j) + \sum_{i=0}^{m}\sum_{j=0}^{n} b\beta_i m(A_i \cap B_j) \\ &= \sum_{i=0}^{m} a\alpha_i m\left(A_i \cap \left[\bigcup_{j=0}^{n} B_j\right]\right) + \sum_{j=0}^{n} b\beta_j m\left(\left[\bigcup_{i=0}^{m} A_i\right] \cap B_j\right) \\ &= a\sum_{i=1}^{m} \alpha_i m(A_i) + b\sum_{j=1}^{n} \beta_j m(B_j) \\ &= a\int\varphi + b\int\psi, \end{aligned}$$

in view of the fact that

$$\begin{cases} A_i \cap \left[\bigcup_{j=0}^{n} B_j\right] = A_i, & i = 0, 1, 2, \cdots, m \\ \left[\bigcup_{i=0}^{m} A_i\right] \cap B_j = B_j, & j = 0, 1, 2, \cdots, n; \end{cases}$$

and that $\alpha_0 = 0 = \beta_0$.

(b) Taking $a = 1$ and $b = -1$ in (a), we get

$$\int \varphi - \int \psi = \int (\varphi - \psi)$$

Since $\varphi - \psi \geq 0$ a.e. is a simple function, by the definition of the elementary integral, we have

$$\int (\varphi - \psi) \geq 0$$

and the result follows. □

Let $f : E \to \mathbb{R}$ be a bounded function and E a measurable set with $m(E) < \infty$. In analogy with the Riemann integral, we consider the numbers

$$\inf_{\psi \geq f} \int_E \psi \text{ and } \sup_{\varphi \leq f} \int_E \varphi,$$

where ψ and φ range over the set of all simple functions defined on the set E. These two numbers do exist and are, respectively, called the **upper Lebesgue integral**, written $\mathcal{L}\overline{\int}_E f(x)dx$, and the **lower Lebesgue integral**, written $\mathcal{L}\underline{\int}_E f(x)dx$.

It is quite clear that every bounded function defined on a set of finite measure possesses both lower and upper Lebesgue integrals of f are equal. The answer is certainly in the affirmative when f is a simple function.

2.5 Theorem *Let $f : E \to \mathbb{R}$ be a simple function. Then*

$$\mathcal{L}\underline{\int}_E f(x)dx = \int_E f = \mathcal{L}\overline{\int}_E f(x)dx.$$

Proof. Since f is a bounded function on E, there exist real numbers α and β such that

$$\begin{cases} \alpha = \inf\{f(x) : x \in E\} \\ \beta = \sup\{f(x) : x \in E\} \end{cases}$$

Consider the set

$$L(f) = \{\varphi : \varphi \text{ is a simple function defined on } E \text{ and } \varphi \le f\}$$
$$U(f) = \{\psi : \psi \text{ is a simple function defined on } E \text{ and } \psi \ge f\}$$

f being a simple function, $f \in L(f)$ and $f \in U(f)$. Hence

$$\int_E f \le \mathcal{L}\underline{\int}_E f(x)dx \le \mathcal{L}\overline{\int}_E f(x)dx \le \int_E f,$$

and the result follows. □

2.6 Definition *A bounded function f defined on a set E of finite measure is said to be* ***Lebesgue integrable*** *over E, if*

$$\mathcal{L}\overline{\int}_E f(x)dx = \mathcal{L}\underline{\int}_E f(x)dx.$$

The common value is called the ***Lebesgue integral*** *of f over E, written $\mathcal{L}\int_E f(x)dx$.*

Remark : Every simple function is Lebesque integrable and its Lebesgue integral is nothing but the same as the elementary integral of f.

Note : Since we deal mostly with Lebesgue integrals, we shall write $\int_E f(x)$ for $\mathcal{L}\int_E f(x)$, and sometimes simply as $\int_E f(x)$. When $E = [a,b]$, the integral $\int_{[a,b]} f$ is traditionally denoted by $\int_a^b f$. If f is a bounded measurable function which vanishes outside a set E of finite measure, we write $\int f$ for $\int_E f$. Obsetve that $\int_E f = \int f \cdot \chi_E$. However, for Riemann integral, we use the notation $\Re\int_a^b f(x)dx$ in order to distinguish it from the Lebesgue integral.

Next, we show the important role played by measurable functions. We shall prove that every bounded measurable function defined on a set whose measure is finite is Lebesgue integrable.

2.7 Theorem *A bounded function f defined on a measurable set E of finite measure is Lebesgue integrable if and only if f is measurable.*

Proof. Let f be Lebesgue integrable over E. Then

$$\inf_{\psi \geq f} \int_E \psi(x)dx = \sup_{\varphi \leq f} \int_E \varphi(x)dx = I(\text{say}),$$

for all simple function φ and ψ.

Given an integer n, $\exists$ simple functions φ_n and ψ_n such that $\varphi_n(x) \leq f(x) \leq \psi_n(x)$ satisfying

$$\int_E \psi_n(x)dx < I + \frac{1}{2n}$$

and

$$\int_E \varphi_n(x)dx > I - \frac{1}{2n}.$$

This gives

$$\int_E \psi_n(x)dx - \int_E \varphi_n(x)dx < \frac{1}{n}. \tag{1}$$

Define the funtions

$$\psi^* = \inf \psi_n \text{ and } \varphi^* = \sup \varphi_n.$$

Since for each n, φ_n and ψ_n are measurable functions, the functions φ^* and ψ^* are measurable (cf. III.4.15) and $\varphi^*(x) \leq f(x) \leq \psi^*(x)$. Now, consider the sets

$$\begin{aligned} \Delta &= \{x : \varphi^*(x) < \psi^*(x)\}, \\ \Delta_v &= \{x : \varphi^*(x) < \psi^*(x) - \frac{1}{v}\}, \\ \Delta_{v,n} &= \{x : \varphi_n(x) < \psi_n(x) - \frac{1}{v}\}. \end{aligned}$$

We note that:

(a) $\Delta = \bigcup_{v=1}^{\infty} \Delta_v$

(b) $\Delta \subset \Delta_{v,n}$, $\forall n$

(c) $m(\Delta_{v,n} < v/n$; for if $m(\Delta_{v,n}) \geq v/n$, then

$$\begin{aligned}\int_{\Delta_{v,n}} \psi_n(x)dx - \int_{\Delta_{v,n}} \varphi_n(x)dx &= \int_{\Delta_{v,n}} \{\psi_n(x) - \varphi_n(x)\}\, dx \\ &> \frac{1}{v} m(\Delta_{v,n}) \\ &\geq \frac{1}{n},\end{aligned}$$

which contradicts (1).

Since n is arbitrary, we, in view of (b) and (c) above, have $m(\Delta_v) = 0$ and hence $m(\Delta) = 0$. This proves that $\varphi^* \geq \psi^*$ a.e. But $\varphi^* \leq \psi^*$. Hence $\varphi^* = \psi^* = f$ a.e., and since each of the functions φ^* and ψ^* is measurable, the function f is measurable.

On the other hand, assume that f is a measurable function on E. Suppose f is bounede by M. Then

$$-M \leq f(x) \leq M, \quad \forall x \in E.$$

Divide the inteval $[-M, M]$ into $2n$ equal parts and consider thd sets

$$E_k = \left\{x \in E : \frac{M}{n} k \geq f(x) > \frac{M}{n}(k-1)\right\}, \quad -n \leq k \leq n.$$

Clearly, $\{E_k : -n \leq k \leq n\}$ is a countable collection of pairwise disjoint measurable sets such that $E = \cup_{k=-n}^{n} E_k$. Therefore,

$$m(E) = \sum_{k=-n}^{n} m(E_k)$$

For each n, if we define simple functions ψ_n and φ_n as

$$\psi_n(x) = \frac{M}{n} \sum_{k=-n}^{n} k \chi_{E_k}(x)$$

and

$$\varphi_n(x) = \frac{M}{n} \sum_{k=-n}^{n} (k-1) \chi_{E_k}(x),$$

then they satisfy $\varphi_n(x) \leq f(x) \leq \psi_n(x)$. Thus,

$$\begin{cases} \inf\limits_{\psi \geq f} \int_E \psi(x)dx \leq \int_E \psi_n(x)dx = \dfrac{M}{n} \sum\limits_{k=-n}^{n} km(E_k) \\ \sup\limits_{\varphi \leq f} \int_E \varphi(x)dx \geq \int_E \varphi_n(x)dx = \dfrac{M}{n} \sum\limits_{k=-n}^{n} (k-1)m(E_k) \end{cases}$$

$$\Rightarrow \qquad \inf_{\psi \geq f} \int_E \psi(x)dx - \sup_{\varphi \leq f} \int_E \varphi(x)dx \leq \frac{M}{n} \sum_{k=-n}^{n} m(E_k) = \frac{M}{n} m(E).$$

Since n is arbitrary, we have

$$0 \leq \inf_{\psi \geq f} \int_E \psi(x)ds - \sup_{\varphi \leq f} \int_E \varphi(x)dx \leq 0.$$

Hence f is Lebesgue integrable on E. □

Remark : The connection between integrability and measurability provides the major justification for introducing the ideas of measurable sets and measurable functions in the preceding chapters.

4.3 Comparison of Riemann Integral and Lebesgue Integral

In this section, we undertake to establish that the Lebesgue integral is a generalization of the Riemann integral.

3.1 Theorem *Let f be a bounded function defined on $[a, b]$. If f is Riemann integrable over $[a, b]$. then it is Lebesgue integrable and*

$$\Re \int_a^b f(x)dx = \int_a^b f(x)dx.$$

Proof. Since f is Riemann integrable over $[a, b]$, we have

$$\inf_{\psi_1 \geq f} \int_a^b \psi_1(x)dx = \sup_{\varphi_1 \leq f} \int_a^b \varphi_1(x)dx = \Re \int_a^b f(x)dx.$$

where φ_1 and ψ_1 vary over all step functions defined on $[a, b]$

But every step function is a simple function, Therefore,

$$\sup_{\varphi_1 \leq f} \int_a^b \varphi_1(x)dx \leq \sup_{\varphi \leq f} \int_a^b \varphi(x)dx$$

$$\inf_{\psi_1 \geq f} \int_a^b \psi_1(x)dx \geq \inf_{\psi \geq f} \int_a^b \psi(x)dx,$$

where φ and ψ very over all the simple functions defined on $[a, b]$. Thus, in view of the above relations, we have

$$\Re \int_a^b f(x)dx \leq \sup_{\varphi \leq f} \int_a^b \varphi(x)dx \leq \inf_{\psi \geq f} \int_a^b \psi(x)dx \leq \Re \int_a^b f(x)dx$$

$$\Rightarrow \quad \sup_{\varphi \leq f} \int_a^b \varphi(x)dx = \inf_{\psi \geq f} \int_a^b \psi(x)dx = \Re \int_a^b f(x)dx$$

$$\Rightarrow \quad \int_a^b f(x)dx = \Re \int_a^b f(x)dx.$$

□

Remark : Though all bounded measurable functions are Lebesgue integrable, there are functions of this kind which are not Riemann integrable.

3.2 Example Consider the Dirichlet function $f : [0, 1] \to \mathbb{R}$ defined by

$$f(x) = \begin{cases} 1 & \text{if } x \text{ is rational} \\ 0 & \text{if } x \text{ is irrational.} \end{cases}$$

The function f is clearly bounded and measurable on $[0, 1]$ and hence Lebesgue integrable. Also $\int_0^1 f(x) = 0$. However, f is not Riemann integrable since

$$\Re \overline{\int_0^1} f(x)dx = 1 \quad \text{and} \quad \Re \underline{\int_0^1} f(x)dx = 0.$$

Note : We have observed that a bounded function defined on an interval may be Lebesgue integrable without being Riemann integrable (cf. Example 3.2). On the other hand, if a function is Riemann integrable, then it is Lebesgue integrable too (cf. Theorem 3.1) and the two integrals are equal. Thus the class of Lebesgue integrable functions is wider than that of Riemann integrable functions. However, the situatiion is entirely different in the case of improper Riemann integrals. There do exist improper Riemann integrable functions which are not Lebesgue integrable.

3.3 Theorem *Let f and g be bounded measurable functions defined on a set E of finite measure. Then:*

(*a*) $\int_E af = a\int_E f$, *for all real numbers a.*

(*b*) $\int_E (f+g) = \int_E f + \int_E g$.

(*c*) *If $f = g$ a.e., then*

$$\int_E f = \int_E g$$

(*d*) *If $f \leq g$ a.e., then*

$$\int_E f \leq \int_E g$$

Hence $\left|\int f\right| \leq \int |f|$.

(*e*) *If $A \leq f(x) \leq B$, then $Am(E) \leq \int_E f \leq Bm(E)$.*

(*f*) *If A and B are disjoint measurable sets of finite measure, then*

$$\int_{A\cup B} f = \int_A f + \int_B f.$$

Proof. (a) The result is trivially true when $a = 0$. Assume that $a \neq 0$. if ψ is a simple function so is $a\psi$, and conversely. Hence for $a > 0$,

$$\int_E af = \inf_{a\psi \geq af} \int_E a\psi = a \inf_{\psi \geq f} \int_E \psi = a\int_E f.$$

If $a < 0$,

$$\int_E af = \inf_{a\varphi \geq af} \int_E a\varphi = a \sup_{\varphi \leq f} \int_E \varphi = a \inf_{\psi \geq f} \int_E \psi = a \int_E f,$$

(b) If ψ_1 and ψ_2 are simple functions such that $\psi_1 \geq f$ and $\psi_2 \geq g$, then $\psi_1 + \psi_2$ is a simple function and $\psi_1 + \psi_2 \geq f + g$. Hence

$$\begin{aligned} & \int_E (f+g) \leq \int_E (\psi_1 + \psi_2) = \int_E \psi_1 + \int_E \psi_2 \\ \Rightarrow \quad & \int_E (f+g) \leq \inf_{\psi_1 \geq f} \int_E \psi_1 + \inf_{\psi_2 \geq f} \int_E \psi_2 = \int_E f + \int_E g. \end{aligned}$$

Similarly, if $\varphi_1 \leq f$ and $\varphi_2 \leq g$ are simple functions, then $\varphi_1 + \varphi_2$ is a simple function with $\varphi_1 + \varphi_2 \leq f + g$, and so

$$\int_E (f+g) \geq \int_E (\varphi_1 + \varphi_2) = \int_E \varphi_1 + \int_E \varphi_2,$$

Since now the supremum of the right side is $\int f + \int g$, we have

$$\int_E f + g \geq \int_E f + \int_E g,$$

Hence the result follows.

(c) It is sufficient to show that

$$\int_E (f - g) = 0.$$

Since $f - g = 0$ a.e., it follows that if $\psi \geq f - g$, then $\psi \geq 0$ a.e. Therefore, in view of Theorem 2.4(b), we have

$$\begin{aligned} & \int_E \psi \geq 0 \\ \Rightarrow \quad & \int_E (f - g) \geq 0 \end{aligned}$$

Similarly, one can prove that

$$\int_E (f - g) \leq 0.$$

This proves the result.

(d) It is similar to that of (c).

(e) It follows from (d) and the fact that $\int_E 1 = m(E)$.

(f) It is an immediate consequence of (b) and the fact that $\chi_{E_1 \cup E_2} = \chi_{E_1} + \chi_{E_2}$, since E_1 and E_2 are disjoint sets. □

3.4 Theorem *If C and D are two disjoint measurable subsets of E with $E = C \cup D$, then*

$$\int_E f = \int_C f + \int_D f$$

holds for every measurable function $f : E \to \mathbb{R}^*$ which is Lebesgue integrable.

Proof. Applying (b) of 3.3 we obtain

$$\begin{aligned} \int_E f &= \int_E (\chi_C + \chi_D) f = \int_E \chi_C f + \int_E \chi_D f \\ &= \int_C f + \int_D f \end{aligned}$$

□

We derive certain simple results from Theorem 3.4 in the form of corollaries.

3.5 Corollary *If $f(x) \geq 0$ on E, then $\int_E f(x)dx \geq 0$; and if $f(x) \leq 0$ on E, then $\int_E f(x)dx \leq 0$.*

3.6 Corollary *If $m(E) = 0$, then $\int_E f = 0$.*

3.7 Corollaty *If $f(x) = k$ a.e. on E, then $\int_E f = km(E)$. In particular, if $f = 0$ a.e. on E, then $\int_E f = 0$;*
and if $f = 1$ a.e. on E, then $\int_E f = m(E)$.

Remarks 1. The converse of Theorem 3.3 (c) and Corollary 3.7 need not be true (cf. Problems 1 and 2)

2. The result in Theorem 3.3 (c) indicates one of the great advantages of the Lebesgue integral over the Riemann integral. If f is a bounded measurable function on $[a, b]$(and hence Lebesgue integrable over $[a, b]$), then changing the values of the function f on a set of measure zero has no effect either on the Lebesgue integrability of or on the value of its integral over $[a, b]$. On the other hand, changing the values of a Riemann integrable function on a set of measure zero may even destroy the Riemann integrability of the function.

3.8 Example Consider the function f defined by $f(x) = 1$, $(0 \le x \le 1)$. Then the characteristic function χ_A of the set A, where A is the set of all irrationals in $[0, 1]$, can be regarded as a function obtained by changing the values of f on a set of measure zero in $[0, 1]$; namely, on the rational points where it takes value zero. Now, one can easily observe that f is Riemann integrable in $[0, 1]$ while χ_A is not.

Problem 1. Show by an example that the converse of Theorem 3.3 (c) need not be true.

Solution. Let the functions $f : [-1, 1] \to \mathbb{R}$ and $g : [-1, 1] \to \mathbb{R}$ be defined by

$$f(x) = \begin{cases} 2 & \text{if } x \le 0 \\ 0 & \text{if } x > 0 \end{cases}$$

and

$$g(x) = 1, \quad \forall x.$$

Then

$$\int_{-1}^{1} f(x)dx = 2 = \int_{-1}^{1} g(x)dx.$$

But $f \neq g$. a.e. In other words, they are not equal even for a single point in $[-1, 1]$.

Problem 2. Show by an example that the converse of Corollary 3.7 need not be true.

Solution. Consider the function $f : [-1, 1] \to \mathbb{R}$ defined by

$$f(x) = \begin{cases} 1 & \text{if } x \geq 0 \\ -1 & \text{if } x < 0. \end{cases}$$

Then

$$\int_{-1}^{1} f(x)dx = \int_{-1}^{0} f(x)dx + \int_{0}^{1} f(x)dx = 0.$$

But $f \neq g$ a.e. on $[-1, 1]$.

In fact, towards the converse of Corollary 3.7 we prove the following.

3.9 Theorem *If* $\int_E f = 0$ *and* $f(x) \geq 0$ *on* E, *then* $f = 0$ *a.e.*

Proof. Suppose $A \subset E$ is the set on which $f(x) > 0$. Then

$$\begin{aligned} A &= \{x \in E : f(x) > 0\} \\ &= \bigcup_{n=1}^{\infty} \left\{x \in E : f(x) > \frac{1}{n}\right\}. \end{aligned}$$

Now it is enough to prove that the measure of each of the sets in the union is zero. Let, if possible, there be a natural number N such that

$$m\left(\left\{x \in E : f(x) > \frac{1}{N}\right\}\right) = \lambda(\neq 0).$$

Write

$$\begin{aligned} E_1 &= \left\{x \in E : f(x) > \frac{1}{N}\right\}, \\ E_2 &= \left\{x \int E : f(x) \leq \frac{1}{N}\right\}. \end{aligned}$$

Thus E_1 and E_2 are two measurable disjoint sets such that $E = E_1 \cup E_2$. Therefore

$$\int_E f = \int_{E_1} f + \int_{E_2} f.$$

But

$$\int_E f > \frac{1}{N} m(E_1) \geq \frac{\lambda}{N} > 0.$$

Consequently, $\int_E f > 0$ which contradicts the hypothesis that $\int_E f = 0$.

This completes the proof of the theorem. □

We now proceed to prove a result concerning the interchange of the limit and integral operations of a convergent sequence of bounded measurable functions.

3.10 Proposition (Bounded Convergence Theorem) *Let $\{f_n\}$ be a sequence of measurable functions defined on a set E of finite measure, and suppose that there is a real number M such that $|f_n(x)| \leq M$ for all n and all x. If $f(x) = \lim f_n(x)$ for each x in E, then*

$$\int_E f = \lim \int_E f_n.$$

Proof. The proof furnishes a nice illustration of the use of Littlewood's 'three principles'. Since the function f is the limit of a sequence of measurable functions, it is measurable (cf. III.4.16) and hence Lebesgue integrable. Futher, by Theorem III.5.15, for a given $\varepsilon > 0$, there is a measurable set $A \subset E$ with $m(A) < \varepsilon/4M$ and a positive integer N such that

$$|f_n(x) - f(x)| < \frac{\varepsilon}{2m(E)},$$

for all $n \geq N$ and $x \in E - A$. Also,

$$\begin{aligned} & |f_n(x)| \leq M, \quad \forall n \in \mathbb{N} \text{ and } x \in E \\ \Rightarrow \quad & |f(x)| \leq M, \quad x \in E \\ \Rightarrow \quad & |f_n(x) - f(x)| \leq 2M, \quad x \in E \text{ and, in particular, for } x \in A. \end{aligned}$$

Therefore

$$\begin{aligned} \left| \int_E f_n - \int_E f \right| &= \left| \int_E (f_n - f) \right| \\ &\leq \int_E |f_n - f| \\ &= \int_{E-A} |f_n - f| + \int_A |f_n - f| \end{aligned}$$

$$< \frac{\varepsilon}{2m(E)} m(E-A) + 2Mm(A)$$
$$< \varepsilon \quad \forall n \geq N.$$

Hence

$$\int_E f_n \to \int_E f. \qquad \square$$

A slightly improved form of Theorem 3.9 is the following.

3.11 Theorem *Let $\{f_n\}$ be a sequence of measurable functions defined on a set E of finite measure, and suppose that $|f_n(x)| \leq M$ for all n and all x in E. If $f_n \to f$ a.e. on E, then f is integrable and*

$$\int_E f = \lim_{n\to\infty} \int_E f_n.$$

Remark : The results in Theorem 3.10 and 3.11 need not be true in case of Riemann integrals.

3.12 Example Let $\{r_i\}$ be an enumeration of all rational numbers in $[a,b]$. Write $S_n = \{r_i : i = 1,2,3,\cdots,n\}$, $n \in \mathbb{N}$. Define, for case $n \in \mathbb{N}$, the function $f_n : [a,b] \to \mathbb{R}$ by

$$f_n(x) = \begin{cases} 1 & \text{if } x \in S_n \\ 0 & \text{if } x \notin S_n. \end{cases}$$

We note that f_n is discontinuous only at n points in $[a,b]$; namely, the points of S_n. Therefore f_n is Riemann integrable over $[a,b]$, and we have

$$\Re \int_a^b f_n(x)dx = 0 \quad \forall n \in N.$$

Futher, we observe that $f_n \to f$ on $[a,b]$, where the limit function f is given by

$$f_n(x) = \begin{cases} 1 & \text{if } x \text{ is rational} \\ 0 & \text{if } x \text{ is irrationl} \end{cases}$$

which is not Riemann integrable although f is Lebesgue integrable,cf. Example 3.2

We find from Example 12 that the result in Theorem 11 does not hold good for Riemann integral simply because the limit function of convergent sequence for Riemann integrable functions can be a function which is not Riemann integrable. However, we prove the following.

3.13 Theorem *Let $\{f_n\}$ be a sequence of Riemann integrable functions defined on $[a, b]$ such that $|f_n(x)| \leq M$, for all n and $x \in [a, b]$.*

If $\{f_n\}$ converges a.e. to a Riemann integrable function f defined on $[a, b]$, then

$$\Re \int_a^b f(x)dx = \lim_{n\to\infty} \Re \int_a^b f_n(x)dx.$$

Proof. It follows in view of Theorems 3.1 and 3.11. □

4.4 Integral of Nonnegative Measurable Functions

In this section, we generalize the definition of the Lebesgue integral to include unbounded functions, particularly nonnegative functions(measurable) defined on sets of finite or infinite measure.

4.1 Definition *If f is a nonnegative measurable function defined on a set E (measurable), we define*

$$\int_E f = \sup_{h \leq f} \int_E h,$$

where h is a bounded measurable function such that $m(\{x \in E : h(x) \neq 0\})$ is finite.

4.2 Proposition *Let f and g be nonnegative measurable functions defined on a set E. Then:*

(a) $\int_E cf = c\int_E f, \quad c > 0$

(b) $\int_E (f+g) = \int_E f + \int_E g.$

(c) *If $f \leq g$ a.e., then* $\int_E f \leq \int_E g.$

Proof. (a) By Definition 4.1, we have

$$\int_E cf = \sup_{ch \le cf} \int_E ch = c \sup_{h \le f} \int_E h = c \int_E f.$$

(b) Let h and k be bounded and measurable functions such that $h \le f$, $k \le g$ and vanish outside sets of finite measure, i.e.

$$\begin{cases} m(\{x \in E : h(x) \ne 0\}) < \infty \\ m(\{x \in E : k(x) \ne 0\}) < \infty. \end{cases}$$

Then, we have $h + k \le f + g$, and so

$$\begin{aligned} & \int_E h + \int_E k = \int_E (h+k) \le \int_E (f+g) \\ \Rightarrow \quad & \sup_{h \le f} \int_E h + \sup_{k \le g} \int_E k \le \int_E (f+g). \end{aligned}$$

This gives

$$\int_E f + \int_E g \le \int_E (f+g).$$

To prove the reverse inequality, let ψ be a bounded measurable function which vanishes outside a set of finite measure and satisfies $\psi \le f + g$. Define functions h and k by setting

$$\begin{cases} h(x) = \min(f(x), \psi(x)) \\ k(x) = \psi(x) - h(x) \end{cases}$$

Now, we claim that $h(x) \le f(x)$ and $k(x) \le g(x)$. The first assertion is obvious from the definition. In order to establish the second relation we note that

$$h(x) + k(x) = \psi(x) \le f(x) + g(x).$$

If $h(x) = \psi(x)$, then $k(x) = 0$ and hence $k(x) \le g(x)$ since g is nonnegative; if $h(x) = f(x)$, then obviously $k(x) \le g(x)$.

Since h and k are bounded by the bounds of ψ, and also vanish where ψ vanishes, we have

$$\begin{aligned} \int_E \psi &= \int_E (h+k) = \int_E h + \int_E k \\ &\le \int_E f + \int_E g \\ \sup_{\psi \le f+g} \int_E \psi &\le \int_E f + \int_E g. \end{aligned}$$

Hence

$$\int_E f + \int_E g \geq \int_E (f+g).$$

This completes the proof of (b).

(c) It follows directly in view of Definition 4.1 and Theorem 3.3 (d). □

Problem 1. Let f be a nonnegative measurable function. Show that $f = 0$ a.e. on E if and only if $\int_E f = 0$.

4.3 Theorem (Fatou's Lemma) *Let $\{f_n\}$ be a sequence of nonnegative measurable functions and $f_n \to f$ a.e. on E. Then*

$$\int_E f \leq \varliminf_{n\to\infty} \int_E f_n.$$

Proof. We may assume, without any loss of generality, that the sequence $\{f_n\}$ converges to f everywhere on E since the integrals over sets of measure zero are zero.

Let h be a bounded measurable function such that $k \leq f$ and vanishes outside a set of finite measure, viz.

$$m(\{x \in E : h(x) \neq 0\}) < \infty$$

Let us denote this set by E'. Define a sequence $\{h(x)\}$ of functions by setting

$$h_n(x) = \min\{h(x), f_n(x)\}.$$

Then each h_n is clearly bounded by the bounds of h and vanishes outside E'. Moreover, we note that

$$\begin{aligned} \lim_{n\to\infty} h_n(x) &= \lim_{n\to\infty} \min\{h(x), f_n(x)\} \\ &= \min\{h(x), f(x)\} \\ &= h(x), \quad x \in E'. \end{aligned}$$

Thus $\{h_n\}$ is a uniformly bounded sequence of measurable functions such that $h_n \to h$ on E'. Therefore, by the Bounded Convergence Theorem 3.9, we have

$$\lim_{n\to\infty}\int_{E'} h_n = \int_{E'} h$$

$$\Rightarrow \qquad \int_E h = \int_{E'} h = \lim_{n\to\infty}\int_{E'} h_n \le \underline{\lim}_{n\to\infty}\int_E f_n.$$

Hence, taking the supremum over all $h \le f$, we get

$$\int_E f \le \underline{\lim}_{n\to\infty}\int_E f_n.$$

□

Remark : Theorem 4.3 need not hold good, even in the presence of uniform convergence, unless the functions f_n are nonnegative,

4.4 Example Let the function $f_n : [0,1] \to \mathbb{R}$ be given by

$$f_n(x) = \begin{cases} -n & \text{if } \frac{1}{n} \le x \le \frac{2}{n} \\ 0 & \text{if otherwise.} \end{cases}$$

Then $f(x) = \lim_{n\to\infty} f_n(x) = 0$ a.e. on $[0,1]$, and

$$\underline{\lim}_{n\to\infty}\int_0^1 f_n(x)dx = -1,$$

while

$$\int_0^1 f(x)dx = 0.$$

Problem 2 Show that the strict inequality exists in Theorem 4.4.

Solution. Consider a sequence $\{f_n\}$ of functions defined on $\mathbb{R}$, where

$$f_n(x) = \begin{cases} 1 & \text{if } n \le x < n+1 \\ 0 & \text{if otherwise .} \end{cases}$$

We note that

$$\lim_{n\to\infty} f_n(x) = 0,$$

and so $f = 0$. Therefore $\int_{\mathbb{R}} f = 0$.

If we write $E_1 = (n, n+1)$ and $E_2 = \mathbb{R} - [n, n+1)$, then

$$\int_{\mathbb{R}} f_n = \int_{E_1} f_n + \int_{E_2} f_n = 1$$

Hence

$$\int_{\mathbb{R}} f < \varliminf_{n\to\infty} \int_{\mathbb{R}} f_n.$$

Note : For another example which ascertains that the strict inequality exist in Theorem 4.3.

The conclusion of Theorem 4.3 is weak as it only asserts that $\int f \leq \varliminf_{n\to\infty} \int f_n$. In order to get equality between $\int f$ and $\varliminf_{n\to\infty} f_n$, we should pose some more conditions on the sequence $\{f_n\}$. In fact, we prove the following.

4.5 Theorem (Monotone Convergence Theorem)
Let $\{f_n\}$ be an increasing sequence of nonnegative measurable functions, and let $f = \lim_{n\to\infty} f_n$. then,

$$\int f = \lim_{n\to\infty} \int f_n.$$

Proof. By Theorem 4.3, we have

$$\int f \leq \varliminf_{n\to\infty} \int f_n.$$

Since for each n, $f_n \leq f$ and so $\int f_n \leq \int f$, c.f. Theorem 3.3(c). This implies

$$\varlimsup_{n\to\infty} \int f_n \leq \int f,$$

and hence the result follows. □

4.6 Corollary *Let $\{u_n\}$ be a sequence of nonnegative measurable functions, and let $f = \sum_{n=1}^{\infty} u_n$. Then*

$$\int f = \sum_{n=1}^{\infty} \int u_n.$$

Proof. Setting $f_n = \sum_{i=1}^{n} u_i$, we get a sequence $\{f_n\}$ satisfying the hypothesis of Theorem 4.5. Since $\lim_{n\to\infty} f_n = \sum_{n=1}^{\infty} u_n = f$, we have

$$\begin{aligned} \int f &= \lim_{n\to\infty} \int f_n \\ &= \lim_{n\to\infty} \int \sum_{i=1}^{n} u_i. \end{aligned}$$

In view of Theorem 3.3(b), we may write

$$\int \sum_{i=1}^{n} u_i = \sum_{i=1}^{n} \int u_i.$$

Hence

$$\int f = \sum_{i=1}^{\infty} \int u_i. \qquad \square$$

4.7 Corollary *Let $\{E_i\}$ be a sequence of disjoint measurable sets and $E = \cup_i E_i$. If f is a nonnegative measurable function defined on E, then,*

$$\int_E f = \sum_i \int_E f.$$

Proof. Set $f_i = f \cdot \chi_{E_i}$. Clearly, each f_i is a nonnegative measurable function. Also

$$\begin{aligned} f \cdot \chi_E &= f \cdot \chi_{\cup_i E_i} \\ &= f(\chi_{E_1} + \chi_{E_2} + \cdots + \chi_{E_n} + \cdots +) \\ &= \sum_i f \cdot \chi_{E_i} \\ &= \sum_i f_i \end{aligned}$$

Threrfore, using Corollary 2.25, we obtain

$$\begin{aligned} & \int f \cdot \chi_E = \sum_{i=1} \int f_i = \sum_i \int f \cdot \chi_{E_i} \\ \Rightarrow \quad & \int_E f = \int \sum_i \int_{E_i} f. \end{aligned} \qquad \square$$

Problem 3. Show that Theorem 4.5 need not hold good for a decreasing sequence of functions.
Hint : Ref [11]

Problem 4. Prove that if $\{f_n\}$ is a sequence of nonnegative measurable functions defined on E, then

$$\int_E \underline{\lim}_{n\to\infty} f_n \leq \underline{\lim}_{n\to\infty} \int_E f_n$$

Hint : Ref [11]

Remark : The result in Preblem 4 is a generalization of Fatou's Lemma (cf. Theorem 4.3).

Unlike the convention in the case of the Riemann integral, we now acknowledge ∞ as a possible value for a Lebesgue integral. It is necessary to distinguish Lebesgue integrals having finite and infinite values.

4.8 Definition *A nonnegative measurable function f defined on a set E (measurable) is said to be* ***integrable*** *if*

$$\int_E f < \infty.$$

Note : Every nonnegative measurable function has an integral (Lebesgue) but only functions having finite integrals are called integrable.

4.9 Theorem *Let f and g be nonnegative measurable functions. If $f > g$ on E and f is integrable over E, then g is also integrable over E and*

$$\int_E (f-g) = \int_E f - \int_E g.$$

Proof. By Theorem 3.3(b), we have

$$\int_E f = \int_E (f-g+g) = \int_E (f-g) + \int_E g, \tag{1}$$

since the functions $f-g$ and g are nonnegative and measurable. Futher, f being integrable over E, $\int_E f < \infty$. Therefore, each integral on the right of (1) is finite. In particular,

$$\int_E g < \infty.$$

This verifies that g is an integrable function over E. The second part of the theorem is now obvious. □

4.10 Proposition *Let f be a nonnegative function which is interable over a set E. Then given $\varepsilon > 0$ there is a $\delta > 0$ such that for every set $A \subset E$ with $mA < \delta$ we have*

$$\int_A f < \varepsilon.$$

Proof. The proposition would be trivial if f were bounded. Set $f_n(x) = f(x)$ if $f(x) \leq n$ and $f_n(x) = n$ otherwise. Then each f_n is bounded and f_n converges to f at each point. By the monotone convergence theorem there is an N such that $\int_E f_N > \int_E f - \varepsilon/2$, and $\int_E f - f_N < \varepsilon/2$. Choose $\delta < \frac{\varepsilon}{2N}$. If $mA < \delta$, we have

$$\begin{aligned} \int_A f &= \int_A (f - f_N) + \int_A f_N \\ &< \int_E (f - f_N) + NmA \\ &< \frac{\varepsilon}{2} + \frac{\varepsilon}{2} = \varepsilon. \end{aligned}$$

□

Problem 5. Evaluate the integral (Lebesgue) of the function $f : [0,1] \to \mathbb{R}$ defined by

$$f(x) = \begin{cases} \frac{1}{\sqrt[3]{x}} & \text{if } 0 < x \leq 1 \\ 0 & \text{if } x = 0. \end{cases}$$

Solution. Here f is an unbounded nonnegative measurable function defined on $[0,1]$. Define a sequence $\{f_n\}$ of functions on $[0,1]$, where

$$f_n(x) = \begin{cases} f(x) & \text{if } f(x) \leq n \\ n & \text{if } f(x) > n. \end{cases}$$

In other words,

$$f_n(x) = \begin{cases} \frac{1}{\sqrt[3]{x}} & \text{if } x \geq \frac{1}{n^3} \\ n & \text{if } x \leq \frac{1}{n^3}. \end{cases}$$

Clearly, as in proof of Theorem 4.10, we observe that $\{f_n\}$ is an increasing sequence of nonnegative measurable function such that $f_n \to f$ on $[0,1]$. Therefore, by the Monotone Convergence Theorem, we get

$$\begin{aligned}\int_0^1 f &= \lim_{n\to\infty}\int_0^1 f_n = \lim_{n\to\infty}\left[\int_0^{n^{-3}} n dx + \int_{n^{-3}}^1 x^{-1/3}dx\right]\\ &= \lim_{n\to\infty}\left[\frac{1}{n^2}+\frac{3}{2}-\frac{3}{2n^2}\right] = \frac{3}{2}.\end{aligned}$$

Hence, by Definition 4.8, f is integrable over $[0,1]$, and the value of its integral is $3/2$.

Remark : We note that the integral

$$\int_0^1 \frac{1}{\sqrt[3]{x}}dx.$$

regarded as a Riemann integral is an improper integral, whereas, regarded as a Lebesgue integral, it is perfectly proper, even though the integrand is unbounded, Also, observe that the value of the integral in each case is the same, i.e. $3/2$.

Problem 6. Evaluate the integral of the function $f : [0,1] \to \mathbb{R}$ defined by

$$f(x) = \begin{cases} \frac{1}{x} & \text{if } 0 < x \le 1 \\ 0 & \text{if } x = 0 \end{cases}$$

and show that f is not Lebesgue integrable on $[0,1]$.
Hint : Ref. [11]

PROBLEMS [4.4]

1. Let $\{f_n\}$ be a sequence of nonnegative functions on $(-\infty,\infty)$ such that $f_n \to f$ a.e., and suppose that $\int f_n \to \int f < \infty$. Then for each measurable set E we have $\int_E f_n \to \int_E f$.

2. (a) Let f be a bounded function on $[a,b]$. and let h be the upper envelope of f. Then $\mathfrak{R}\int_a^b f = \int_a^b h$ (If $\varphi \ge f$ is

a step function, then $\varphi \geq h$ except at a finite number of points and so $\int_a^b h \leq \Re\overline{\int_a^b} f$. But there is a sequence $\{\varphi_n\}$ of step functions such that $\varphi_n \downarrow h$. By Proposition 3.9 we have $\int_a^b h = \lim \int_a^b \varphi_n \geq \Re\overline{\int_a^b} f)$.

(b) Use part (a) to prove *Lebesgue's Theorem*: A bounded function f on $[a, b]$ is Riemann integrable if and only if the set of points at which f is discontinuous has measure zero.

3. Let f be a nonnegative measurable function. Show that $\int f = 0$ implies $f = 0$ a.e.

4. Let f be a nonnegative measurable function.

(a) Show that there is an increasing sequence $\{\varphi_n\}$ of nonnegative simple functions each of which vanishes outside a set of finite measure such that $f = \lim \varphi_n$.

(b) Show that $\int f = \sup \int \varphi$ over all simple functions $\varphi \leq f$.

5. Let $\{f_n\}$ be a sequence of nonnegative measurable functions which converges to f, and suppose $f_n \leq f$ for each n, then

$$\int f = \lim \int f_n.$$

4.5 The Lebesgue Integral of General Functions

By the positive part f^+ of a function f we mean the function $f^+ = f \vee 0$; that is,

$$f^+(x) = \max\{f(x), 0\}.$$

Similarly, we define the negative part f^- by $f^- = (-f) \vee 0$. If f is measurable, so are f^+ and f^- and non-negative. Also We have

$$f = f^+ - f^-$$

and

$$|f| = f^+ + f^-.$$

With these notions in mind we make the following definition:

5.1 Definition *A measurable function f is said to be **integrable** over E if f^+ and f^- are both integrable over E. In this case we define the Lebesgue integral, by taking*

$$\int_E f = \int_E f^+ - \int_E f^-.$$

5.2 Theorem *A function f is integrable over E if and only if $|f|$ is integrable over E. And $\left|\int_E f\right| \leq \int_E |f|$ hold.*

Proof. If f is integrable over E, it follows that f^+ and f^- both are integrable over E, By Theorem 4.2 (b), we have

$$\int_E |f| = \int_E f^+ + \int_E f^-.$$

Hence the function $|f|$ is integrable over E. Conversely, if $\int_E |f|$ is finite, then so are $\int_E f^+$ and $\int_E f^-$, and, hence, the result follows. The proof of rest part is left to the reader. (Problem 1) □

Problem 1. If f is an integrable function, then prove that

$$\left|\int f\right| \leq \int |f|.$$

When does equality occur?

Solution. Since $|f| - f \geq 0$ and $|f| + f \geq 0$, we have

$$\int |f| \geq \int f \text{ and } \int |f| \geq -\int f.$$

Thus

$$\int |f| \geq \left| \int f \right|.$$

Futher, if $\int f \geq 0$, then $\int |f| = \int f$ giving $\int_E (|f| - f) = 0$ so that, in view of Problem 3, $|f| = f$ a.e.,; if $\int f \leq 0$, then $\int |f| = \int (-f)$ which, again in view of Problem 3, gives $|f| = -f$ a.e. Hence equality occurs when either $f \geq 0$ a.e. or $f \leq 0$ a.e.

Remark : Lebesgue integration is absolute integration in the sense that f is Lebesgue integrable if and only if $|f|$ is so. In other words, the Lebesgue integral integrates only those functions whose absolute value functions are also integrable. However, this is not always true in the case of Riemann integrals.

5.3 Example Consider a function $f : [0,1] \to \mathbb{R}$ given by

$$f(x) = \begin{cases} 1 & \text{if } x \text{ rational} \\ -1 & \text{if } x \text{ irrational.} \end{cases}$$

Observe that $|f| = 1$ on $[0,1]$ and hence Riemann integrable while f is not.

Problem 2. Show that the function $f : [0, \infty] \to \mathbb{R}$ defined by

$$f(x) = \begin{cases} \frac{\sin x}{x} & \text{if } x \neq 0 \\ 0 & \text{if } x = 0 \end{cases}$$

is not Lebesgue integrable over $[0, \infty]$.

Solution. In view of Theorem 5.2, it is sufficient to prove that

$$\int_0^\infty \left| \frac{\sin x}{x} \right| = \infty.$$

For this, consider the integral

$$\begin{aligned}
\int_0^{n\pi} \frac{|\sin x|}{x} dx &= \sum_{k=1}^{n} \int_{(k-1)\pi}^{k\pi} \frac{|\sin x|}{x} dx \\
&= \sum_{k=1}^{n} \int_0^{\pi} \frac{|\sin\{z+(k-1)\pi\}|}{z+(k-1)\pi} dz \\
&\geq \sum_{k=1}^{n} \int_0^{\pi} \frac{|\sin\{z+(k-1)\pi\}|}{k\pi} dz \\
&= \frac{1}{\pi} \sum_{k=1}^{n} \frac{1}{k} \int_0^{\pi} |\sin z| dz \\
&= \sum_{k=1}^{n} \frac{2}{k\pi}.
\end{aligned}$$

Thus

$$\begin{aligned}
&\lim_{n\to\infty} \int_0^{\pi} \frac{|\sin x|}{x} dx \geq \sum_{k=1}^{\infty} \frac{2}{k\pi} = \infty \\
\Rightarrow \quad &\int_0^{\infty} \frac{|\sin x|}{x} dx = \infty.
\end{aligned}$$

5.4 Theorem *Let f be a measurable function over a set E. If there is an integrable function g such that $|f| \leq g$. then f is integrable over E.*

Proof. Form $f^+ \leq g$, it follows that $\int_E f^+ \leq \int_E g$, and so f^+ is integrable on E. Similarly $f^- \leq g$ implies the integrability of f^-. Hence f is integrable over E. □

5.5 Proposition *If a measurable function $f : E \to \mathbb{R}^*$ is Lebesgue integrable, then it is finite almost everywhere in E, that is, the subsets $f^{-1}(\infty)$ and $f^{-1}(-\infty)$ are of measure 0.*

Proof. Let $A = f^{-1}(\infty)$ and $B = f^{-1}(-\infty)$. To prove $m(A) = 0$, let n denote any natural number. Since $n\chi_A \leq f^+$, we have

$$\int_E f^+ \geq nm(A).$$

Since f^+ is Lebesque integrable and this holds for all $n \in \mathbb{N}$, we must have $m(A) = 0$. Similarly, by considering f^-, we can also prove $m(B) = 0$. □

The following proposition is an immediate consequence of the definition.

5.6 Proposition *If $m(E) = 0$, then every function $f : E \to \mathbb{R}^*$ is measurable and Lebesgue integrable with*

$$\int_E f = 0.$$

Let $BM(E)$ be the set of all bounded measurable functions and let $LI(E)$ denote the set of all measurable functions $f : E \to \mathbb{R}^*$ which are Lebesgue integrable. In case $m(E) < \infty$, $LI(E)$ contains $BM(E)$ as a subset. The assignment $f \to \int_E f$ defines a function

$$\int_E LI(E) \to \mathbb{R}$$

which is an extension of the function $\int_E BM(E) \to \mathbb{R}$ if $m(E) < \infty$.

The elementary properties of this function are given by the following theorem which can be easily verified by means of the definitions.

5.7 Proposition *Let f and g integrable over E. Then:*

(a) The function cf is integrable over E, and $\int_E cf = c\int_E f$.

(b) The function $f + g$ is integrable over E, and

$$\int_E f + g = \int_E f + \int_E g.$$

(c) If $f \leq g$ a.e., then $\int_E f \leq \int_E g$.

(d) If A and B are disjoint measurable sets contained in E, then

$$\int_{A \cup B} f = \int_A f + \int_B f.$$

Proof. (a) If $c \geq 0$, then

$$(cf)^+ = cf^+, \quad (cf)^- = cf^-,$$

and if $c < 0$, then

$$(cf)^+ = (-c)f^-, \quad (cf)^- = (-c)f^+.$$

The result now follows since the integrability of f implies that of f^+ and f^-; and conversely.

(b) We first prove that if f_1 and f_2 are nonnegative functions with $f = f_1 - f_2$, then

$$\int_E f = \int_E f_1 - \int_E f_2. \tag{1}$$

Since $f = f^+ - f^-$, it follows that

$$\begin{aligned} & f^+ - f^- = f_1 - f_2 \\ \Rightarrow \quad & f^+ + f_2 = f_1 + f^- \end{aligned}$$

This, by Theorem 3.3 (b), gives

$$\int_E f^+ + \int_E f_2 = \int_E f_1 + \int_E f^-$$

Hence, by Theorem 4.9 we have

$$\int_E f = \int_E f^+ - \int_E f^- = \int_E f_1 - \int_E f_2$$

This establishes (1) above.

Now, if f and g are integrable, then so are $f^+ + g^+$ and $f^- + g^-$, and

$$f + g = (f^+ + g^+) - (f^- + g^-).$$

Therefore

$$\begin{aligned} \int_E (f+g) &= \int_E f^+ + \int_E g^+ - \int_E f^- - \int_E g^- \\ &= \int_E f + \int_E g. \end{aligned}$$

Alternative Proof (b) Ref[11]

(c) We note that $f \leq g$ a.e. implies $g - f \geq 0$ a.e. Therefore

$$\int_E (g-f) \geq 0.$$

Hence

$$\int_E g = \int_E (g-f) + \int_E f \geq \int_E f,$$

by using part(b).

(d)

$$\begin{aligned} \int_{A \cup B} f &= \int f \cdot \chi_{A \cup B} \\ &= \int f \cdot \chi_A + \int f \cdot \chi_B = \int_A f + \int_B f. \end{aligned}$$ □

Note : Observe that $f + g$ is not defined at points where $f = \infty$ and $g = -\infty$ or where $f = -\infty$ and $g = \infty$. However, the set of such points must have measure zero, since f and g are integrable functions. As such the integrability and the value of $f + g$ are independent of the choice of the valuee in these ambiguous cases.

Problem 3. If f is an integrable function such that $f = 0$ a.e., then show that $\int f = 0$ [Hint: Apply Problem 3 of 4.2 to functions f^+ and f^-.]

It is the restriction to nonnegative functions in Fatou's Lemma and Monotone Convergence Theorem that ensures the integrability of $\lim f_n$. If it is desired to relax the condition of nonnegativeness of the functions, some other conditions are to be imposed. One such concerns the notion of dominant function of the sequence.

5.8 Theorem (Lebesgue Convergence Theorem) *Let g be an integrable function on E and let $\{f_n\}$ be a sequence of measurable functions such that $|f_n| \leq g$ on E and $\lim_{n\to\infty} f_n = f$ a.e. on E. Then*

$$\int_E f = \lim_{n\to\infty} \int_E f_n. \tag{2}$$

Proof. Clearly each f_n is integrable over E (cf. Theorem 5.6). Further, it follows from $\lim_{n\to\infty} f_n = f$ a.e. on E and $|f_n| \le g$ a.e. on E that $|f| \le g$ a.e. on E. Hence f is integrable over E. Consider a sequence $\{h_n\}$ of functions defined by $h_n = f_n + g$. Clearly, h_n is a nonnegative and integrable function for each n. Therefore, by Fatou's Lemma, we get

$$\int_E (f+g) \le \underline{\lim}_{n\to\infty} \int_E (f_n + g)$$

$$\Rightarrow \qquad \int_E f \le \underline{\lim}_{n\to\infty} \int_E f_n.$$

Similarly, consider a sequence $\{k_n\}$ defined by $k_n = g - f_n$, and observe that k_n is a nonnegative and integrable function for each n. So, again by Fatou's Lemma, we have

$$\int_E (f+g) \le \underline{\lim}_{n\to\infty} \int_E (f_n + g)$$

$$\Rightarrow \qquad \int_E f \le \overline{\lim}_{n\to\infty} \int_E f_n.$$

Hence the theorem follows. □

5.9 Corollary *Let $\{u_n\}$ be a sequence of integrable function on E such that $\sum_{n=1}^{\infty} u_n$ converges a.e. on E. Let g be a function which is integrable on E and satisfy $|\sum_{i=1}^{n} u_i| \le g$ a.e. on E for each n, Then $\sum_{n=1}^{\infty} u_n$ is integrable on E and*

$$\int_E \sum_{n=1}^{\infty} u_n = \sum_{n=1}^{\infty} \int_E u_n.$$

Proof. Let $\sum_{i=1}^{n} u_i = f_n$. Applying Theorem 3.8 for the sequence $\{f_n\}$, we get

$$\int_E \sum_{n=1}^{\infty} u_n = \sum_{n=1}^{\infty} \int_E u_n.$$

□

5.10 Corollary *If f is integrable over E and $\{E_i\}$ is a sequence of disjoint measurable sets such that $\cup_{i=1}^{\infty} E_i = E$, then*

$$\int_E f = \sum_{i=1}^{\infty} \int_E f_i.$$

Proof. Since $\{E_i\}$ is a sequence of disjoint measurable sets, we may write

$$f = \sum_{i=1}^{\infty} f \cdot \chi_{E_i}.$$

The function $f \cdot \chi_{E_i}$ is integrable over E since $|f \cdot \chi_{E_i}| \leq |f|$ and $|f|$ is integrable over E. Morever,

$$|\sum_{i=1}^{n} f \cdot \chi_{E_i}| \leq |f|, \quad \forall n \in \mathbb{N}$$

Thus the conditions of Corollary 5.9 are satisfied, and, hence,

$$\begin{aligned} \int_E f &= \int_E \sum_{i=1}^{\infty} f \cdot \chi_{E_i} \\ &= \sum_{i=1}^{\infty} \int_E f \cdot \chi_{E_i} = \sum_{i=1}^{\infty} \int_E f \end{aligned}$$

□

Note : The result in Corollary 5.10 concerns a sequence of integrals, rather than a sequence of functions, based on a decomposition of the domain of an integrable function. This, in fact, extends the result in Theorem 5.7(d) to any countable collection of disjoint measurable sets.
Remark : The condition of integrability of f on E in Corollary 5.10 cannot be relaxed.

5.11 Example Let $E_n = [n - 1/n, n]$ and $E = \cup_{i=1}^{\infty} E_i$. Define a function $f : E \to \mathbb{R}$ given by

$$f(x) = (-1)^n n, \quad x \in E_n, \forall n \in \mathbb{N}.$$

Clearly,

$$\begin{aligned} \int_{E_n} f &= \int f \cdot \chi_{E_n} = (-1)^n n m(E_n) \\ &= (-1)^n, \quad \forall n \in \mathbb{N} \\ \Rightarrow \quad \sum_{n=1}^{\infty} \int_{E_n} f &= \sum_{n=1}^{\infty} (-1)^n. \end{aligned}$$

On the other hand

$$\int_E |f| = \infty$$

$\Rightarrow f$ if not integrable on E.

If we replace the function g in Theorem 5.8 by certain appropriate g_n's, we get the following generalization of the theorem.

5.12 Theorem *Let $\{g_n\}$ be a sequence of integrable function which converge a.e. to an integrable function g. Let $\{f_n\}$ be a sequence of measurable functions such that $|f_n| \leq g_n$ and $\{f_n\}$ converge to f a.e. If*

$$\int g = \lim_{n\to\infty} \int g_n,$$

then

$$\int f = \lim_{n\to\infty} \int f_n.$$

Proof. It follows on the lines of the proof of Theorem 3.8 by using Fatou's Lemma for the sequence $\{g_n - f_n\}$ and $\{g_n + f_n\}$. □

5.13 Corollary *Let $\{f_n\}$ be a sequence of integrable functions such that $f_n \to f$ a.e. with f integrable. Then $\int_E |f - f_n| \to 0$ if and only if*

$$\int_E |f_n| \to \int_E |f|$$

Proof. Suppose first that

$$\int_E |f_n| \to \int_E |f|$$

Then, since $|f_n - f| \leq |f_n| + |f|$ and

$$\lim_{n\to\infty} \int (|f_n| + |f|) = 2\int |f| < \infty$$

we have

$$\int |f - f_n| \to 0.$$

in view of Theorem 5.12. The other part follows trivially. □

Remark : Theorem 5.8 is false without the assumption that an integrable function g exist. First, a sequence of integrable functions, although convergent everywhere, can have a nonintegrable limit function and, secondly, even if the limit function is integrable, relation (2) may be false. In this direction, we now work out some examples.

5.14 Example

1. Let $f_n(x) = nx^n$, $0 \le x \le 1$. Then $\lim_{n\to\infty} f_n = 0$ a.e. and

$$\lim_{n\to\infty}\int_0^1 f_n(x)dx = \lim_{n\to\infty}\int_0^1 nx^n dx = \lim_{n\to\infty}\frac{n}{n+1} = 1$$

$$\Rightarrow \qquad \lim_{n\to\infty}\int_0^1 f_n \ne \int_0^1 \lim_{n\to\infty} f_n.$$

2. Let $f_n(x) = n^2x^n$, $0 \le x \le 1$. Then $\lim_{n\to\infty} f_n = 0$ a.e. and

$$\begin{aligned}\lim_{n\to\infty}\int_0^1 f_n(x)dx &= \lim_{n\to\infty}\int_0^1 n^2x^n dx \\ &= \lim_{n\to\infty}\frac{n^2}{n+1} = \infty\end{aligned}$$

$$\Rightarrow \qquad \lim_{n\to\infty}\int_0^1 f_n = \infty \ne 0 = \int_0^1 \lim_{n\to\infty} f_n.$$

P R O B L E M S [4.5]

1. Considering the sequence $\{f_n\}$ defined by

$$f_n(x) = \begin{cases} n^2x & \text{if } 0 < x \le \frac{1}{n} \\ 0 & \text{if } x > \frac{1}{n} \end{cases}$$

Show that strict inequality holds in Theorem 4.3.

2. Let g be an integrable function on a set E and suppose $\{f_n\}$ is a sequence of measurable functions such that $|f_n(x)| \le g(x)$ a.e. on E. Then

$$\int_E \underline{\lim} f_n \le \underline{\lim}\int_E f_n \le \overline{\lim}\int_E f_n \le \int_E \overline{\lim} f_n.$$

3. Let h be an integrable function and $\{f_n\}$ a sequence of measurable functions with $f_n \geq -h$ and $\lim f_n = f$. Show that $\int f_n$ and $\int f$ have a meaning and that $\int f \leq \underline{\lim} \int f_n$.

4. Establish the Riemann-Lebesgue Theorem : If f is an integrable function on $(-\infty, \infty)$, then $\lim_{n \to \infty} \int_{-\infty}^{\infty} f(x) \cos nx dx = 0$

   (Hint : The theorem is easy if f is a step function. Use problem 5)

5. (a) Let f be integrable over E. Then given $\varepsilon > 0$, there is a simple function φ such that

   $$\int_E |f - \varphi| < \varepsilon.$$

   [Apply Problem 4.2(4) to the positive and negatice parts of f.]

   (b) Under the same hypothesis there is a step function ψ such that

   $$\int_E |f - \psi| < \varepsilon.$$

   [Combine part (a) with Proposition 5.11 of III.]

Chapter 5

GENERAL MEASURE AND INTEGRATION

The purpose of the present chapter is to abstract the most important properties of Lebesgue measure and Lebesgue integration. We shall do this by giving certain axioms which Lebesgue measure satisfies and base our integration theory on these axioms. As a consequence our theory will be vaild for every system satisfying the given axioms. Next the Radon-Nikodym Theorem and the theory of signed measure will be studied.

5.1 Measure Space

We begin by recalling that a σ-algebra $\mathfrak{M}$ is a fammily of subsets of a given set X which contains $\emptyset$ and is closed with respect to complements and with respect to countable unions. By a set function μ we mean a function which assigns an extended real number to certain sets. With this in mind we make the following definitions:

Consider an arbitrarily given set X and let $\mathfrak{M}$ denote any subset of the set $\wp(X)$ of all subsets of X. A function

$$\mu : \mathfrak{M} \to \mathbb{R}^*$$

from $\mathfrak{M}$ into the extended real line $\mathbb{R}^*$, usually called a **set function**, is said to be **additive** iff

$$\mu(A \cup B) = \mu(A) + \mu(B)$$

holds for any two members A and B of $\mathfrak{M}$ satisfying

$$A \cup B \in \mathfrak{M}, \quad A \cap B = \emptyset$$

The function $\mu : \mathfrak{M} \to \mathbb{R}^*$ is said to be **finitely addition** iff

$$\mu\left(\bigcup_{i=1}^{n} E_i\right) = \sum_{i=1}^{n} \mu(E_i)$$

holds for every finite subset $\{E_1, \cdots, E_n\}$ of $\mathfrak{M}$ satisfying.

$$\bigcup_{i=1}^{n} E_i \in \mathfrak{M}, \quad E_i \cap E_j = \emptyset(j \neq j).$$

The function $\mu : \mathfrak{M} \to \mathbb{R}^*$ is saed to be **countably additive** iff

$$\mu\left(\bigcup_{i=1}^{\infty} E_i\right) = \sum_{i=1}^{\infty} \mu(E_i)$$

holds for every sequence $\{E_n : n \in \mathbb{N}\}$ in $\mathfrak{M}$ satislying.

$$\bigcup_{i=1}^{\infty} E_i \in \mathfrak{M}, \quad E_i \cap E_j = \emptyset(j \neq j).$$

1.1 Definition *Let $\mathfrak{M}$ be a σ-algebra on arbitrary set X. A* ***measure*** *on $\mathfrak{M}$ is a nonnegative set function*

$$\mu : \mathfrak{M} \to \mathbb{R}^*$$

defined on $\mathfrak{M}$ such that the following conditions are satisfied;

$(M1)$ If $E \in \mathfrak{M}$, then $\mu(E) \geq 0$
$(M2)$ For empty set $\emptyset$, $\mu(\emptyset) = 0$
$(M3)$ μ is countably additive, that is,

$$\mu\left(\bigcup_{i=1}^{\infty} E_i\right) = \sum_{i=1}^{\infty} \mu E_i$$

for any sequense E_i of disjoint measurable sets.

In this case, the triple $(X, \mathfrak{M}, \mu)$ is called a **measure space**. In other words, a measure space we mean a set X furnished with a Boolean σ-algebra $\mathfrak{M}$ of subsets of X and a measure $\mu : \mathfrak{M} \to \mathbb{R}^*$ on $\mathfrak{M}$. Likewise, if X is a set and if $\mathfrak{M}$ is a σ-algebra on X, then the pair $(X, \mathfrak{M})$ is often called a **measurable space**. The members of $\mathfrak{M}$ will be called the measurable sets of the measure space.

Note that a measute

$$\mu : \mathfrak{M} \to \mathbb{R}^*$$

on $\mathfrak{M}$ is always finitely additive because of the identity

$$\bigcup_{i=1}^{n} E_i = E_1 \cup \cdots \cup E_n \cup \emptyset \cup \emptyset \cdots$$

and the condition $\mu(\emptyset) = 0$. In the literature, nonnegative functions $\mu : \mathfrak{M} \to \mathbb{R}^*$ which are finitely additive and satisfy $\mu(\emptyset) = 0$ are called **finitely additive measures** on $\mathfrak{M}$.

Finite additivity might at first seem to be a more natural property than countable additivity. However countably additive measures on the one hand seem to be sufficient for almost all applications, and on the other hand support a much more powerful theory of integration than do finitely additive measures.

1.2 Example of measure spaces

(1) **The trivial measures** : Let X be an arbitrary set and let $\mathfrak{M} = 2^X$. Define two functions $\mu_0, \mu_m : \mathfrak{M} \to \mathbb{R}^*$ by taking

$$\mu_0(E) = 0, \quad \mu_\infty(E) = \infty$$

for every nonempty subset E of X and

$$\mu_0(\emptyset) = 0 = \mu_\infty(\emptyset)$$

Then μ_0 and μ_∞ are measures on $\mathfrak{M}$ and hence $(X, \mathfrak{M}, \mu_0)$ and $(X, \mathfrak{M}, \mu_\infty)$ are measure spaces.

(2) **The counting measure** : Let X be an arbitrary set and let $\mathfrak{M} = \wp(x)$. Define a function $\mu : \mathfrak{M} \to \mathbb{R}^*$ as follows. Let

E denote any subset of X. If E is infinite, We set $\mu(E) = \infty$; otherwise, we define $\mu(E)$ to be the number of elements in E. One can easily verify that μ is a measure on $\mathfrak{M}$ called the **conuting measure** in the set X. Thus $(X, \mathfrak{M}, \mu)$ is a measure space.

(3) **The Lebesgue measure** : Let X be the real line $\mathbb{R}$ and let $\mathfrak{M}$ denote the σ-algebra of all Lebesgue measurable subsets of $\mathbb{R}$. Define a function $m : \mathfrak{M} \to \mathbb{R}^*$ by taking, for every $E \in \mathfrak{M}$, the Lebesgue measure $m(E)$ of E. According to (III.2), m is a measure on $\mathfrak{M}$, and hence $(X, \mathfrak{M}, m)$ is a measure space. We can also take $\mathcal{B}$ to be the σ-algebra of all Borel sets of $\mathbb{R}$ and m to be the restriction of the Lebesgue measure to Borel sets.

Then the Lebesgue measure $m : \beta \to \mathbb{R}^*$ is called the **Lebesgue-Borel measure** and $(\mathbb{R}, \beta, m)$ is called the Lebesgue-Borel measure space.

(4) Let X be a non-empty set, and let $\mathcal{A}$ be a σ-algebra on X. Let x be a member of X. Define a function $\delta_x : \mathcal{A} \to [0, +\infty]$ by letting $\delta_x(A)$ be 1 if $x \in A$, and letting $\delta_x(A)$ be 0 if $x \notin A$. Then δ_x is a measure; it is called a **point mass** concentrated at x.

(5) Let X be the set of all positive integers, and let $\mathcal{A}$ be the collection of all subsets A of X such that either A or A^c is finite. Then $\mathcal{A}$ is an algebra. but not a σ-algebra (see Example 4 in Section 1.1). Define a function $\mu : \mathcal{A} \to [0, +\infty]$ by letting $\mu(A)$ be 1 if A infinite, and letting $\mu(A)$ be 0 if A is finite. It is easy to check that μ is a finitely additive measure; however it is impossible to extend μ to a countably additive measure on the σ-algebra generated by $\mathcal{A}$ if $A_k = \{k\}$ for each k, then $\mu(\cup_{k=1}^{\infty} A_k) = \mu(X) = 1$, while $\sum_{k=1}^{\infty} \mu(A_k) = 0$.

(6) Let X be an arbitrary set, and let $\mathcal{A}$ be an arbitrary σ-algebre on X. Define a function $\mu : \mathcal{A} \to [0, +\infty]$ by letting $\mu(A)$ be $+\infty$ if $A \neq \emptyset$, and letting $\mu(A)$ be 0 if $A = \emptyset$. Then μ is a measure.

(7) Let X be a set that has at least two members, and let $\mathcal{A}$ be the σ-algebra consisting of all subsets of X. Define a function $\mu : \mathcal{A} \to [0, +\infty]$ by letting $\mu(A)$ be 1 if $A \neq \emptyset$, and letting $\mu(A)$ be 0 if $A = \emptyset$. Then μ is not a measure, nor even a finitely additive measure, for if A_1 and A_2 are disjoint non-empty subsets

of X, then $\mu(A_1 \cup A_2) = 1$, while $\mu(A_1) + \mu(A_2) = 2$.

Remark : As above example, it is clear that many different σ-algebra can be defined on the same set, and so many different measure space can be defined on X.

Throughout the remainder of the present section. let $(X, \mathfrak{M}, \mu)$ denote an arbitrarily given measure space. We shall give a few of its elementary properties.

<Further properties of measurable sets>

1.3 Proposition *Let $(X, \mathfrak{M}, \mu)$ be a measure space, and let A and B be subsets of X that belong to $\mathfrak{M}$ and satisfy $A \subset B$. Then $\mu(A) \leq \mu(B)$. If in addition A satisfies $\mu(A) < +\infty$, then $\mu(B - A) = \mu(B) - \mu(A)$.*

Proof. The sets A and $B - A$ are disjoint and satisfy $B = A \cup (B - A)$; thus the additivity of μ implies that

$$\mu(B) = \mu(A) + \mu(B - A).$$

Since $\mu(B - A) \geq 0$, it follows that $\mu(B) \geq \mu(A)$. In case $\mu(A) < +\infty$, the relation $\mu(B) - \mu(A) = \mu(B - A)$ also follows. □

1.4 Theorem *Let $\{E_i\}$ be an infinite decreasing sequence of measurable sets; that is, a sequence with $E_{i+1} \subset E_i$ for each $i \in \mathbb{N}$. Let $m(E_i) < \infty$ for at least one $i \in \mathbb{N}$. Then*

$$m\left(\bigcap_{i=1}^{\infty} E_i\right) = \lim_{n \to \infty} m(E_n)$$

Proof. Set $E = \cap E_i$ Then

$$E_1 = E \cup \bigcup_{i=1}^{\infty} (E_i \sim E_{i+1}),$$

and this is a disjoint union. Hence

$$\mu E_1 = \mu E + \sum_{i=1}^{\infty} \mu(E_i \sim E_{i+1}).$$

Since

$$E_i = E_{i+1} \cup (E_i \sim E_{i+1})$$

is a disjoint union, we have

$$\mu(E_i \sim E_{i+1}) = \mu E_i - \mu E_{i+1}.$$

Hence

$$\begin{aligned} \mu E_1 &= \mu E + \sum_{i=1}^{\infty}(\mu E_i - \mu E_{i+1}) \\ &= \mu E + \lim_{n\to\infty} \sum_{i=1}^{n-1}(\mu E_i - \mu E_{i+1}) \\ &= \mu E + \mu E_1 - \lim \mu E_n, \end{aligned}$$

whence the proposition follows. □

1.5 Proposition *For every sequence $\{E_i : i \in \mathbb{N}\}$ in $\mathfrak{M}$, we have*

$$\mu\left(\bigcup_{i=1}^{\infty} E_i\right) \leq \sum_{i=1}^{\infty} \mu E_i.$$

Proof. Let $G_n = E_n \sim [\cup_{i=1}^{n-1} E_i]$. Then $G_n \subset E_n$ and the sets G_n are disjoint, Hence

$$\mu G_n \leq \mu E_n,$$

while

$$\mu(\cup E_i) = \sum_{n=1}^{\infty} \mu G_n \leq \sum_{n=1}^{\infty} \mu E_n.$$

1.6 Theorem *Let $\{E_i\}$ be an infinete increasing sequence of sets (not necessarily measurable). Then*

$$\mu\left(\bigcup_{i=1}^{\infty} E_i\right) = \lim_{n\to\infty} \mu(E_n).$$

Proof. If $\mu(E_n) = \infty$ for some $n \in \mathbb{N}$, then it follows that

$$\mu\left(\bigcup_{i=1}^{\infty} E_i\right) = \lim_{n\to\infty} \mu(E_n).$$

Hereafter, we assume $\mu(E_n) < \infty$ for every $n \in \mathbb{N}$. Let $E_0 = \emptyset$. Then we have

$$\bigcup_{i=1}^{\infty} E_i = \bigcup_{i=1}^{\infty}(E_i - E_{i-1}).$$

Since the subsets $\{E_i - E_{i+1} | i \in \mathbb{N}\}$ are disjoint, we have

$$\begin{aligned} \mu\left(\bigcup_{i=1}^{\infty} E_i\right) &= \mu\left[\bigcup_{i=1}^{\infty}(E_i - E_{i-1})\right] = \sum_{i=1}^{\infty} \mu(E_i - E_{i-1}) \\ &= \sum_{i=1}^{\infty}[\mu(E_i) - \mu(E_{i-1})] = \lim_{n\to\infty} \mu(E_n). \end{aligned}$$

This completes the proof. □

A set E is said to be of **finite measure** if $E \in \mathfrak{M}$ and $mE < \infty$. A set E is said to be of **σ-finite measure** if E is the union of a coumtable collection of measurable sets of finite measure. Any measurable set contained in a set of σ-finite measure is itself of σ-finite measure, and the union of a countable collection of sets of σ-finite measure is again of σ-finite measure. If m is σ-finite, then every measurable set is of σ-finite measure.

A measure μ is called **finite** if $m(X) < \infty$. It is called **σ-finite measure** if X is **σ-finite** that is, there is a sequence $\{x_n\}$ of sets in $\mathfrak{M}$ such that

$$X = \bigcup_{n=1}^{\infty} X_n$$

and $m(X_n) < \infty$.

If m is σ-finite, then every measurable set is of σ-finite measure.

By virtue of Proposition 1.5 we may always take $\{x_n\}$ to be a disjoint sequence of sets. Lebesgue measure on $[0, 1]$ is an example of a finite measure, Also the trivial measure m_0 in any set and the conting measure in a finite set are finite. While Lebesgue measure on $(-\infty, \infty)$ and the conting measure in a countable set are σ-finite.

On the other hand the counting measure on a countable set is σ-finite, but on an uncountable set is a measure which is not σ-finite. Other examples are given in problems

1.7 Example Let X be an arbitrary uncountable set and let $\mathfrak{M} = \{E \subset X : E \text{ is a countable or } \sim E \text{ is uncountable}\}$. Then $\mathfrak{M}$ is a σ-algebra on X(section 3.1). Define a set function $\mu : \mathfrak{M} \to \mathbb{R}^*$ by

$$\mu(E) = \begin{cases} 0, & \text{if } E \text{ is a countable set.} \\ 1, & \text{if } \sim E \text{ is a countable set.} \end{cases}$$

Then μ is a measure.

In fact, i) $\mu(E) \geq 0$ and ii) It is clear that $\mu(\emptyset) = 0$. It remains to show the countably additivity. iii) Let (E_i) be a pairwise disjoint sequence. If $\{E_i\}$ is a countable set, then $\cup_{i=1}^{\infty} E_i$ is also countable. and so

$$\mu\left(\bigcup_{i=1}^{\infty} E_i\right) = \sum_{i=1}^{\infty} \mu(E_i) = 0$$

If at least one E_i is uncountable, that is, there is n for which E_n is uncountable, then $\sim E_n$ is a countable set by the definition of $\mathfrak{M}$.

Now since $\{E_i\}$ are pairwise disjoint, $E_i \subset E_j$ for $i \neq j$. And since $\sim E_n$ is countable, E_i $(i \neq j)$ are all countable sets.

Hence $\mu(E_n) = 1$ and $\mu(E_i) = 0$ for $i \neq n$. And since $\cup_{i=1}^{\infty} E_i \in \mathfrak{M}$ and $\cup_{i=1}^{\infty} E_i$ is not countable, we obtain

$$\mu\left(\bigcup_{i=1}^{\infty} E_i\right) = 1 = \mu(E_n) = \sum_{i=1}^{\infty} \mu(E_i)$$

Thus μ is a measure and furthermore finite measure.

1.8 Example The Lebesgue measure and Lebesgue-Borel measure in Example 1.2 are σ-finite measure.

A measure space $(X, \mathfrak{M}, \mu)$ is said to be **complete** iff $\mathfrak{M}$ contains all subsets of every set of measure zero. In other words, $(X, \mathfrak{M}, \mu)$ is complete iff the conditions

$$E \in \mathfrak{M},\ \mu(E) = 0,\ D \subset E$$

imply $D \in \mathfrak{M}$ and hence $\mu(D) = 0$ by (1.3).

1.9 Example The trivial measure space (Example 1.2) and the counting measure space (Example 1.2) are complete because $\mathfrak{M} = P(x)$. As to Example 1.2(3), the Lebesgue measure space is complete if $\mathfrak{M}$ is the family of all Lebesgue measurable subset of $\mathbb{R}$. While Lebesgue measure restricted to the σ-algebra of Borel sets is not complete.

The following proposition, whose proof is left to the reader shows that each measure space can be completed by the addition of subsets of measure zero. The measure space $(X, \mathcal{B}_0, \mu_0)$ given in the proposition is called the **completion** of $(X, \mathcal{B}_0, \mu_0)$.

1.10 Proposition *If $(X, \mathfrak{M}, \mu)$ is a measure space, then we can find a complete measure space $(X, \mathfrak{M}_0, \mu_0)$ such that*

i. $\mathfrak{M} \subset \mathfrak{M}_0$

ii. $E \in \mathfrak{M} \Rightarrow \mu E = \mu_0 E$.

iii. $E \in \mathfrak{M}_0 \Leftrightarrow E = A \cup B$ *where* $B \in \mathfrak{M}$ *and* $A \subset C$, $C \in \mathfrak{M}$, $\mu C = 0$.

If $(X, \mathfrak{M}, \mu)$ is a measure space, we say that a subset E of X is locally measurable if $E \cap B \in \mathfrak{M}$ for each $B \in \mathfrak{M}$ with $\mu B < \infty$. The collection $\mathcal{A}$ of all locally measurable sets is σ-algebra containing $\mathfrak{M}$. The measure μ is called **saturated** if every locally measurable set is measurable, i.e., is in $\mathfrak{M}$. Every σ-finite measure is saturated. A measure can always be extended to a saturated measure, but, unlike the process of completion, the process of saturation is not uniquely determined: There is no defficulty with the σ-algebra, for we simply take the σ-algebra $\mathcal{A}$ of all locally measurable sets, but the measure μ can often be extended to $\mathcal{A}$ in different ways. Some details are given in Problem 10, and we shall return to this topic in the next chapter.

PROBLEMS [5.1]

1. Let $\{(X_\alpha, \mathfrak{M}_\alpha, \mu_\alpha)\}$ be a collection of measure spaces, and suppose that the sets $\{X_\alpha\}$ are disjoint. Then we can form

a new measure space (called their union) by letting $X = \cup X_\alpha$, $\mathcal{A} = \{B : (\alpha)[B \cap X_\alpha \in \mathfrak{M}_\alpha]\}$, and defining $\mu(B) = \sum \mu_\alpha(B \cap X_\alpha)$.

(a) Show that $\mathfrak{M}$ is a σ-algebra.

(b) Show that μ is a measure.

(c) Show that μ is σ-finite if and only if all but a countable number of the μ_α are zero and the remainder are σ-finite.

2. (a) Show that $\mu(E_1 \triangle E_2) = 0$ implies $\mu E_1 = \mu E_2$ provided that E_1 and $E_2 \in \mathfrak{M}$.

   (b) Show that if μ is complete, then $E_1 \in \mathfrak{M}$ and $\mu(E_1 \triangle E_2) = 0$ imply $E_2 \in \mathfrak{M}$.

3. For any two measurable sets A and B of a measure space $(X, \mathfrak{M}, \mu)$ prove:

$$\mu(A) + \mu(B) = \mu(A \cup B) + \mu(A \cap B).$$

4. For an arbitrary uncountable set X, consider the family $\mathfrak{M}$ of subsets $E \subset X$ such that either E or $X \sim E$ is countable. Prove that $\mathfrak{M}$ is a Boolean σ-algebra. Define a function $\mu : \mathfrak{M} \to \mathbb{R}^*$ by taking

$$\mu(E) = \begin{cases} 0, & \text{if } E \text{ is countable} \\ 1, & \text{otherwise} \end{cases}$$

Prove that μ is a finite measure on $\mathfrak{M}$ and $(X, \mathfrak{M}, \mu)$ is a complete measure space.

5. Let $(X, \mathfrak{M}, \mu)$ be an arnitrary measure space. Define a family $\mathfrak{M}^*$ of subsets of X as follows: an arbitrary subset E of X is in $\mathfrak{M}^*$ iff E is the union of a set $M \in \mathfrak{M}$ and a set D which is contained in a set $C \in \mathfrak{M}$ with $\mu(C) = 0$. Prove that $\mathfrak{M}^*$ is a Boolean σ-algebra of subsets of X. For an arbitrary $E \in \mathfrak{M}^*$, assume $E = M \cup D$, $M \in \mathfrak{M}$, $D \subset C$,

$C \in \mathfrak{M}$ and $\mu(C) = 0$. Prove that the measure $\mu(M)$ is uniquely determined by the set $E \in \mathfrak{M}^*$. Hence the assignment $E \to \mu^*(E) = \mu(M)$ defines a function $\mu^* : \mathfrak{M}^* \to \mathbb{R}^*$. Prove that μ^* is a measure on $\mathfrak{M}^*$ and that $(X, \mathfrak{M}^*, \mu^*)$ is a complete measure space satisfying

$$\mathfrak{M} \subset \mathfrak{M}^* \quad \mu = \mu^*|_{\mathfrak{M}}$$

This complete measure space $(X, \mathfrak{M}^*, \mu^*)$ is called the *completion* of the given measure space $(X, \mathfrak{M}, \mu)$. If the latter is complete, then we have $\mathfrak{M}^* = \mathfrak{M}$ and hence $\mu^* = \mu$.

6. Let $(X, \mathfrak{M}, \mu)$ be a measure space and $Y \in \mathfrak{M}$. Let $\mathfrak{M}_Y$ consist of those sets of $\mathfrak{M}$ which are contained in Y. Set $\mu_Y E = \mu E$ if $E \in \mathfrak{M}_Y$. Then $(Y, \mathfrak{M}_Y, \mu_Y)$ is a measure space. The measure μ_Y is called the **restriction** of μ to Y.

7. Let $(X, \mathfrak{M})$ be a measurable space.

   (a) If μ and ν are measures defind on $\mathfrak{M}$, then the set function λ definded on $\mathfrak{M}$ by $\lambda E = \mu E + \nu E$ is also a measure. We denote λ by $\mu + \nu$.

   (b) If μ and ν are measures on $\mathfrak{M}$ and $\mu \geq \nu$, then these is a measure λ on $\mathfrak{M}$ such that $\mu = \nu + \lambda$.

   (c) If ν is σ-finite, the measure λ in (b) is unique.

   (d) Show that in general the measure λ need not be unique but that these is always a smallest such λ.

8. Prove Proposition 1.10 [First show that the family $\mathfrak{M}_0$ defind by (iii) is a σ-algebra. If $E \in \mathfrak{M}_0$, show that μA is the same for all sets $A \in \mathfrak{M}$ such that $E = A \cup B$ with $\mathfrak{M}$ a set of measure zero. Use this fact to define μ_0 and show μ_0 is a measure.]

9. (a) Show that each σ-finite measure is saturated.

   (b) Show that the collection of locally measurable sets is a σ-algebra.

(c) Let $(X, \mathfrak{M}, \mu)$ be a measure space and $\mathscr{C}$ the σ-algebra of locally measurable sets. For $E \in \mathscr{C}$ set $\overline{\mu}E = \mu E$ if $E \in \mathfrak{M}$ and $\overline{\mu}E = \infty$ if $E \notin \mathfrak{M}$. Show that $(X, \mathscr{C}, \overline{\mu})$ is a saturated measure space.

(d) Give an example to show that $\overline{\mu}$ and μ may be different.

10. **σ-Rings and σ-algebras.** Some authors prefer to define a measure on a σ-ring $\mathfrak{R}$, that is, a collection of subsets of X such that if A and B are in $\mathfrak{R}$ so is $A \sim B$, and if $< A_n >$ is a sequence from $\mathfrak{R}$ then $\cup A_n$ is in $\mathfrak{R}$. Thus a σ-ring $\mathfrak{R}$ is a σ-algebra if and only if $X \in \mathfrak{R}$. Some of the relations between σ-rings and σ-algebras are given below:

(a) Let $\mathfrak{R}$ be a σ-ring which is not a σ-algebra, let $\mathscr{B}$ be the smallest σ-algebra containing $\mathfrak{R}$, and set $\mathfrak{R}' = \{E : \tilde{E} \in \mathfrak{R}\}$. Then $\mathscr{B} = \mathfrak{R} \cup \mathfrak{R}'$ and $\mathfrak{R} \cap \mathfrak{R}' = \emptyset$.

(b) If μ is a measure on $\mathfrak{R}$, define $\overline{\mu}$ on $\mathscr{B}$ by $\overline{\mu}E = \mu E$ if $E \in \mathfrak{R}$ and $\overline{\mu}E = \infty$ if $E \in \mathfrak{R}'$. Then $\overline{\mu}$ is a measure on $\mathscr{B}$.

(c) Define $\underline{\mu}$ on $\mathscr{B}$ by $\underline{\mu}E = \mu E$ if $E \in \mathfrak{R}$ and $\underline{\mu}E = \sup\{\mu A | A \subset E, A \in \mathfrak{R}\}$ if $E \in \mathfrak{R}'$. Then $\overline{\mu}$ is also a measure on $\mathscr{B}$.

(d) More generally, for any nonnegative extended real number r define μ_r on $\mathscr{B}$ by $\mu_r E = \mu E$ if $E \in \mathfrak{R}$ and $\mu_r E = \underline{\mu}E + r$ if $E \in \mathfrak{R}'$. Then μ_r is a measure on $\mathscr{B}$.

(e) If ν is any measure on $\mathscr{B}$ which agrees with μ on $\mathfrak{R}$, then $\nu = \mu_r$ for some $r \geq 0$.

5.2 Measurable Function

The concept of a measurable function on an abstract measurable space is almost identical with that for function of a real variable. Consequently, those propositions and theorems whose proofs are essentially the same as those in Section 1 of Chapter 4 are stated without proof. The reader should verify that the proofs in Chapter

4 can be carried out in the abstract case. Throughout this section we assume that a fixed measurable space $(X, \mathfrak{M})$ is given.

2.1 Proposition *Let f be an extended real-valued function defind on X. Then the following statements are equivalent :*

i. $\{x : f(x) < \alpha\} \in \mathfrak{M}$ *for all* $\alpha \in \mathbb{R}$.

ii. $\{x : f(x) \leq \alpha\} \in \mathfrak{M}$ *for all* $\alpha \in \mathbb{R}$.

iii. $\{x : f(x) > \alpha\} \in \mathfrak{M}$ *for all* $\alpha \in \mathbb{R}$.

iv. $\{x : f(x) \geq \alpha\} \in \mathfrak{M}$ *for all* $\alpha \in \mathbb{R}$.

(*See Proposition III* 1.1)

2.2 Definition *The extended real-valued function f definded on X is called measurable (or measurable with respect to $\mathfrak{M}$) if any one of the statements of Proposition* 2.1 *holds.*

Note : The measure of the set $E(f > \alpha)$ may be finite or infinite.

Problem 1. Show that a constant function with a measurable domain is measurable.

2.3 Theorem *If c is a constant and the functions f and g are measurable, then so are the functions $f + c$, cf, $f + g$, $f \cdot g$, and $f \vee g$. Moreover, if $\{f_n\}$ is a sequence of measurable functions, then* $\sup f_n$, $\inf f_n$, $\overline{\lim} f_n$, *and* $\underline{\lim} f_n$ *are all measurable.*
(*See Theorem* 1.4 *and* 1.16 *of III.*)

Next we shall generalize the notion of simple functions introduced in Chapter III.

By a **simple function** on a measurable subset E of X, we mean a measurable function

$$f : E \to \mathbb{R}$$

such that the image $f(E)$ is a finite subset of $\mathbb{R}$. And the properties and their proofs are still valid.

By a simple function we mean as before a finite linear combination

$$\varphi(x) = \sum_{i=1}^{n} c_i \chi_{E_i}(x)$$

of characteristic functions of measurable sets E_i.

2.4 Proposition *Let f be a nonnegative measurable function. Then there is a sequence $\{\varphi_n\}$ of simple functions with $\varphi_{n+1} \geq \varphi_n$ such that $f = \lim \varphi_n$ at each point of X. If f is defind on σ-finite measure space, then we may choose the functions φ_n so that each vanishes outside a set of finite measure.*

2.5 Proposition *If μ is a complete measure and f is a measurable function, then $f = g$ a.e. implies g is measurable.*
(See Proposition 4.19 of Chapter III.)

The sets $\{f : f(x) < \alpha\}$ are somtimes called ordinate sets for f. They increase with α. The following two lemmas will be useful later. They state that, given a countable collection $\{B_\alpha\}$ of measurable sets which increase with α, we can find a measurable function f which nearly has these for ordinate sets in the sense that

$$\{x : f(x) < \alpha\} \subset B_\alpha \subset \{x : f(x) \leq \alpha\}$$

2.6 Lemma *Suppose that to each α in a countable set D of real numbers these is assigned a set $B_\alpha \in \mathfrak{M}$ such that $B_\alpha \subset B_\beta$ for $\alpha < \beta$. Then there is a measurable extended real-valued function f on X such that $f \leq \alpha$ on B_α and $f \geq \alpha$ on $X \sim B_\alpha$.*

Proof. For each $x \in X$, let $f(x) = \inf\{\alpha : x \in B_\alpha\}$, where as usual $\inf \emptyset = \infty$. If $x \in B_\alpha$, $f(x) \leq \alpha$. If $x \notin B_\alpha$, $x \notin B_\beta$ for each $\beta < \alpha$ and so $f(x) \geq \alpha$. Thus we have only to show that f is measurable. But for any real α, $\{x : f(x) < \alpha\}$. For if $f(x) < \alpha$, then $x \in B_\beta$ for some $\beta < \alpha$ by the definition of f. On the other hand, if $x \in B_\beta$ for some $\beta < \alpha$, then $f(x) \leq \beta < \alpha$.

2.7 Lemma *Suppose that for each α in a countable set D of real numbers there is assigned a set $B_\alpha \in \mathfrak{M}$ such that $\mu(B_\alpha \sim B_\beta) = 0$ for $\alpha < \beta$. Then there is a measurable function f such that $f \leq \alpha$ on B_α and $f \geq \alpha$ a.e. on $X \sim B_\alpha$.*

Proof. Let $C = \cup_{\alpha<\beta} B_\alpha \sim B_\beta$. Then $\mu C = 0$. Let $B'_\alpha = B_\alpha \cup C$. If $\alpha < \beta$, we have $B'_\alpha \sim B'_\beta = (B_\alpha \sim B_\beta) \sim C = \emptyset$. Thus by

Lamma 2.6 there is a measurable f such that $f \leq \alpha$ on B'_α and $f \geq \alpha$ on $X \sim B'_\alpha$. Thus $f \leq \alpha$ on B_α and $f \geq \alpha$ on $X \sim B_\alpha$ except for $x \in C$. □

PROBLEMS [5.2]

1. Prove Proposition 2.4. [For each pair $< n, k >$ of integers let $E_{n,k} = \{x : k2^{-n} \leq f(x) < (k+1)2^{-n}\}$, and set $\varphi_n = 2^{-n} \sum_{k=0}^{2^{2n}} k\chi_{E_{n,k}}$.]

2. Prove Proposition 2.5 and show that it is false if the word 'complete' is omitted.

3. Let $(X, \mathfrak{M}, \mu)$ be a measure space and $(X, \mathfrak{M}_0, \mu_0)$ its completion. Then a function f is measurable with respact to $\mathfrak{M}_0$ if and only if there is a function g measurable with respect to $\mathfrak{M}_0$ such that $f = g$ almost everywhere in the sense that there is a set $E \in \mathfrak{M}$ with $\mu E = 0$ and $f = g$ on $X \sim E$. Note that this does not require the set $\{x : f(x) \neq g(x)\}$ to be in $\mathfrak{M}$.

4. Show that the requirement that D be countable in Lemma 6 and 7 is unnecessaey. [Replace an uncountable D by a countable subset E which is dense in D.]

5. Show that if D is dense in $\mathbb{R}$, then the function f in Lemma 6 is unique and the function f in Lamma 7 is unique to within equivalense; that is, any other function satisfying the conclusion of the lemma is equal to f almost everywhere.

6. Let D be the rationals and f the function in Lemma 6. Express the sets $\{x : f(x) < \alpha\}$, $\{x : f(x) \leq \alpha\}$, $\{x : f(x) = \alpha$ in terms of the sets $\{\mathfrak{M}_\beta\}$.

5.3 Integration

Many definrtions and proofs of Chapter III depend on only those properties of Lebesgue measure which are also true for an arbitrary measure in an abstract measure space and carry over to this case.

In the present section, we shall generaliz the construction of Lebesgue integerals to an arbitrary complete measure space. The completeness assumption is not a severe restriction because every measure space can be completed by means of Proposition 1.11 of Section 1.

If E is a measurable set and φ a nonnegative simple function, we define

$$\int_E \varphi d\mu = \sum_{i=1}^{n} c_i \mu(E_i \cap E),$$

Where

$$\varphi(x) = \sum_{i=1}^{n} c_i \chi_{E_i}(x).$$

It is easily seen that the value of this integral is independent of the representation of φ which we use. If a and b are positive numbers and φ and ψ nonnegative simple functions, then $\int a\varphi + b\psi = a\int\varphi + b\int\psi$.

In Chapter 4 we first defined the integral for bounded measurable functions and then defined the integral of a nonnegative measurable function f to be the supremum of $\int g d\mu$ as g ranged over all bounded measurable functions which vanished outside a set of finite measure. Unforturnately, this definition is not quite suitable in the case of general measures, for, if $\mathfrak{M} = \{X, \emptyset\}$ and $\mu\emptyset = 0$, $\mu X = \infty$, then we certainly want $\int 1 d\mu = \infty$. But the only measurable function g which vanishes outside a set of finite measure is $g = 0$, and hence $\sup \int g d\mu = 0$. To avoid this difficulty, we define the integral of a nonnegative measurable function directly in terms of integrals of nonnegative simple functions.

3.1 Definition *Let f be a nonnegative extended real-valued measurable function on the measure space $(X, \mathfrak{M}, \mu)$. Then $\int f d\mu$ is the supremum of the integrals $\int \varphi d\mu$ as φ ranges over all simple function with $0 \leq \varphi \leq f$.*

It follows immediately from this definition that $f < g$ implies $\int f \leq \int g$ and that $\int cf = c \int f$. In other respects, however, this definition is somewhat awkward to apply, since we are taking a supremum over a large collection of simple functions, and it is not apparent from the definition that $\int (f+g) = \int f + \int g$. Consequently, we begin our treatment of the integral by first establishing the convergence theorems. These then enable us to determine the value of $\int f$ by taking the limite of $\int \varphi_n$ for any increasing sequence $\{\varphi_n\}$ of simple functions which converge to f.

Returning to the general case, we consider for each function $f : E \to \mathbb{R}^*$ its **trivial extension** $f^* : X \to \mathbb{R}^*$ defined by

$$f^*(x) = \begin{cases} f(x), & \text{if } x \in E \\ 0, & \text{if } x \in X \backslash E. \end{cases}$$

Obviously, f^* is integrable iff f is so. In this case, we also have

$$\int f^* = \int_E f.$$

Thus we may consider only integrals over the whole set X.

Now let D denote an arbitrary measurable subset of E. According to (IV. 4.7), the restriction

$$g = f|_D : D \to \mathbb{R}^*$$

of any measurable function $f : E \to \mathbb{R}^*$ is also measurable. If f is integrable, then so is g and we have

$$\int_D g = \int_E \chi_D f.$$

For convenience, this will be denoted by

$$\int_D f = \int_D f d\mu$$

and (IV. 3.4) is still vaild.

Motivated by the convergence theorems on Lebesgue integrals given in the text of (IV.2), we shall establish these important theorems for an arbitrary measure space $(X, \mathfrak{M}, \mu)$.

3.2 Lebesgue Bounded Convergence Theorem *Let $\{f_n\}$ be a sequence of measurable functions defined on a set E of finite measure, Suppose there is a real number M such that $|f_n(x)| \leq M$, for all x and all n. If $f(x) = \lim_{n\to\infty} f_n(x)$ for each $x \in E$, then*

$$\int_E f = \lim_{n\to\infty} \int_E f_n.$$

Proof. The conclusion of the proposition would be trivial if $\{f_n\}$ converged to f uniformly. Littlewood's third principle states that, if $\{f_n\}$ converges to f pointwise, then $\{f_n\}$ is "nearly" uniformly convergent to f. A precise version of this principle is given by Proposition 3.10 of IV-3. The rest of proof is omitted. □

3.3 Theorem (Fatou's Lemma) *Let $\{f_n\}$ be a sequence of nonnegative measurable functions which converge almost everywhere on a set E to a function f. Then*

$$\int_E f \leq \underline{\lim} \int_E f_n.$$

Proof. Without loss of generality we may assume that $f_n(x) \to f(x)$ for each $x \in E$. From the definition of $\int f$ it suffices to show that, if φ is any nonnegative simple function with $\varphi \leq f$, then $\int_E \varphi \leq \underline{\lim} \int_E f_n$.

If $\int_E \varphi = \infty$, then there is a measurable set $A \subset E$ with $\mu A = \infty$ such that $\varphi > a > 0$ on A. Set $A_n = \{x \in E : f_k(x) > a \text{ for all } k \geq n\}$. Then $\{A_n\}$ is an increasing sequence of measurable sets whose union contains A, since $\varphi \leq \lim f_n$. Thus $\lim \mu A_n = \infty$. Since $\int_E f_n \geq a\mu A_n$, We have $\lim \int_E f_n = \infty = \int_E \varphi$.

If $\int_E \varphi < \infty$, then there is a measurable set $A \subset E$ with $\mu A < \infty$ such that φ vanishes identically on $E \sim A$. Let M be the maximum of φ, let ε be a given positive number, and set

$$A_n = \{x \in E : f_k(x) > (1-\varepsilon)\varphi(x) \text{ for all } k \geq n\}$$

Then $\{A_n\}$ is an increasing sequence of sets whose union contains A, and so $\{A \sim A_n\}$ is a decreasing sequence of sets whose intersection is empty. By Theorem 1.4 $\lim(A \sim A_n) = 0$, and so we can find an n such that $\mu(A \sim A_k) < \varepsilon$ for all $k \geq n$. Thus for $k \geq n$

$$\begin{aligned}\int_E f_k \geq \int_{A_k} f_k &\geq (1-\varepsilon)\int_{A_k} \varphi \\ &\geq (1-\varepsilon)\int_E \varphi - \int_{A-A_k} \varphi \\ &\geq \int_E \varphi - \varepsilon\left[\int_E \varphi + M\right].\end{aligned}$$

Hence

$$\underline{\lim}\int_E f_n \geq \int_E \varphi - \varepsilon\left[\int_E \varphi + M\right].$$

Since ε is arbitrary,

$$\underline{\lim}\int_E f_n \geq \int_E \varphi.$$

□

3.4 Monotone Convergence Theorem *Let $\{f_n\}$ be a sequence of nonnegative measurable functions which converge almost everywhere to a function f and suppose that $f_n \leq f$ for all n. Then*

$$\int f = \lim \int f_n.$$

Proof. Since $f_n \leq f$, We have $\int f_n \leq \int f$. Thus by Fatou's Lemma we have

$$\int f \leq \underline{\lim} \int f_n \leq \overline{\lim} \int f_n \leq \int f.$$

□

Problem 1. Show that Theorem 3.4 need not hold good for a decreasing sequence of functions.

Solution. Consider a sequence $\{f_n\}$ of functions each defined on $\mathbb{R}$, where

$$f(x) = \begin{cases} 0 & \text{if } x < n \\ 1 & \text{if } x \geq n. \end{cases}$$

Clearly, $\{f_n\}$ is a decreasing sequence of nonnegative measurable functions and converging to zero function, f. Thus

$$\int_{-\infty}^{\infty} f(x)dx = 0.$$

But

$$\int_{-\infty}^{\infty} f_n(x)dx = \int_{-\infty}^{n} f(x)dx + \int n^{\infty} f(x)dx$$

$$\Rightarrow \qquad \lim_{n\to\infty} \int_{-\infty}^{\infty} f_n(x)dx = \infty.$$

We are now able to establish some of the standard properties of the integral.

3.5 Corollary *Let $\{E_i\}$ be a sequence of disjoint measurable sets and $E = \bigcup_i E_i$. If f is a nonnegative measurable function defined on E, then*

$$\int_E f = \sum_i \int_E f.$$

Proof. Set $f_i = f \cdot \chi_{E_i}$. Clearly, each f_i is a nonnegative measurable function. Also

$$\begin{aligned} f \cdot \chi_E &= f \cdot \chi_{\bigcup\limits_i E_i} \\ &= f(\chi_{E_1} + \chi_{E_2} + \cdots + \chi_{E_n} + \cdots) \\ &= \sum_i f \cdot \chi_{E_i} = \sum_i f_i. \end{aligned}$$

Therefore, we obtain

$$\int f \cdot \chi_E = \sum_{i=1} \int f_i = \sum_i \int f \cdot \chi_{E_i}$$

$$\Rightarrow \int_E f = \sum_i \int_{E_i} f.$$

□

Problem 2. Prove that if $\{f_n\}$ is sequence of nonnegative measurable functions defined on E, then

$$\int_E \underline{\lim}_{n\to\infty} f_n \le \underline{\lim}_{n\to\infty} \int_E f_n.$$

Solution. Corresponding to each positive integer n, define $h_n = \inf\{f_v : v \ge n\}$. Each h_n is measurable. Since $h_n \le f_v$, for $v > n$

we have

$$\int_E h_n \leq \int_E f_v, \qquad \text{for } v \geq n$$
$$\Rightarrow \int_E h_n \leq \underline{\lim}_{v\to\infty} \int_E f_v, \quad \forall n \in \mathbb{N}.$$

One may verify that $\{h_n\}$ is an increasing sequence which converges to the limit function $h(= \underline{\lim}_{n\to\infty} f_n)$. Therefore, in view Theorem 3.3, we have

$$\int_E \underline{\lim}_{n\to\infty} f_n = \int_E h = \lim_{n\to\infty} \int_E f_n \leq \underline{\lim}_{n\to\infty} \int_E f_n.$$

Remark : The result in Problem 2 is a generalzation of Fatou's Lemma (cf. Theorem 3.3)

3.6 Proposition *If f and g are nonnegative measurable functions and a and b nonnegative constants, then*

$$\int af + bg = a\int f + b\int g$$

We have

$$\int f \geq 0$$

with equality only if $f = 0$ a.e.

Proof. To prove the first statement, let $\{\varphi_n\}$ and $\{\psi_n\}$ be increasing sequence of simple functions which converge to f and g. Then $\{a\varphi + b\psi\}$ is an increasing sequence of simple functions which converge to $af + bg$. By the Monotone Convergence Theorem

$$\begin{aligned} \int af + bg &= \lim \int a\varphi_n + b\psi_n \\ &= \lim\left(a\int \varphi_n + b\int \psi_n\right) \\ &= a\int f + b\int g. \end{aligned}$$

Clearly $\int f \geq 0$, if $\int f = 0$, let $A_n\{x : f(x) \geq 1/n\}$. Then $f \geq \frac{1}{n}\chi_{A_n}$, and so $\mu A_n = \int \chi_{A_n} = 0$. Since the set where $f > 0$ is the union of the sets A_n, it has measure zero. □

Taking this proposition together with the Monotone Convergence Theorem gives us the following corollary:

3.7 Corollary *Let $\{f_n\}$ be a sequence of nonnegative measurable functions, Then*

$$\int \sum_{n=1}^{\infty} f_n = \sum_{i=1}^{\infty} \int f_n.$$

3.8 Definition *A nonnegative function f defined on a set E(measurable) is said to be* ***integrable****(over a measurable set E with respect to μ) if it is measurable and*

$$\int_E f d\mu < \infty$$

Note : Every nonnegative measurable function has an integral(Lebesgue) but only functions having finite integrals are called integrable.

Let us recall that if f is a real-valued function defined on a set E, its positive and negative parts are defined as

$$f^+ = \max(f, 0) = f \vee 0,$$

and

$$f^- = \min(-f, 0) = -(f \wedge 0).$$

and that f is measurable if and only if both f^+ and f^- are measurable,

One can easily verify that

$$f = f^+ - f^-$$

and

$$|f| = f^+ + f^-.$$

3.9 Definition *A measurable function f is said to be* ***integrable*** *over E if f^+ and f^- both are integrable over E. In this case, we define*

$$\int_E f = \int_E f^+ - \int_E f^-$$

3.10 Theorem *Let f and g be integrable functions over E. then:*
(a) *The function of cf(c is finite) is integrable over E, and*

$$\int_E cf = c\int_E f.$$

(b) *The function $f+g$ is integrable over E, and*

$$\int_E (f+g) = \int_E f + \int_E g.$$

(c) *If $f \le g$ a.e., then $\int_E f \le \int_E g$.*
(d) *If E_1 and E_2 are disjoint measurable sets in E, then*

$$\int_{E_1 \cup E_2} f = \int_{E_1} f + \int_{E_2} f.$$

(e) *If $|h| \le |f|$ and h is measurable, then h is integrable.*

Proof. Since the proof is similar as that of Proposition 3.3 of IV-3, proof is left to the reader. □

3.11 Theorem *Let f be a measurable function over a set E. If there is an integrable function g such that $|f| \le g$. then f is integrable over E.*

Proof. From $f^+ \le g$, it follows that $\int_E f^+ \le \int_E g$, and so f^+ is integrable on E. Similarly $f^- \le g$ implies the integrability of f^-. Hence f is integrable over E. □

3.12 (Lebesgue Dominated Convergence Theorem) *Let g be an ingegrable function on E and let $\{f_n\}$ be a sequence of measurable functions such that $|f_n| \le g$ on E and $\lim_{n\to\infty} f_n = f$ a.e. on E. Then*

$$\int_E f = \lim_{n\to\infty} \int_E f_n.$$

Proof. Since the proof is similar as that of Theorem 3.10 of IV-3, the proof is omitted. Hint : Apply Fatou's Lemma to the sequences $g+f_n$ and $g-f_n$. □

Because of (IV. 3.10) the following remark is obvious.

Remark : If we replace "*almost everywhere*" by "*pointwise*" in 3.12 the statements are still true.

A slightly improved form of Theorem 3.12 is the following.

3.13 Theorem *Let $\{f_n\}$ be a sequence of measurable functions defined on a set E of finite measure, and suppose that $|f_n(x)| \le M$ for all n and all x in E. If $f_n \to f$ a.e. on E, then f is integrable and*

$$\int_E f = \lim_{n\to\infty} \int_E f_n.$$

Remark : The results in Theorem 3.12 and 3.13 need not be true in case of Riemann integrals.

3.14 Lemma *If a sequence $\{f_n : E \to \mathbb{R} | n \in \mathbb{N}\}$ of measurable functions on a measurable subset E of X with $m(E) < \infty$ converges pointwise to a function $f : E \to \mathbb{R}$, then for arbitrary positive real number ε and δ, there exists a measurable subset D of E with $m(D) < \delta$ and a natural number k such that*

$$|f_n(x) - f(x)| < \varepsilon$$

holds for every point $x \in E \sim D$ and every natural number $n \ge k$.

Proof. For each $i \in \mathbb{N}$, let

$$E_n = \{x \in E : |f_n(x) - f(x)| \ge \varepsilon\}$$

and consider the subset

$$F_n = \bigcup_{i=n}^{\infty} E_i$$

of E for every $n \in \mathbb{N}$. For any $x \in E$, there exists an $n \in \mathbb{N}$ with $x \notin F_n$ since σ converges to f at the point x. This implies

$$\bigcap_{n=1}^{\infty} F_n = \emptyset.$$

By definition of F_n, we have $F_{n+1} \subset F_n$ for every $n \in \mathbb{N}$. We have

$$\lim_{n\to\infty} \mu(F_n) = \mu\left(\bigcap_{n=1}^{\infty} F_n\right) = \mu(\emptyset) = 0.$$

Hence there exists a natural number k with $\mu(F_k) < \delta$. Let $D = F_k$. Then we have $\mu(D) < \delta$ and $|f_n(x) - f(x)| < \varepsilon$ for every $x \in E \sim D$ and every $n \geq k$. □

3.15 Theorem *if a nonnegative measurable function $f : E \to \mathbb{R}^*$ is integrable, then, for every positive real number r, the subset*

$$E_r = \{x \in E | f(x) \geq r\}$$

is of finite measure and the subset

$$D = \{x \in E | f(x) > 0\}$$

is of σ-finite measure.

Proof. Consider the simple function $\phi : E \to \mathbb{R}$ define by

$$\phi(x) = \begin{cases} r & \text{if } x \in E_r \\ 0 & \text{if } x \in E - E_r. \end{cases}$$

Then $0 \leq \phi \leq f$. This implies

$$\int_E f \geq r\mu(E_r).$$

Since $\int_E f < \infty$, this proves E_r is of finite measure. Finally, since

$$D = \cup_{n=1}^{\infty} E_{1/n},$$

it follows that D is of σ-finite measure. □

PROBLEMS [5.3]

1. Shows that if f is integrable then the set $\{x : f(x) \neq 0\}$ is of σ-finite measure.

2. (a) Let $(X, \mathfrak{M}, \mu)$ be a measure space and g a nonnegative measurable function on X. Set $\nu E = \int_E g d\mu$. Show that ν is measure on $\mathfrak{M}$.

   (b) Let f be a nonnegative measurable function on X. Then
$$\int f d\nu = \int f g d\mu.$$
[Hint: First establish this for the case that g is simple and then use the Monotone Convergence Theorem]

3. A function f on a measure space $(X, \mathfrak{M}, \mu)$ is called locally measurable if the restriction of f to each E in $\mathfrak{M}$ with $\mu E < \infty$ is measurable, that is, if $f\chi_E$ is measurable.

   (a) Show that f is locally measurable if and only if it is measurable with respect to the σ-algebra of locally measurable sets.

   (b) If we define integration for nonnegative locally measurable functions f by taking $\int f$ to the supremum of $\int \varphi$ as φ ranges over all simple functions less than f, then $\int f = \int f d\underline{\mu}$, where $\underline{\mu}$ is the extension of μ.

4. Let $(X, \mathfrak{M}, \mu)$ be a complete measure space and $f : X \to \mathbb{R}^*$ be any integrable function. Prove that, for every positive real number ε, there exists a positive real number δ such that
$$\left|\int_E f d\mu\right| \leq \int_E |f| d\mu < \varepsilon$$
for every $E \in \mathfrak{M}$ satisfying $\mu(E) < \delta$

5. Let f be an integrable function on the measure space $(X, \mathfrak{M}, \mu)$. Show that given $\varepsilon > 0$, there is a $\delta > 0$ such that for

each measurable set E with $\mu E < \delta$ we have

$$\left|\int_E f\right| < \varepsilon.$$

6. Consider any nonnegative measurable function $f : E \to \mathbb{R}^*$ on a measurable subset E of complete measure space $(X, \mathfrak{M}, \mu)$. Let $B(f)$ denote the set of all bounded measurable functions $\xi : E \to \mathbb{R}$ satisfying

$$\xi \leq f, \quad \mu[\xi^{-1}(\mathbb{R}\backslash 0)] < \infty.$$

Prove that

$$\int_E f = \sup\left\{\int_E \xi : \xi \in B(f)\right\}$$

holds in case the subset $f^{-1}(R^*\backslash 0$ of E is of σ-finite measure.

7. Let $(X, \mathfrak{M}, \mu)$ be a σ-finite measure space. Prove that for any nonnegative measurable function $f : X \to \mathbb{R}^*$, there exists a nondecreasing sequence $\sigma = \{\phi_n | n \in \mathbb{N}\}$ of simple functions $\phi_n : X \to \mathbb{R}$ with $\mu[\phi^{-1}(\mathbb{R}\backslash 0)] < \infty$ for every $n \in \mathbb{N}$ such that σ converges (or diverges) pointwise to f.

8. Let $f : X \to \mathbb{R}^*$ be a nonnegative measurable function on a measure space $(X, \mathfrak{M}, \mu)$ with

$$\int f d\mu < \infty.$$

Prove that there exists a nondecreasing sequence $\sigma = \{\phi_n | n \in \mathbb{N}\}$ of simple functions $\phi_n : X \to \mathbb{R}$ with $\mu[\phi_n^{-1}(\mathbb{R}\backslash 0)] < \infty$ for every $n \in \mathbb{N}$ such that σ converges(or diverges) pointwise to f.

9. For an arbitrary complete measure space $(X, \mathfrak{M}, \mu)$ and any real number $p \geq 1$, let $\mathcal{L}_\nu(\mu)$ denote the set of all measurable functions $f : X \to \mathbb{R}$ such that

$$\int |f|^p d\mu < \infty.$$

Prove that $\mathcal{L}_p(\mu)$ is a linear space. Define a function $\nu : \mathcal{L}_p(\mu) \to \mathbb{R}$ by taking

$$\nu(f) = ||f||_p = \left| \int |f|^p d\mu \right|^{1/p},$$

for every $f \in \mathcal{L}_p(\mu)$. As in section 8.2, prove that ν is a pseudo-norm in $\mathcal{L}_p(\mu)$ and makes $\mathcal{L}_p(\mu)$ a complete pseudo-normed linear space. Hence its reduced normed linear space $\mathcal{L}_p^*(\mu)$ is a Banach space. Prove that the assignment $f \to \int f$ defines a continuous linear functional

$$\int : \mathcal{L}_p(\mu) \to \mathbb{R}$$

with $|| \int || = 1$. Also show that $\mathcal{L}_p(E)$ and l_p in section 9.2 are special cases of $\mathcal{L}_p(\mu)$.

5.4 Signed Measure

Throughout the present section, let X denote an arbitrary set and let $\mathfrak{M}$ be any given Boolean σ-algebra of subsets of X.

The measures on $\mathfrak{M}$ do not form a linear space; in fact, for any measure $\mu : \mathfrak{M} \to \mathbb{R}^*$ different from the trivial measure μ_0, $-\mu$ is never a measure. To overcome this difficulty, we have to allow a measure to take on negative values as well as positive values. This leads to the notion of signed measures which will be defined as follows.

We first note that if μ_1 and μ_2 are two measures defined on the same measurable space $(X, \mathfrak{M})$, then we may define a new measure μ_3 on $(X, \mathfrak{M})$ by setting

$$\mu_3(E) = c_1\mu_1(E) + c_2\mu_2(E), \quad c_1, c_2 \geq 0.$$

What happens if we try to define a measure by

$$\nu E = \mu_1 E - \mu_2 E$$

The first thing that may occur is that ν is not always nonnegative, and this leads to the consideration of signed measures which we shall define later. A more serious difficulty comes from the fact that ν is not defined when $\mu_1 E = \mu_2 E = \infty$. For this reason we should have either μ_1 or μ_2 finite. With these considerations in mind we make the following definition;

4.1 Definition *By a signed measure on $\mathfrak{M}$, we mean a function*

$$\mu : \mathfrak{M} \to \mathbb{R}^*$$

which satisfies the following three conditions:

($SM1$) μ assumes at most one of the values ∞ and $-\infty$;

($SM2$) $\mu(\phi) = 0$;

($SM3$) μ is countably additive. that is,

$\mu\left(\bigcup\limits_{i=1}^{\infty} E_i\right) = \sum\limits_{i=1}^{\infty} \mu E_i$, for any sequence E_i of disjoint measurable sets, the equality taken to mean that series on the right converges absolutely if $\mu(\cup E_i)$ is finite and that it properly diverges otherwise.

Thus a measure is a special case of a signed measure, but a signed measure is not in gerneral a measure. That is every measure on $\mathfrak{M}$ is also a signed measure on $\mathfrak{M}$, but the converse is false in general.

A signed measure $\mu : \mathfrak{M} \to \mathbb{R}^*$ is said to be **finite** iff $\mu(\mathfrak{M}) \subset \mathbb{R}$. The set of all finite signed measures clearly forms a linear space.

Now let us consider an arbitrarily given signed measure $\mu : \mathfrak{M} \to \mathbb{R}^*$ on $\mathfrak{M}$. We recall that the members of $\mathfrak{M}$ are called the measurable subsets of X.

4.2 Definition *Let $(X, \mathfrak{M})$ be a measurable space and ν be a signed measure on $\mathfrak{M}$. If A is measurable and for every measurable subset E of A we have $\nu(E) \geq 0$ $[\nu(E) \leq 0]$, then set A is called a* ***positive set**[**negative set**]* *with respect to a signed measure ν. A set which is both positive and negative with respect to ν is called a* ***null set****.*

Every measurable subset of a positive set is again positive, and if we take the restriction of ν to a positive set we obtain a measure.

A measurable set is a null set if and only if every measurable subset of it has ν measure zero. The reader should carefully note the distinction between a null set and a set of measure zero: while every null set must have measure zero, a set of measure zero may well be a union of two sets whose measures are not zero but are negatives of each other. Similarly, a positive set is not to be confused with a set which merely has positive measure. We have the following lemmas concerning positive sets. Similar statements hold, of course, for negative sets.

4.3 Lemma *Every measurable subset of a positive set is itself positive. The union of a countable collection of positive sets is positive.*

Proof. The first statement is trivially true by the definition of a positive set. To prove the second statement, let A be the union of a sequence $\{A_n\}$ of positive sets. If E is any measurable subset of A, set

$$E_n = E \cap A_n \cap A_{n-1}^c \cap \cdots \cap A_1^c.$$

Then E_n is a measurable subset of A_n and so $\nu E_n \geq 0$. Since the E_n are disjoint and $E = \cup E_n$, we have

$$\nu E = \sum_{n=1}^{\infty} \nu E_n \geq 0.$$

Thus A is a positive set. □

Similarly, one can establish the following proposition.

4.4 Proposition *Every measurable subset of a negative set is itself negative, and the union of any countable collection of negative sets is negative.*

The following Lemmas will be used in the proof of the Han Decomposition Theorem.

4.5 Lemma *Let E be a measurable set such that $0 < \nu E < \infty$. Then there is a positive set A contained in E with $\nu A > 0$.*

Proof. Either E itself is a positive set or it contains sets of negative measure. In the latter case let n_1 be the smallest positive integer such that there is a measurable set $E_1 \subset E$ with

$$\nu E_1 < -\frac{1}{n_1}.$$

Proceeding inductively, let n_k be the smallest positive integer for which there is a measurable set E_k such that

$$E_k \subset E \sim \left[\bigcup_{j=1}^{k-1} E_j\right]$$

and

$$\nu E_k < -\frac{1}{n_k}.$$

If we set

$$A = E \sim \bigcup_{k=1}^{\infty} E_k,$$

then

$$E = A \cup \left[\bigcup_{k=1}^{\infty} E_k\right].$$

Since this is a disjoint union, we have

$$\nu E = \nu A + \sum_{i=1}^{\infty} \nu E_k$$

with the series on the right absolutely convergent, since νE is finite. Thus $\sum \frac{1}{n_k}$ converges, and we have $n_k \to \infty$. Since $\nu E_k \leq 0$ and $\nu E > 0$, we must have $\nu A > 0$.

To show that A is a positive set, let $\varepsilon > 0$ be given. Since $n_k \to \infty$, we may choose k so large that $(n_k - 1)^{-1} < \varepsilon$. Since

$$A \subset E \sim \left[\bigcup_{j=1}^{k} E_j\right],$$

A can contain no measurable sets with measure less than $-(n_k - 1)^{-1}$, which is greater than $-\varepsilon$. Thus A contains no measurable sets of measure less than $-\varepsilon$. Since ε is an arbitrary positive number, it follows that A can contain no sets of negative measure and so must be a positive set. □

Similarly, one can establish the following proposition.

4.6 Theorem *Let E be a measurable set with finite negative measure, that is, $-\infty < \nu(E) < 0$. Then there is a negative subset A contained in E with $\nu(A) < 0$.*

4.7 Proposition (Hahn Decomposition Theorem) *Let ν be a signed measure on the measurable space $(X, \mathfrak{M})$. Then there is a positive set A and negative set B such that $X = A \cup B$ and $A \cap B = \emptyset$.*

Proof. Without loss of generality we may assume that ν does not take the infinite value $-\infty$ by passing to the signed measure $-\nu$ if necessary ($SM1$). Let λ be the supremum of νA over all sets A which are positive with respect to ν. Since the empty set is positive, $\lambda \geq 0$. Let $\{A_i\}$ be a sequence of positive sets such that

$$\lambda = \lim_{i \to \infty} \nu A_i,$$

and set

$$A = \bigcup_{i=1}^{\infty} A_i.$$

By Lemma 4.3 the set A is itself a positive set, and so $\lambda \geq \nu A$. But $A \sim A_i \subset A$ and so $\nu(A \sim A_i) \geq 0$. Thus

$$\nu A = \nu A_i + \nu(A \sim A_i) \geq \nu A_i.$$

Hence $\nu A \geq \lambda$, and so $\nu A = \lambda$, and $\lambda < \infty$.

Let $B = \sim A$, and suppose E is a positive subset of B. Then E and A are disjoint and $E \cup A$ is a positive set. Hence

$$\lambda \geq \nu(E \cup A) = \nu E + \nu A = \nu E + \lambda,$$

whence $\nu E = 0$, since $0 \leq \lambda < \infty$. Thus B contains no positive subsets of positive measure and hence no subsets of positive measure by Lemma 4.5 Conseqently, B is a negative set. □

A decomposition of X into two disjoint sets A and B such that A is positive for ν and B negative is called a **Hahn decomposition of X** for ν. Proposition 4.7 states the existence of a Hahn decomposition for each signed measure. Unfortunately, a Hahn decomposition is not unique.

For any nonempty null set in one of the components may be moved to the other component without destroying its being a Hahn decomposition. However, these are *almost nuique* in the sence of the following proposition.

4.8 Proposition *If $\{A, B\}$ and $\{A', B'\}$ are any two Hahn decompositions of X with respect to a signed measure $\nu : \mathfrak{M} \to \mathbb{R}^*$, then we have*

$$\nu(E \cap A) = \nu(E \cap A'), \quad \nu(E \cap B) = \nu(E \cap B')$$

for every $E \in \mathfrak{M}$.

Proof. Let $E \in \mathfrak{M}$ be arbitrarily given. Since the set $E \cap A \cap B'$ is a subset of both A and B', we must have $\nu(E \cap A \cap B') = 0$. Similarly, we also have $\nu(E \cap A' \cap B) = 0$. By (SM3), these imply

$$\begin{aligned} \nu(E \cap A) &= \nu[E \cap (A \cup A')] = \nu(E \cap A'), \\ \nu(E \cap B) &= \nu[E \cap (B \cup B')] = \nu(E \cap B'). \end{aligned}$$ □

Because of 4.6 and 4.7, we obtain, for an arbitrarily given signed measure $\nu : \mathfrak{M} \to \mathbb{R}^*$, three uniquely defined functions

$$\nu^+, \nu^-, |\nu| : \mathfrak{M} \to \mathbb{R}^*$$

as follows. Let $\{A, B\}$ be any Hahn decomposition of X with respect to ν. Then these functions are defined by setting

$$\begin{aligned} \nu^+(E) &= \nu(E \cap A), \\ \nu^-(E) &= -\nu(E \cap B), \\ |\nu|(E) &= \nu^+(E) + \nu^-(E) \end{aligned}$$

for every $E \in \mathfrak{M}$.

By means of the conditions (SM1-3), one can easily verify that these functions $\nu^+, \nu^-, |\nu|$ are measures on the Boolean σ-algebra $\mathfrak{M}$ of subsets of X in the sence of 5.4. These three measures $\nu^+, \nu^-, |\nu|$ are called the **positive variation**, the **negative variation**, and the **total variation** of the signed measure ν respectively. Furthermore, we clearly also have $\nu = \nu^+ - \nu^-$, that is,

$$\nu(E) = \nu^+(E) - \nu^-(E)$$

holds for every $E \in \mathfrak{M}$.

4.9 Definition *Let $(X, \mathfrak{M})$ be a measurable space, and μ, ν be measurable on $\mathfrak{M}$. If there are disjoint measurable sets A and B with $X = A \cup B$ such that $\nu_1(A) = \nu_2(B) = 0$, then two measures $\nu_1, \nu_2 : \mathfrak{M} \to \mathbb{R}^*$ are said to* ***mutually singular*** *in symbols*

$$\nu_1 \perp \nu_2$$

Thus the measures ν^+ and ν^+ defined above are mutually singular. We have thus etablished the existence part of the following proposition. The uniqueness part is left to the reader.

4.10 Proposition *Let ν be a signed measure on the measurable space $(X, \mathfrak{M})$. Then there are two mutually singular measures ν^+ and ν^- on $(X, \mathfrak{M})$ such that $\nu = \nu^+ - \nu^-$. Moreover, there is only one such pair of mutually singular measures.*

The decomposition $\{\nu^+, \nu^-\}$ of ν given by the proposition is called the **Jordan decomposition** of ν. That is, by a Jordan decomposition of signed measure $\nu : \mathfrak{M} \to \mathbb{R}^*$, we mean a pair $\{\nu^+, \nu^-\}$ of mutually singular measures ν_1 and ν_2 satisfying $\nu = \nu_1 - \nu_2$. Obviously, $\{\nu^+, \nu^-\}$ is a Jordan decomposition of ν. The measures ν^+ and ν^- are called the positive and negative parts(or variations) of ν. Since ν assumes at most one of the values $+\infty$ and $-\infty$, either ν^+ or ν^- must be finite. If they are both finite, we call ν a finite signed measure.

The measure $|\nu|$ defined by

$$|\nu|(E) = \nu^+ E + \nu^- E$$

is called the **absolute value** or **total variation** of ν. A set E is positive for ν if $\nu^-(E) = 0$. It is a null set if $|\nu|(E) = 0$. Note that $|\nu(E)| \leq |\nu|(E)$.

As to the uniqueness, we have the following proposition.

4.11 Proposition *Every signed measure $\nu : \mathfrak{M} \to \mathbb{R}^*$ has a unique Jordan decomposition.*

Proof. It remains to prove the uniqueness. For this purpose, let $\{\nu_1, \nu_2\}$ denote any Jordan decomposition of ν. Since $\nu_1 \perp \nu_2$, there exists a measurable subset Q of X satisfying

$$\nu_1(Q) = \nu_2(X \backslash Q).$$

Let $P = (X \backslash Q)$. Since ν_1, ν_2 are measures on $\mathfrak{M}$ and since $\nu = \nu_1 - \nu_2$, it is obvious that P is positive and Q is negative with respect to ν. Hence $\{P, Q\}$ is a Hahn decomposition of X with respect to ν. Consequently, we have

$$\nu_1(E) = \nu_1(E \cap P) + \nu_1(E \cap Q) = \nu_1(E \cap P) = \nu(E \cap P) = \nu^+(E),$$
$$\nu_2(E) = \nu_2(E \cap P) + \nu_2(E \cap Q) = \nu_2(E \cap Q) = \nu(E \cap Q) = \nu^-(E)$$

for every $E \in \mathfrak{M}$. Hence $\{\nu_1, \nu_2\} = \{\nu^+, \nu^-\}$. This completes the proof of 4.9 $\square$

PROBLEMS [5.4]

1. (a) Give an example to show that the Hahn decomposition need not be unique.

   (b) Show that the Hahn decomposition is unique except for null sets

2. Show that there is only one pair of mutually singular measure ν^+ and ν^- such that $\nu = \nu^+ - \nu^-$.

   [Hint: Show that any such pair determines a Hahn decomposition and apply the results of Problem 1.b.]

3. Show that if E is any measurable set, then

$$-\nu^- E \leq \nu E \leq \nu^+ E$$

and

$$|\nu E| \leq |\nu|(E).$$

4. Show that if ν_1 and ν_2 are any two finite signed measures, then so is $\alpha\nu_1 + \beta\nu_2$, where α and β are real numbers. Show that

$$|\alpha\nu| = |\alpha||\nu|$$

and

$$|\nu_1 + \nu_2| \leq |\nu_1| + |\nu_2|,$$

where $\nu \leq \mu$ means $\nu E \leq \mu E$ for all measurable sets E.

5. We define integration with respect to a signed measure ν by defining

$$\int f d\nu = \int f d\nu^+ - \int f d\nu^-.$$

If $|f| \leq M$,

$$\left|\int_E f d\nu\right| \leq M|\nu|(E).$$

Moreover, there is a measurable function f with $|f| \leq 1$ such that

$$\int_E f d\nu = |\nu|(E).$$

6. (a) Let μ and ν be finite signed measures. Show that there is a signed measure $\nu \wedge \mu$ which is smaller than μ and ν but larger than any other signed measure which is smaller than μ and ν.
[Hint: $\mu \wedge \nu = \frac{1}{2}(\mu + \mu - |\mu - \nu|)$.]

   (b) Show that there is a measure $\mu \vee \nu$ which is larger than μ and ν but smaller than any other measure which is larger than μ and ν. Also $\mu \vee \nu + \mu \wedge \nu = \mu + \nu$.

   (c) If μ and ν are positive measures, then they are mutually singular if and only if $\mu \wedge \nu = 0$.

7. Consider an arbitrary measure space $(X, \mathfrak{M}, \mu)$ and any integrable function $f : X \to \mathbb{R}^*$. Define a function $\nu : \mathfrak{M} \to \mathbb{R}$ by taking

$$\nu(E) = \int_E f$$

for every $E \in \mathfrak{M}$. Verify that ν is a signed measure on $\mathfrak{M}$. Let $f^+ = f \vee 0$ and $f^- = f \wedge 0$. Prove

$$\nu^+(E) = \int_E f^+, \quad \nu^-(E) = \int_E f^-, \quad |\nu|(E) = \int_E |f|$$

for every $E \in \mathfrak{M}$, ν^+, ν^-, $|\nu|$ are finite measure on $\mathfrak{M}$.

5.5 Absolute Continuity of a Measure

Throughout the present section, let $(X, \mathfrak{M}, \mu)$ denote an arbitrarily given measure space.

Let $(X, \mathfrak{M})$ be a fixed measurable space. If μ and ν are two measures defined on $(X, \mathfrak{M})$, we say that μ and ν are **mutually singular**(and write $\mu \perp \nu$) if there are disjoint sets A and B in $\mathfrak{M}$ such that $X = A \cup B$ and $\nu A = \mu B = 0$. Despite the fact that the notion of singularity is symmetric in ν and μ, we sometimes say that ν is singular with respect to μ. The notion antithetical to singularity is absolute continuity.

5.1 Definition *Let μ, ν be sign measures on a measurable space $(X, \mathfrak{M})$. An arbitrary measure $\nu : \mathfrak{M} \to \mathbb{R}^*$ is said to be* ***absolutely continuous*** *with respect to the measure μ, in symbols,*

$$\nu \ll \mu,$$

iff, for every $E \in \mathfrak{M}$, $\mu(E) = 0$ implies $\nu(E) = 0$.

If ν is defined by a nonnegative measurable function $f : X \to \mathbb{R}^*$ and μ, then it follows that $\nu \ll \mu$.

In the case of signed measures μ and ν, we say $\nu \ll \mu$ if $|\nu| \ll |\mu|$ and $\nu \perp \mu$ if $|\nu| \perp |\mu|$.

5.2 Example Let $(X, \mathfrak{M})$ be a measurable space. Then the followings are equivalent.

(1) $\nu \ll \mu$

(2) $\nu^+ \ll \mu, \quad \nu^- \ll \mu$

(3) $|\nu| \ll |\mu|$

Solution. (1) $\Rightarrow$ (2) Suppose that $\nu \ll \mu$, if $|\mu|(E) = 0$, then $\nu(E) = 0$.

Now let $\{A, B\}$ be a Hahn-Decomposition of X with respect to ν, then

$$0 \le |\mu|(E \cap A) \le |\mu|(E) = 0$$
$$0 \le |\mu|(E \cap B) \le |\mu|(E) = 0$$

Hence
$$\nu^+(E) = \nu(E \cap A) = 0$$
$$\nu^-(E) = \nu(E \cap B) = 0$$

(2) $\Rightarrow$ (3) Suppose that $\nu^+ \ll \mu$ and $\nu^- \ll \mu$. Then as $|\mu|(E) = 0$, $\nu^+(E) = \nu^-(E) = 0$.

Hence $|\nu|(E) = (\nu^+ + \nu^-)(E) = \nu^+(E) + \nu^-(E) = 0$.
That is, $|\nu| \ll |\mu|$

(3) $\Rightarrow$ (1) It is clear by $0 \le |\nu(E)| \le |\nu|(E)$.

5.3 Theorem *Let $(X, \mathfrak{M})$ be a measurable space, and μ, ν be signed measurable on $\mathfrak{M}$. Then*

(1) $\nu \perp \mu, \ \nu \ll \mu \quad \Rightarrow \nu = 0$

(2) $\nu_1 \perp \mu, \ \nu_2 \perp \mu$

$\Rightarrow \ (c_1\nu_1 + c_2\nu_2) \perp \mu \quad {}^\forall c_1, c_2 \in \mathbb{R}$ *with* $c_1, c_2 \ge 0$

(3) $\nu_1 \ll \mu, \ \nu_2 \ll \mu$

$\Rightarrow \ c_1\nu_1 + c_2\nu_2 \ll \mu \quad {}^\forall c_1, c_2 \in \mathbb{R}$

Proof. (1) ν : signed measure with $\nu \perp \mu$

$\Rightarrow \ {}^\exists A, B \in \mathfrak{M}$ such that $X = A \cup B, \ A \cap B = \emptyset$ and $\nu(A) = \nu(B) = 0$

$\nu \ll \mu \Rightarrow \ \nu(A) = 0$ as $|\mu|(A) = 0$

Since ${}^\forall E \in \mathfrak{M}$ $E = (E \cap A) \cup (E \cap B)$, and $E \cap A \subset A$,

$|\nu|(E \cap A) \le |\nu|(A) = 0$

i.e. $|\nu|(E \cap A) = 0$.
Since $E \cap B \subset B$ and $|\nu|(E \cap B) \leq |\nu|(B) = 0$
$|\nu|(E \cap B) = 0$.
And so
$$\begin{aligned} |\nu|(E) &= |\nu|\{(E \cap A) \cup (E \cap B)\} \\ &= |\nu|(E \cap A) + |\nu|(E \cap B) = 0 + 0 = 0 \end{aligned}$$

Hence $|\nu|(E) = \nu^+(E) + \nu^-(E) = 0$
That is, $\nu^+(E) = \nu^-(E) = 0$
Thus, $\nu(E) = \nu^+(E) - \nu^-(E) = 0$

(2) $\nu_1 \perp \mu \Rightarrow {}^\exists E \in \mathfrak{M}$ such that $\mu(E) = \nu_1(E^c) = 0$
$\nu_2 \perp \mu \Rightarrow {}^\exists F \in \mathfrak{M}$ such that $\mu(F) = \nu_1(F^c) = 0$
now $E \cup F \in \mathfrak{M}$ and

$$\begin{aligned} 0 &\leq \mu(E \cup F) \leq \mu(E) + \mu(F) = 0 \\ 0 &\leq (c_1\nu_1 + c_2\nu_2)\{(E \cup F)^c\} = (c_1\nu_1 + c_2\nu_2)(E^c \cap F^c) \\ &= c_1\nu_1(E^c \cap F^c) + c_2\nu_2(E^c \cap F^c) \\ &\leq c_1\nu_1(E^c) + c_2\nu_2(F^c) = 0 \end{aligned}$$

Hence $\mu(E \cup F) = (c_1\nu_1 + c_2\nu_2)\{(E \cup F)^c\} = 0$
Thus $(c_1\nu_1 + c_2\nu_2) \perp \mu$

(3) $\nu_1 \ll \mu \Rightarrow |\nu_1| \ll |\mu|$
$\nu_2 \ll \mu \Rightarrow |\nu_2| \ll |\mu|$
Since ${}^\forall E \in \mathfrak{M}$, $|\mu|(E) = 0$,

$$|\nu_1|(E) = 0, \quad \text{and} \quad |\nu_2|(E) = 0.$$

Hence
$$\begin{aligned} |c_1\nu_1 + c_2\nu_2|(E) &\leq |c_1\nu_1|(E) + |c_2\nu_2|(E) \\ &= |c_1||\nu_1|(E) + |c_2||\nu_2|(E) \\ &= 0 \end{aligned}$$

Thus $c_1\nu_1 + c_2\nu_2 \ll \mu$. □

Whenever we are dealing with more than one measure on a measurable space $(X, \mathfrak{M})$, the term 'almost everywhere' becomes ambiguous, and we must specify almost everywhere with respect to μ or almost everywhere with respect to ν, etc. These are abbreviated a.e. $[\mu]$ and a.e. $[\nu]$. If $\nu \ll \mu$ and a property holds a.e. $[\mu]$, then it holds a.e. $[\nu]$.

As in Chapter 4, let us consider any nonnegative measurable function

$$f : X \to \mathbb{R}^*.$$

Define a function $\nu : \mathfrak{M} \to \mathbb{R}^*$ by taking

$$\nu(E) = \int_E f d\mu$$

for every $E \in \mathfrak{M}$.

Then ν is a set function defined on $\mathfrak{M}$, and it follows from Corollary 3.7 that ν is countably additive and hence a measure. The measure ν will be finite if and only if f is integrable. Since the integral over a set of μ-measure zero is zero, we have ν absolutely continuous with respect to μ.

In case μ is σ-finite, then the converse is also true; in fact, we have the following theorem. The next theorem shows that, subject to σ-finiteness restrictions, every absolutely continuous measure ν is obtained in this fashion.

5.4 Theorem (Radon-Nikodym) *If the measure*

$$\mu : \mathfrak{M} \to \mathbb{R}^* \textit{ of a measure space } (X, \mathfrak{M}, \mu)$$

is σ-finite, and let ν be a measure defined on $\mathfrak{M}$ which is absolutely continuous with respect to μ. Then there is a nonnegative measurable function $f : X \to \mathbb{R}^*$ such that for each set E in $\mathfrak{M}$ we have

$$\nu E = \int_E f d\mu.$$

holds for every $E \in \mathfrak{M}$. The function f is almost unique in the sense that if any measurable function $g : X \to \mathbb{R}^*$ has this property then we have $f(x) = g(x)$ a.e. in X with respect to μ.

Proof. The extension from the finite to the σ-finite case is not difficult and is left to the reader. Thus we shall assume that μ is finite. Then $\nu - \alpha\mu$ is a signed measure for each rational number α. Let (A_α, B_α) be a Hahn decomposition for $\nu - \alpha\mu$. and take $A_0 = X$, $B_0 = \emptyset$.

Now $B_\alpha \sim B_\beta = B_\alpha \cap A_\beta$. Thus $(\nu - \alpha\mu)(B_\alpha \sim B_\beta) \leq 0$ and $(\nu - \beta\mu)(B_\alpha \sim B_\beta) \geq 0$. If $\beta > \alpha$, these imply $\mu(B_\alpha \sim B_\beta) = 0$, and so by Lemma 2.7 of 5.2 there is an measurable function f such that for each rational α we have $f \geq \alpha$ a.e. on A_α and $f \leq \alpha$ a.e. on B_α. Since $B_0 = \emptyset$, we may take f to be nonnegative.

Let E be an arbitrary set in $\mathfrak{M}$, and set

$$E_k = E \cap \left(B_{\frac{k+1}{N}} \sim B_{\frac{k}{N}}\right), \quad E_\infty = E \sim \cup B_{\frac{k}{N}}.$$

Then $E = E_\infty \cup \bigcup\limits_{k=0}^{\infty} E_k$, and this a disjoint union. Thus

$$\nu E = \nu E_\infty + \sum_{k=0}^{\infty} \nu E_k.$$

Since $E_k \subset B_{\frac{k+1}{N}} \cap A_{\frac{k}{N}}$, We have $\frac{k}{N} \leq f \leq \frac{k+1}{N}$ on E_k and so

$$\frac{k}{N}\mu E_k \leq \int_{E_k} f d\mu \leq \frac{k+1}{N}\mu E_k.$$

Since $\frac{k}{N}\mu E_k \leq \nu E_k \leq \frac{k+1}{N}\mu E_k$, we have

$$\nu E_k - \frac{1}{N}\mu E_k \leq \int_{E_k} f d\mu \leq \nu E_k + \frac{1}{N}\mu E_k.$$

On E_∞ we have $f = \infty$ a.e. If $\mu E_\infty > 0$, we must have $\nu E_\infty = \infty$, since $(\nu - \alpha\mu)E_\infty$ is positive for each α. If $\mu E_\infty = 0$, we must have $\nu E_\infty = 0$, since $\nu \ll \mu$. In either case

$$\nu E_\infty = \int_{E_\infty} f d\mu.$$

Adding this equality and our previous inequalities give;

$$\nu E - \frac{1}{N}\mu E \leq \int_E f d\mu \leq \nu E + \frac{1}{N}\mu E.$$

Since μE is finite and N arbitrary, we must have

$$\nu E = \int_E f d\mu.$$

□

The function f given by Theorem 5.4 is called the **Radon-Nikodym derivative** of ν with respect to μ. It is sometimes denoted by $\left[\frac{d\nu}{d\mu}\right]$.

5.5 Proposition (Lebesgue Decomposition) *Let $(X, \mathfrak{M}, \mu)$ be a σ-finite measure space and ν a σ-finite measure defined on $\mathfrak{M}$. Then we can find a measure ν_0 which is singular with respect to μ and a measure ν_1 which is absolutely continuous with respect to μ such that $\nu = \nu_0 + \nu_1$. The measures ν_0 and ν_1 are unique.*

Proof. Since μ and ν are σ-finite measures, so is the measure $\lambda = \mu + \nu$. Since both μ and ν are absolutely continuous with respect to λ, the Radon-Nikodym Theorem asserts the existence of nonnegative measurable functions f and g such that for each $E \in \mathfrak{M}$

$$\mu E = \int_E f d\lambda, \quad \nu E = \int_E g d\lambda.$$

Let $A = \{x : f(x) > 0\}$ and $B = \{x : f(x) = 0\}$. Then X is the disjoint union of A and B, while $\mu B = 0$. If we define ν_0 by

$$\nu_0 E = \nu(E \cap B),$$

we have $\nu_0(A) = 0$ and so $\nu_0 \perp \mu$. Let

$$\nu_1(E) = \nu(E \cap A) = \int_{E \cap A} g d\lambda.$$

Then $\nu = \nu_0 + \nu_1$, and we have only to show that $\nu_1 \ll \mu$. Let E be a set of μ-measure zero. Then

$$0 = \mu E = \int_E f d\lambda,$$

and $f = 0$ a.e. $[\lambda]$ on E. Since $f > 0$ on $A \cap E$, we must have $\lambda(A \cap E) = 0$. Hence $\nu(A \cap E) = 0$,and so $\nu_1(E) = \nu(A \cap E) = 0$. This establishes the proposition except for the uniqueness, which is left to the reader. □

The decomposition $\nu = \nu_0 + \nu_1$ in (5.2) is called the **Lebesgue decomposition** of the σ-finite measure $\nu : \mathfrak{M} \to \mathbb{R}^*$ with respect to the σ-finite measure $\mu : \mathfrak{M} \to \mathbb{R}^*$.

PROBLEMS [5.5]

1. Consider any two signed measures $\mu, \nu : \mathfrak{M} \to \mathbb{R}^*$ and their total variations $|\mu|$ and $|\nu|$. Define $\nu \perp \mu$ and $\nu \ll \mu$ by $|\nu| \perp |\mu|$ and $|\nu| \ll |\mu|$ respectively. Prove:

   (a) $\nu \ll \mu$ implies $\nu^+ \ll \mu$ and $\nu^- \ll \mu$.

   (b) $\nu \perp \mu$ and $\nu \ll \mu$ imply $\nu = 0$.

2. Let $\mu, \nu : \mathfrak{M} \to \mathbb{R}^*$ be two measures with $\nu \ll \mu$. Prove that
$$\int f d\nu = \int f \frac{d\nu}{d\mu} d\mu$$
holds for every measurable function $f : X \to \mathbb{R}^*$ which is integrable with respect to ν.

3. Generalize the Radon-Nikodym theorem to the case where
$$\nu : \mathfrak{M} \to \mathbb{R}^*$$
is a signed measure.

4. Let $(X, \mathfrak{M}, \mu)$ be a σ-finite measure space and $p \geq 1$, $q \geq 1$ be real numbers satisfying $p^{-1} + q^{-1} = 1$. Prove that, for an arbitrary continuous linear functional $f : \mathcal{L}_p(\mu) \to \mathbb{R}$, there exists a function $\psi \in \mathcal{L}_q(\mu)$ satisfying $||\psi||_q = ||f||$ and
$$f(\phi) = \int (\phi\psi) d\mu$$
for every $\phi \in \mathcal{L}_p(\mu)$.

Chapter 6

MEASURE THEORY

In the case of Lebesgue measure, we defined outer measure by open sets and used this to define Lebesgue measure. Such a procedure is feasible in general. In the first section we discuss the process of deriving a measure from an outer measure, and in the second section we derive an outer measure from a measure which is defined only on an algebra of sets. The remainder of the chapter is devoted to some application of this process.

6.1 Outer Measure and Measurability

Motivated by the Lebesgue outer measure in the real line R,(III–3) we define outer measures in an arbitrarily given set X as follows.

1.1 Definition. *Let X be set, and let $\wp(X)$ be the collection of all subsets of X. An* ***outer measure on*** *X is a set function.*

$$\mu^* : \wp(X) \to R^* = [0, +\infty] \quad \text{such that}$$

(OM1) $\mu^*(\emptyset) = 0$

(OM2) μ^* is monotonic. that is,
if $A \subset B \subset X$, then $\mu^*(A) \leq \mu^*(B)$, and

(OM3) μ^* is countably subadditive, that is,
if $\{E_i\}$ is an infinite sequence of subsets of X, then
$E \subset \bigcup_{i=1}^{\infty} E_i \Longrightarrow \mu^* E \leqslant \sum_{i=1}^{\infty} \mu^* E_i.$

In view of (OM1) finite subadditivity follows from (OM3). Because of (OM2), property (OM3) can be replaced by

$$\text{(OM3)}': \; E = \bigcup_{i=1}^{\infty} E_i, E_i \text{ disjoint } \Longrightarrow \mu^* E \leq \sum_{i=1}^{\infty} \mu^* E_i$$

The outer measure μ^* is called finite if $\mu^* X < \infty$.

1.2 Example. Let X be a space of at least two points and $x_0 \in X$. For each $A \subset X$, define

$$\mu^*(A) = \begin{cases} 0 & if \quad x_0 \notin A \\ 1 & if \quad x_0 \in A. \end{cases}$$

Then μ^* is an outer measure on X.

1.3 Example. Let $X = R^2$. The Lebseque outer plane measure $m^*(A)$ of a set $A \subset R^2$ is defined by

$$m^*(A) = \begin{cases} 0 & if \quad A = \emptyset \\ \inf\{y | y = \sum |\lambda_n| \; with \; A \subset (\cup \lambda_n)\} & if \quad A \neq \emptyset \end{cases}$$

where each λ_n is a rectangle in R^2 and $|\lambda_n|$ is its area.

Then Lebseque outer plane measure is an outer measure.

Also it is clear that Lebsque outer measuer is an outer measure. But the Lebseque measure in the real line R is not an outer measure in R, since it is not defined for every subset of R.

By analogy with the case of Lebesgue measure we define a set E to be **measurable** with respect to m^* if for every set A we have

$$\mu^*(A) = \mu^*(A \cap E) + \mu^*(A \cap \tilde{E}).$$

Since μ^* is subadditive, it is only necessary to show that

$$\mu^*(A) \geq \mu^*(A \cap E) + \mu^*(A \cap \tilde{E})$$

for every A in order to prove that E is measurable. This inequality is trivially true when $\mu^* A = \infty$, and so we need only establish it for sets A with $\mu^* A$ finite.

While outer measure has the advantage that it is defined for all sets, it is not countably additive. It becomes countably additive, however, if we suitably reduce the family of sets on which it is defined.

1.4 Theorem. *The class $\mathfrak{M}$ of μ^* measurable subset of X is a σ-algebra.*

Proof. Trivially, the empty set is measurable. The symmetry of the definition of measurability in E and $\widetilde{E}$ shows that $\widetilde{E}$ is measurable whenever E is.

Let E_1 and E_2 be measurable sets. From the measurability of E_2

$$\mu^*(A) = \mu^*(A \cap E_2) + \mu^*(A \cap \widetilde{E}_2)$$

and

$$\mu^*(A) = \mu^*(A \cap E_2) + \mu^*(A \cap \widetilde{E}_2 \cap E_1) + \mu^*(A \cap \widetilde{E}_1 \cap \widetilde{E}_2)$$

by the measurability of E_1 . Since

$$A \cap [E_1 \cup E_2] = [A \cap E_2] \cup [A \cap E_1 \cap \widetilde{E}_2],$$

we have

$$\mu^*(A \cap [E_1 \cup E_2]) \leq \mu^*(A \cap E_2) + \mu^*(A \cap \widetilde{E}_2 \cap E_1)$$

by subadditivity, and so

$$\mu^* A \geq \mu^*(A \cap [E_1 \cup E_2]) + \mu^*(A \cap \widetilde{E}_1 \cap \widetilde{E}_2).$$

This means that $E_1 \cup E_2$ is measurable, since

$$\sim (E_1 \cup E_2) = \widetilde{E}_1 \cap \widetilde{E}_2.$$

Thus the union of two measurable sets is measurable, and by induction the union of any finite number of measurable sets is measurable, showing that $\mathfrak{M}$ is an algebra of sets.

Assume that $E = \bigcup E_i$, where $\{E_i\}$ is a disjoint sequence of measurable sets, and set

$$G_n = \bigcup_{i=1}^{n} E_i$$

Then G_n is measurable, and

$$\mu^*(A) = \mu^*(A \cap G_n) + \mu^*(A \cap \widetilde{G}_n) \geq \mu^*(A \cap G_n) + \mu^*(A \cap \widetilde{E}),$$

since $\widetilde{E} \subset \widetilde{G}_n$. Now $G_n \cap E_n = E_n$ and $G_n \cap \widetilde{E}_n = G_{n-1}$, and by the measurability of E_n we have

$$\mu^*(A \cap G_n) = \mu^*(A \cap E_n) + \mu^*(A \cap G_{n-1}).$$

By induction

$$\mu^*(A \cap G_n) = \sum_{i=1}^{n} \mu^*(A \cap E_i)$$

and so

$$\begin{aligned} \mu^*(A) &\geq \mu^*(A \cap \widetilde{E}) + \sum_{i=1}^{\infty} \mu^*(A \cap E_i) \\ &\geq \mu^*(A \cap \widetilde{E}) + \mu^*(A \cap E), \end{aligned}$$

since

$$A \cap E \subset \bigcup_{i=1}^{\infty} (A \cap E_i).$$

Thus E is measurable. Since the union of any sequence of sets in an algebra can be replaced by a disjoint union of sets in the algebra, it follows that $\mathfrak{M}$ is a σ-algebra.

1.5 Theorem. *Let $\mathfrak{M}$ be the family of all μ^*-measurable subset of X. Then the restriction $\mu = \mu^* \mid \mathfrak{M}$ is a measure on $\mathfrak{M}$. Furthermore, the measure space $(X, \mathfrak{M}, \mu)$ is complete.*

Proof. If E is the disjoint union of the measurable sets $\{E_i\}$, then

$$\overline{\mu}(E) \geq \overline{\mu}\left(\bigcup_{i=1}^{n} E_i\right) = \sum_{i=1}^{n} \overline{\mu}(E_i),$$

and so

$$\overline{\mu}(E) \geq \sum_{i=1}^{\infty} \overline{\mu}(E_i).$$

But

$$\overline{\mu}(E) \leq \sum_{i=1}^{\infty} \overline{\mu}(E_i)$$

by the subadditivity of μ^*. Hence $\overline{\mu}$ is countably additive and thus a measure since it is nonnegative and $\overline{\mu}\phi = \mu^*\phi = 0$.

We next demonstrate the finite additivity of $\overline{\mu}$. Let E_1 and E_2 be disjoint measurable sets. Then the measurability of E_2 implies that

$$\begin{aligned}
\mu(E_1 \cup E_2) &= \mu^*(E_1 \cup E_2) \\
&= \mu^*([E_1 \cup E_2] \cap E_2) + \mu^*([E_1 \cup E_2] \cap \widetilde{E}_2) \\
&= \mu^* E_2 + \mu^* E_1
\end{aligned}$$

Finite additivity follows by induction.
Now let $\overline{\mu}(E) = 0$ and $D \subset E$. Then we have

$$0 \leq \mu^*(D) \leq \mu^*(E) = \overline{\mu}(E) = 0$$

This implies $\mu^*(D) = 0$. By (III.3.4) and its proof with $\mathbb{R}$, X replaced by X, A, respectively, we obtain $D \subset \mathfrak{M}$

Hence $(X, \mathfrak{M}, \mu)$ is complete.

This complete measure space $(X, \mathfrak{M}, \overline{\mu})$ will be referred to as the measure space defined by the given outer measure μ^* in X. Since $X \in \mathfrak{M}$ and $\mu = \mu^*|\mathfrak{M}$ it follows that the measure $\mu : \mathfrak{M} \to R^*$ is finite or σ-finite according as the given outer measure, μ^* is so respectively. □

PROBLEMS [6.1]

1. Prove that every function $\mu^* : \wp(X) \to R^*$ which satisfies (OM1) and (OM2) must be non-negative.

2. Show that $\overline{\mu}$ is saturated.

3. Assume that $\{E_i\}$ a sequence of disjoint measurable sets and $E = \bigcup E_i$. Then for any set A we have
$$\mu^*(A \cap E) = \sum \mu^*(A \cap E_i).$$

6.2 Extension of a measure

Traditionly, the Lebesgue outer measure μ^* was first defined on the Boolean algebra of all open subsets and all closed subsets of the real line R, and then was extended to all subsets of R.

The present section is devoted to the generalization of this process. For this purpose, first we defined a measure on an algebra.

2.1 Definition. *A* ***measure on an algebra*** *is a nonnegative extended real valued set function μ defined on an algebra $\mathcal{A}$ of sets such that:*

i. $\mu(\phi) = 0$;

ii. if $\{A_i\}$ is a disjoint sequence of sets in $\mathcal{A}$ whose union is also in $\mathcal{A}$, then
$$\mu\left(\bigcup_{i=1}^{\infty} A_i\right) = \sum_{i=1}^{\infty} \mu A_i.$$

Thus a measure on an algebra $\mathcal{A}$ is a measure if and only if $\mathcal{A}$ is a σ-algebra.

The purpose of this secton is to show that, if we start with a measure on an algebra $\mathcal{A}$ of sets, we may extend it to be a measure defined on a σ-algebra $\mathcal{B}$ containing $\mathcal{A}$. We shell do this by using the measure on the algebra to construct an outer measure μ^* and show that the measure $\overline{\mu}$ induced by μ^* is the desired extension of μ. The process by which we construct μ^* from μ is analogous to that by which we constructed Lebesgue outer measure from the lengths of intervals:

We define set function $\mu^* : \wp(X) \to R^*$ defined by

$$\mu^* E = \inf \sum_{i=1}^{\infty} \mu A_i \tag{1}$$

where $\{A_i\}$ ranges over all sequences from $\mathfrak{a}$ such that $E \subset \bigcup_{i=1}^{\infty} A_i$. We first establish some lemmas concerning μ^*.

2.2 Lemma. *If $A \in \mathfrak{a}$ and if $\{A_i\}$ is any sequence of sets in $\mathfrak{a}$ such that $A \subset \bigcup_{i=1}^{\infty} A_i$, then $\mu A \leq \sum_{i=1}^{\infty} \mu A_i$.*

Proof. Set

$$B_n = A \cap A_n \cap \tilde{A}_{n-1} \cap \cdots \cap \tilde{A}_1.$$

Then $B_n \in \mathfrak{a}$ and $B_n \subset A_n$. But A is the disjoint union of the sequence $\{B_n\}$, and so by countable additivity

$$\mu A = \sum_{i=1}^{\infty} \mu B_n \leq \sum_{i=1}^{\infty} \mu A_n. \qquad \square$$

2.3 Corollary. *If $A \in \mathfrak{a}$, $\mu^* A = \mu A$.*

2.4 Lemma. *The set function μ^* is an outer measure.*

Proof. Since μ^* is clearly a monotone nonnegative set function defined for all sets and $\mu^* \phi = 0$, we have only to show that it is countably subadditive. Let $E \subset \bigcup_{i=1}^{\infty} E_i$. If $\mu^* E_i = \infty$ for any i, we have $\mu^* E \leq \sum \mu^* E_i = \infty$. If not, given $\epsilon > 0$, there is for each i a sequence $\{A_{ij}\}_{j=1}^{\infty}$ of sets in $\mathfrak{a}$ such that $E_i \subset \bigcup_{j=1}^{\infty} A_{ij}$ and

$$\sum_{j=1}^{\infty} \mu A_{ij} < \mu^* E_i + \epsilon/2^i.$$

Then

$$\mu^* E \le \sum_{ij} \mu A_{ij} < \sum_{i=1}^{\infty} \mu^* E_i + \epsilon.$$

Since ϵ was an arbitrary positive number,

$$\mu^* E \le \sum_{i=1}^{\infty} \mu^* E_i \ ,$$

and μ^* is subadditive. □

2.5 Lemma. *If $A \in \mathfrak{a}$ then A is measurable with respect to μ^*.*

Proof. Let E be an arbitrary set of finite outer measure and ϵ a positive nember. Then there is a sequence $\{A_i\}$ from $\mathfrak{a}$ such that $E \subset \bigcup A_i$ and

$$\sum \mu A_i < \mu^* E + \epsilon.$$

By the additivity of μ on $\mathfrak{a}$ we have

$$\mu(A_i) = \mu(A_i \cap A) + \mu(A_i \cap \overline{A}).$$

Hence

$$\mu^* E + \epsilon > \sum_{i=1}^{\infty} \mu(A_i \cap A) + \sum_{i=1}^{\infty} \mu(A_i \cap \tilde{A})$$

$$> \mu^*(E \cap A) + \mu^*(E \cap \tilde{A}),$$

since

$$E \cap A \subset \bigcup (A_i \cap A)$$

and

$$E \cap \tilde{A} \subset \bigcup (A_i \cap \tilde{A}).$$

Since ϵ was an arbitray positive number,

$$\mu^* E \ge \mu^*(E \cap A) + \mu^*(E \cap \tilde{A}),$$

and A is measurable. □

The outer measure μ^* which we have defined is called the **outer measure induced by** μ. For a given algebra $\mathcal{A}$ of sets we use $\mathcal{A}_\sigma$ to denote those sets which are countable unions of sets of $\mathcal{A}$ and use $\mathcal{A}_\sigma$ to denote those sets which are countable intersections of sets in $\mathcal{A}_\sigma$.

2.6 Proposition. *Let μ be a measure on an algebra $\mathcal{A}$, μ^* the outer measure induced by μ, and E any set. Then for $\epsilon > 0$, there is a set $A \in \mathcal{A}_\sigma$ with $E \subset A$ and*

$$\mu^* A \leq \mu^* E + \epsilon.$$

There is also a set $B \in \mathcal{A}_{\sigma\delta}$ with $E \subset B$ and $\mu^* E = \mu^* B$.

Proof. By the definition of μ^* there is a sequence $\{A_i\}$ from $\mathcal{A}$ such that $E \subset \bigcup A_i$ and

$$\sum_{i=1}^{\infty} \mu A_i \leq \mu^* E + \epsilon.$$

Set $A = \bigcup A_i$. Then $\mu^* A \leq \sum \mu^* A_i = \sum \mu A_i$.

To prove the second statement, we note that for each positive integer n there is a set A_n in $\mathcal{A}_\sigma$ such that $E \subset A_n$ and $\mu^* A_n < \mu^* E + \frac{1}{n}$. Let $B = \bigcap A_n$. Then $B \in \mathcal{A}_{\sigma\delta}$ and $E \subset B$. Since $B \subset A_n$, $\mu^* B \leq \mu^* A_n < \mu^* E + \frac{1}{n}$. Since n is arbitrary, $\mu^* B \leq \mu^* E$. But $E \subset B$, and so $\mu^* B \geq \mu^* E$ by monotonicity. Hence $\mu^* B = \mu^* E$. □

If we apply this proposition in the case that E is a measurable set of finite measure, we see that E must be the difference of a set B in $\mathcal{A}_{\sigma\delta}$ and a set of measure zero. This gives us the structure of the measurable sets of finite measure, and the next proposition extends this to the σ-finite case. It can be considered a generalization of the first principle of Littlewood.

2.7 Proposition. *Let μ be a σ-finite measure on an algebra $\mathcal{A}$, and let μ^* be the outer measure generated by μ. A set E is μ^* measurable if and only if E is the proper difference $A \sim B$ of a set A in $\mathcal{A}_{\sigma\delta}$ and a set B with $\mu^* B = 0$. Each set B with $\mu^* B = 0$ is contained in a set C in $\mathcal{A}_{\sigma\delta}$ with $\mu^* C = 0$.*

Proof. The "if" part of the proposition follows from the fact that each set in $\mathcal{A}_{\sigma\delta}$ must be measurable, since the measurable sets form a σ-algebra, while each set of μ^*-measure zero must be measurable, since $\overline{\mu}$ is complete.

To prove the "only if" part of the proposition, let $\{X_i\}$ be a countable disjoint collection of sets in $\mathcal{A}$ with μX_i finite and $X = \bigcup X_i$. If E is measurable, then E is the disjoint union of the measurable sets $E_i = X_i \cap E$. By proposition 2.6 we can find for each positive integer n, a set A_{ni} in $\mathcal{A}_\sigma$ such that $E_i \subset A_{ni}$ and

$$\overline{\mu} A_{ni} \leq \overline{\mu} E_i + \frac{1}{n2^i}.$$

Set

$$A_n = \bigcup_{i=1}^{\infty} A_{ni}.$$

Then $E \subset A_n$, and $A_n \sim E \subset \bigcup_{i=1}^{\infty} [A_{ni} \sim E_i]$. Hence

$$\begin{aligned}\overline{\mu}(A_n \sim E) &\leq \sum_{i=1}^{\infty} \overline{\mu}(A_{ni} \sim E_i) \\ &\leq \sum_{i=1}^{\infty} \frac{1}{n2^i} = \frac{1}{n}.\end{aligned}$$

Since $A_n \in \mathcal{A}_\sigma$, the set $A = \bigcap_{n=1}^{\infty} A_n$ is in $\mathcal{A}_{\sigma\delta}$, and for each n

$$A \sim E \subset A_n \sim E.$$

Hence

$$\overline{\mu}(A \sim E) \leq \overline{\mu}(A_n \sim E) \leq \frac{1}{n}.$$

Since this holds for each positive n, we must have

$$\overline{\mu}(A \sim E) = 0.$$

□

In this section we are concerned the extension of a measure on algebra. We summarize the results of this section in the following theorem.

2.8 Theorem. *(Carathéodory) Let μ be a measure on an algebra $\mathfrak{a}$, and μ^* the outer measure induced by μ. Then*

(a) *The family β of all μ^*-measurable sets is a σ-algebra containing $\mathfrak{a}$.*

(b) *The restriction $\overline{\mu}$ of μ^* to the μ^*-measurable sets is an extemntion of μ.*
Futhermore, the measure μ^* is complete.

(c) *If μ is finite, so is $\overline{\mu}$.*

(d) *If μ is σ-finite, so is $\overline{\mu}$.*

(e) *If μ is σ-finite, then $\overline{\mu}$ is the only measure on the smallest σ-algebra containing $\mathfrak{a}$ which is an extension of μ.*

Proof. (a),(b) They are direct consequence of Theorem 1.5 of sec. 6.1 and Corollary 2.3 and Lemma 2.5 of sec. 6.2

(c) If μ is finite, $\overline{\mu}(X) = \mu(X) < \infty$ obviously. Hence $\overline{\mu}$ is finite.

(d) Similar as (c)

(e) The fact that $\overline{\mu}$ is an extension of μ from $\mathfrak{a}$ to be a measure on a σ-algebra containing $\mathfrak{a}$ follows directly from Corollary 3, Lemma 5, and Theorem 1.5, and it is readily verified that $\overline{\mu}$ is finite or σ-finite whenever μ is.

To show the unicity of $\overline{\mu}$ when μ is σ-finite, we let β be the smallest σ-algebra containing $\mathfrak{a}$ and $\overline{\mu}$ some measure on β which agrees with μ on $\mathfrak{a}$.

Since each set in $\mathfrak{a}_\sigma$ can be expressed as a disjoint countable union of sets in $\mathfrak{a}$, the measure $\tilde{\mu}$ must agree with $\overline{\mu}$ on $\mathfrak{a}_\sigma$. Let B be any set in $\mathcal{B}$ with finite outer measure. Then by Proposition 6 there is an A in $\mathfrak{a}_0$ such that $B \subset A$ and

$$\mu^* A \leq \mu^* B + \epsilon$$

Since $B \subset A$,

$$\tilde{\mu} B \leq \tilde{\mu} A = \mu^* A \leq \mu^* B + \epsilon.$$

Since ϵ is an arbitrary positive number, we have

$$\tilde{\mu} B \leq \mu^* B$$

for each $B \in \mathcal{B}$.

Since the class of sets measurable with respect to μ^* is a σ-algebra containing $\mathcal{A}$, each B in $\mathcal{B}$ must be measurable. If B is measurable and A is in $\mathcal{A}_\sigma$ with $B \subset A$ and $\mu^* A \leq \mu^* B + \epsilon$, then

$$\mu^* A = \mu^* B + \mu^*(A \sim B),$$

and so

$$\mu^*(A \sim B) \leq \epsilon,$$

if $\mu^* B < \infty$. Hence

$$\begin{aligned} \mu^* B \leq \mu^* A &= \tilde{\mu} A \\ &= \tilde{\mu} B + \tilde{\mu}(A \sim B) \\ &\leq \tilde{\mu} B + \epsilon. \end{aligned}$$

Since ϵ is arbitrary, we have

$$\mu^* B \leq \tilde{\mu} B$$

and so

$$\mu^* B = \tilde{\mu} B.$$

If μ is a σ-finite measure, let $\{X_i\}$ be a countable disjoint collection of sets in $\mathcal{A}$ with $X = \bigcup X_i$ and μX_i finite. If B is any set in $\mathcal{B}$, then

$$B = \bigcup (X_i \cap B)$$

and this is a countable disjoint union of sets in $\mathcal{B}$, and so we have

$$\tilde{\mu} B = \sum \tilde{\mu}(X_i \cap B)$$

and

$$\overline{\mu} B = \sum \overline{\mu}(X_i \cap B).$$

Since $\mu^*(X_i \cap B) < \infty$, we have

$$\tilde{\mu}(X_i \cap B) = \overline{\mu}(X_i \cap B).$$

□

This extension procedure not only extends μ to a measure on the smallest σ-algebra $\mathcal{B}$ containing $\mathcal{A}$, but also completes and saturates the measure. If μ is σ-finite, the extension to $\mathcal{B}$ is already saturated, and the extension to the μ^*-measurable sets is merely the completion of μ. If μ is not σ-finite, then the extension to μ^*-measurable sets also saturates $\overline{\mu}$. It should be observed that in this case the extension of μ to $\mathcal{B}$ need not be unique (Problem 1), although any extension $\tilde{\mu}$ must agree with $\overline{\mu}$ for each set B of $\mathcal{B}$ for which $\tilde{\mu}B < \infty$, and we always have $\overline{\mu}B \leq \mu^* B$.

It is often convenient to start with a set function on a collection $\mathcal{C}$ of sets having less structure than an algebra of sets.

2.9 Definition. *A collection $\mathcal{C}$ of subsets of X is said to be a semialgebra of sets if*

(a) $A,\ B \in \mathcal{C} \Longrightarrow A \cap B \in \mathcal{C}$

(b) $A \in \mathcal{C} \Longrightarrow A^c = \bigcup_{i=1}^{n} B_i$, *with* $B_i \in \mathcal{C}$ *and* $B_i \cap B_j = \phi$, *for* $i \neq j$.

If $\mathcal{C}$ is any semialgebra of sets, then the collection $\mathcal{A}$ consisting of the empty set and all finite disjoint unions of sets in $\mathcal{C}$ is an algebra of sets which is called the algebra generated by $\mathcal{C}$. If μ is a set function defined on $\mathcal{C}$, it is natural to attempt to define a finitely additive set function on $\mathcal{A}$ by setting

$$\mu A = \sum_{i=1}^{n} \mu E_i$$

whenever A is the disjoint union of the sets E_i in $\mathcal{C}$. Since a set A in $\mathcal{A}$ may possibly be represented in several ways as a disjoint union of sets in $\mathcal{C}$, we must be certain that such a procedure leads to a unique value for μA. The following proposition gives conditions under which this procedure can be carried out and will give a measure on the algebra $\mathcal{A}$.

2.10 Proposition. *Let $\mathcal{C}$ be a semialgebra of sets and μ a nonnegative set function defined on $\mathcal{C}$ with $\mu\phi = 0$ (if $\phi \in \mathcal{C}$). Then μ has a unique extension to a measure on the algebra $\mathcal{A}$ generated by $\mathcal{C}$ if the following conditions are satisfied:*

i. If a set C in $\mathcal{C}$ is the union of a finite disjoint collection $\{C_i\}$ of sets in $\mathcal{C}$, then $\mu C = \sum \mu C_i$.

ii. If a set C in $\mathcal{C}$ is the union of a countable disjoint collection $\{C_i\}$ of sets in $\mathcal{C}$, then $\mu C \leq \sum \mu C_i$.

PROBLEMS [6.2]

1. Let X be the set of rational numbers and $\mathcal{A}$ the algebra of finite unions of intervals of form $(a, b]$ with $\mu(a, b] = \infty$ and $\mu\phi = 0$. The extension of μ to the smallest σ-algebra containing $\mathcal{A}$ is not unique.

2. Prove Proposition 10 by showing:

   (a) Condition (i) implies that, if A is the union of each of two finite disjoint collections $\{C_i\}$ and $\{D_j\}$ of sets in $\mathcal{C}$, then $\sum \mu C_i = \sum \mu D_j$. [Hint:$\mu C_i = \sum_j \mu(C_i \cap D_j)$]

   (b) Condition (ii) implies that μ is countably additive on $\mathcal{A}$ [for finite additivity and monotonicity already imply the reverse inequality].

3. Let $\mathcal{A}$ be a collection of sets which is closed under finite union and finite intersections; an algebra of sets, for example.

   (a) Show that $\mathcal{A}_\sigma$ is closed under countable unions and finite intersections.

   (b) Show that each set in $\mathcal{A}_{\sigma\delta}$ is the intersection of a *decreasing* sequence of sets in $\mathcal{A}_\sigma$.

4. Let μ be a finite measure on an algebra $\mathcal{A}$, and μ^* the induced measure. Show that a set E is measurable if and only if for each $\epsilon > 0$ there is a set $A \in \mathcal{A}_\delta$, $A \subset E$, such that $\mu^*(E \sim A) < \epsilon$.

5. If we start with an outer measure μ^* on X and form the induced measure $\overline{\mu}$ on the μ^*-measurable sets, we can use $\overline{\mu}$ to induce an outer measure μ^+.

   (a) Show that for each set E we have $\mu^+ E \geq \mu^* E$.

   (b) For a given set E we have $\mu^+ E = \mu^* E$ if and only if there is a μ^*-measurable set $A \supset E$ with $\mu^* A = \mu^* E$.

   (c) An outer measure which satisfies the criterion in (5b) for each set E is called a **regular outer measure**. Show that every outer measure induced by a measure on an algebra is regular.

   (d) Let X be a set consisting of two points. Costruct an outer measure on X which is not regular.

6.3 Product Measure Space

Let X and Y denote any two sets and consider the Cartesian product $X \times Y$ as defined in (1.2). For each $A \subset X$ and each $B \subset Y$, the subset

$$A \times B = \{(x, y) \in X \times Y | x \in A \quad \textit{and} \quad y \in B\}$$

of $X \times Y$ is called a **rectangle** in $X \times Y$ with A and B as its *sides*.

The following two propositions are direct consequences of the definition.

3.1 Proposition. *A rectangle $A \times B$ in $X \times Y$ is empty iff at least one of its sides A and B is empty.*

3.2 Proposition. *For any two non-empty rectangles $E_1 = A_1 \times B_1$ and $E_2 = A_2 \times B_2$, we have*

(a) $E_1 = E_2 \iff A_1 = A_2$ *and* $B_1 = B_2$.

(b) $E_1 \subset E_2 \iff A_1 \subset A_2$ *and* $B_1 \subset B_2$.

(c) $E_1 \cap E_2 = (A_1 \cap A_2) \times (B_1 \cap B_2)$.

$$\text{(d) } E_1 \sim E_2 = [(A_1 \sim A_2) \times B_1] \cup [(A_1 \cap A_2) \times (B_1 \sim B_2)].$$

It follows from(c) that E_1 and E_2 are disjoint iff $A_1 \cap A_2 = \phi$ or $B_1 \cap B_2 = \phi$. In particular, the rectangles $(A_1 \sim A_2) \times B_2$ and $(A_1 \cap A_2) \times (B_1 \sim B_2)$ are disjoint.

Throughout the remainder of the present section, we are concerned with two arbitrarily given complete measure spaces$(X, \mathfrak{M}, \mu)$ and $(Y, \mathfrak{N}, v)$.

By a **measurable rectangle** in $X \times Y$, we mean a rectangle $A \times B$ in $X \times Y$ with $A \in \mathfrak{M}$ and $B \in \mathfrak{N}$. The collection $\mathfrak{R}$ of measurable rectangles is a semialgebra, since

$$(A \times B) \cap (C \times D) = (A \cap C) \times (B \cap D)$$

and

$$\sim (A \times B) = (\tilde{A} \times B) \cup (A \times \tilde{B}) \cup (\tilde{A} \times \tilde{B}).$$

If $A \times B$ is a measurable rectangle, we set

$$\lambda(A \times B) = \mu A \cdot \nu B.$$

Furthermore, the set of finite disjoint unions of measurable rectangles is a Boolean algebra.

That is, in other words let $\mathfrak{R}$ denote the set of all measurable rectangles in $X \times Y$ and let $\mathcal{B}$ denote the set of all finite disjoint unions of measurable rectangles in $X \times Y$. Then we have the following proposition.

3.3 Proposition. *The set $\mathcal{B}$ is a Boolean algebra of subsets of $X \times Y$.*

Proof. First, $X \times Y$ and its empty subset ϕ are obviously in $\mathcal{B}$. Next, according to (c) and (d) of 3.2 the intersection and the difference of any two measurable rectangles are in $\mathcal{B}$. Now, let

$$E = \bigcup_{i=1}^{m} E_i \qquad F = \bigcup_{j=1}^{n} F_j$$

denote any two members of $\mathcal{B}$. Then we have

$$E \cap F = \bigcup_{i=1}^{m} \left[\bigcup_{j=1}^{n} (E_i \cap F_j) \right],$$

$$E \sim F = \bigcup_{i=1}^{m} \left[\bigcap_{j=1}^{n} (E_i \sim F_j) \right],$$

These prove $E \cap F \in \mathcal{B}$ and $E \sim F \in \mathcal{B}$. Since $X \times Y \in \mathcal{B}$, it follows that we also have $E \cup F \in \mathcal{B}$. This completes the proof of (3.3). □

Now define a function $\lambda' : \mathfrak{R} \to R^*$ by taking

$$\lambda'(A \times B) = \mu(A)\nu(B)$$

for every measurable rectangle $A \times B$ in $X \times Y$. We have the following lemma.

3.4 Lemma. *Let $\{(A_i \times B_i)\}$ be a countable disjoint collection of measurable rectangles whose union is a measurable rectangle $A \times B$. Then*

$$\lambda(A \times B) = \sum \lambda(A_i \times B_i).$$

Proof. Let $y \in B$ be arbitrarily given. For every $x \in A$, the point (x, y) belongs to a unique rectangle $A_i \times B_i$ of the collection. Hence A is the disjoint union of those A_i such that $y \in B_i$. This implies

$$\mu(A)\chi_{\text{B}}(y) = \sum_{i \in M} \mu(A_i)\chi_{\text{B}_i}(y)$$

for every $y \in B$. It follows from (3.3) that we have

$$\int \mu(A)\chi_{\text{B}}\, d\nu = \sum_{i \in \text{M}} \int \mu(A_i)\chi_{\text{B}_i}\, d\nu.$$

Consequently we obtain

$$\lambda'(A \times B) = \mu(A)\nu(B) = \sum_{i \in M} \mu(A_i)\nu(B_i) = \sum_{i \in M} \lambda'(A_i \times B_i).$$

This completes the proof of (3.4). □

Because of (3.4) the function $\lambda' : \mathfrak{R} \to R^*$ can be extended over the Boolean algebra $\mathcal{B}$ by taking

$$\lambda'(E) = \sum_{i=1}^{n} \lambda'(A_i \times B_i)$$

for every finite disjoint union

$$E = \bigcup_{i=1}^{n} (A_i \times B_i)$$

of measurable rectangles in $X \times Y$. Furthermore, by means of (3.4), one can easily verify that $\lambda' : \mathcal{B} \to R^*$ is countably additive and hence is a measure on $\mathcal{B}$. According to section 6.2 $\lambda' : \mathcal{B} \to R^*$ induces an outer measure

$$\lambda^* : 2^{X \times Y} \to R^*$$

which defines a complete measure $\lambda : \mathfrak{L} \to R^*$ on a Boolean σ-algebra $\mathfrak{L}$ containing $\mathcal{B}$. This measure λ is called the **product measure** of μ and ν and will be denoted by

$$\lambda = \mu \times \nu : \mathfrak{L} \to R^*.$$

The complete measure space $(X \times Y, \mathfrak{L}, \mu \times \nu)$ is called the **product measure space** of the given complete measure spaces $(X, \mathfrak{M}, \mu)$ and $(Y, \mathfrak{N}, \nu)$. If μ and ν are finite or σ-finite, then clearly so is $\mu \times \nu$.

3.5 Example. (1) If $X = Y$ is the real line R and $\mu = \nu$ is the Lebesgue measure, then the product measure $\mu^2 = \mu \times \mu$ is called the two-dimensional Lebesgue measure in the plane R^2.

Let $A = [a, b]$, $B = [c, d]$. Then $A \times B$ is a measurable rectangle of the plane R^2 and $\mu \times \nu(A \times B) = \mu(A) \times \nu(B) = (b - a)(d - c)$

Inductively, we may define the n-dimensional Lebesgue measure in the n-space R^n as the product measure.

$$\mu^n = \mu \times \mu^{n-1}$$

(2) Let $X = Y = N$ and $\alpha = \beta = (N)$
Define a counting measure on $\wp(N)(V-1)$. then the product measure $\mu \times \nu$ is a measure defining as follows.
$\mu \times \nu : \wp(N) \longrightarrow R^*$ by

$$s \longrightarrow \mu^*(s) = \begin{cases} \text{the} & \text{number of elements of } S; \\ & \text{if } S \text{ is a finite set.} \\ \infty : & \text{if } S \text{ is an infinite set.} \end{cases}$$

Then $\mu \times \nu$ is a counting measure on $\wp(N)$.

The purpose of the next few lemmas is to describe the structure of the sets which are measureable with respect to the product measure $\mu \times \nu$. If E is any subset of $X \times Y$ and x a point of X, we define the x **cross-section** E_x by

$$E_x = \{y :< x, y >\in E\}$$

and similarly for the y cross-section for y in Y. The characteristic function of E_x is related to that of E by

$$\chi_{E_x}(y) = \chi_E(x, y).$$

We also have $(\widetilde{E})_x = \sim (E_x)$ and $(\bigcup E_\alpha)_x = \bigcup (E_\alpha)_x$ for any collection $\{E_\alpha\}$.

3.6 Lemma. *Let x be a point of X and E a set in $\mathfrak{R}_{\sigma\delta}$. Then E_x is a measurable subset of Y.*

Proof. The lemma is trivially true if E is in the class $\mathfrak{R}$ of measurable rectangles. We next show it to be true for E in $\mathfrak{R}_\sigma$. Let $E = \bigcup_{i=1}^{\infty} E_i$, where each E_i is a measurable rectangle. Then

$$\begin{aligned} \chi_{E_x}(y) &= \chi_E(x, y) \\ &= \sup_i \chi_{E_i}(x, y) \\ &= \sup_i \chi_{(E_i)_x}(y). \end{aligned}$$

Since each E_i is a measurable rectangle, $\chi_{(E_i)_x}(y)$ is a measurable function of y, and so χ_{E_x} must also be measurable, whence E_x is measurable.

Suppose now that $E = \bigcap_{i=1}^{\infty} E_i$ with $E_i \in \mathfrak{R}_\sigma$. Then

$$\begin{aligned}\chi_{E_x} &= \chi_E(x,y)\\ &= \inf_i \chi_{E_i}(x,y)\\ &= \inf_i \chi_{(E_i)_x}(y),\end{aligned}$$

and we see that χ_{E_x} is measurable. Thus E_x is measurable for any $E \in \mathfrak{R}_{\sigma\delta}$. □

3.7 Lemma. *Let E be a set in $\mathfrak{R}_{\sigma\delta}$ with $\mu \times \nu(E) < \infty$. Then the function g defined by $g(x) = \nu E_x$ is a measurable function of x and*

$$\int g \, d\mu = \mu \times \nu(E).$$

Proof. The lemma is trivially true if E is a measurable rectangle. We first note that any set in $\mathfrak{R}_\sigma$ is a disjoint union of measurable rectangles. Let $< E_i >$ be a disjoint sequence of measurable rectangles, and let $E = \bigcup E_i$ Set

$$g_i(x) = \nu\left[(E_i)_x\right].$$

Then each g_i is a nonnegative measurable function, and

$$g = \sum g_i.$$

Thus g is measurable, and by the corollary of the Monotone Convergence Theorem 3.4 of Chapter V. we have

$$\begin{aligned}\int g \, d\mu &= \sum \int g_i \, d\mu\\ &= \sum \mu \times \nu(E_i)\\ &= \mu \times \nu(E).\end{aligned}$$

Consequently, the lemma holds for $E \in \mathfrak{R}_\sigma$.

Let E be a set of finite measure in $\mathfrak{R}_{\sigma\delta}$. Then there is a sequence $< E_i >$ of sets in $\mathfrak{R}$ such that $E_{i+1} \subset E_i$ and $E = \bigcap E_i$. It follows from Proposition 3.5 that we may take $\mu \times \nu(E_1) < \infty$. Let $g_i(x) = \nu[(E_i)_x]$. Then $g(x) = \lim g_i(x)$, and so g is measurable. Since

$$\int g_1 \, d\mu = \mu \times \nu(E_1) < \infty,$$

we have $g_1(x) < \infty$ for almost all x. For an x with $g_1(x) < \infty$, we have $< (E_i)_x >$ a decreasing sequence of measurable sets of finite measure whose intersection is E_x.

Thus by Proposition 3.22 of Chapter III. we have

$$\begin{aligned} g(x) &= \nu(E_x) = \lim \nu[(E_i)_x] \\ &= \lim g_i(x). \end{aligned}$$

Hence

$$g_i \longrightarrow g \ a.e.,$$

and so g is measurable. Since $0 \leq g_i \leq g_1$, the Lebesgue Convergence Theorem implies that

$$\begin{aligned} \int g \, d\mu &= \lim \int g_i \, d\mu \\ &= \lim \mu \times \nu(E_i) \\ &= \mu \times \nu(E), \end{aligned}$$

the last equality following from Proposition 1.4 of V □

3.8 Lemma. *Let E be a set for which $\mu \times \nu(E) = 0$. Then for almost all x we have $\nu(E_x) = 0$.*

Proof. By Proposition 3.6 there is a set F in $\wp_{\sigma\delta}$ such that $E \subset F$ and $\mu \times \nu(F) = 0$. It follows from Lemma 3.7 that for almost all x we have $\nu(F_x) = 0$. But $E_x \subset F_x$, and so $\nu E_x = 0$ for almost all x since ν is complete. □

3.9 Proposition. *Let E be a measurable subset of $X \times Y$ such that $\mu \times \nu(E)$ is finite. Then for almost all x the set E_x is a measurable subset of Y. The function g defind by*

$$g(x) = \nu(E_x)$$

is a measurable function defined for almost all x and

$$\int g \, d\mu = \mu \times \nu(E).$$

Proof. By Proposition 2.6 there is a set F in $\mathfrak{R}_{\sigma\delta}$ such that $E \subset F$ and $\mu \times \nu(F) = \mu \times \nu(E)$. Let $G = F \sim E$. Since E and F are measurable, so is G, and

$$\mu \times \nu(F) = \mu \times \nu(E) + \mu \times \nu(G).$$

Since $\mu \times \nu(E)$ is finite and equal to $\mu \times \nu(F)$, we have $\mu \times \nu(G) = 0$. Thus by lemma 3.8 we have $\nu G_x = 0$ for almost all x. Hence

$$g(x) = \nu E_x = \nu F_x \quad a.e,$$

and so g is a measurable function by Lemma 3.7. Again by Lemma 3.7

$$\int g \, d\mu = \mu \times \nu(F) = \mu \times \nu(E). \qquad \square$$

The following two therems enable us to interchange the order of integration and to calculate integrals with respect to product measures by iteration.

3.10 Theorem. ***(Fubini)*** *Let $(X, \mathfrak{a}, \mu)$ and (Y, β, ν) be two complete measure spaces. For an arbitrary measurable function*

$$f : X \times T \longrightarrow R^*$$

which is integrable with respect to $\mu \times \nu$, we have:

(a) for almost all x, the function $f_x : Y \to \mathbb{R}^*$ defined by $f_x(y) = f(x, y)$ is an integrable function on Y with respect to ν.

(b) *for almost all y, the function $f_y : X \to \mathbb{R}^*$ defined by $f_y(x) = f(x, y)$ is an integrable function on X with respect to μ.*

(c) $\int_Y f(x, y)\, d\nu(y)$ *is an integrable function on X; Simdilarly, The function $\xi : X \to R$ defined by*

$$\xi(x) = \begin{cases} \int f_x \, d\nu, & \text{if } f_x \text{ is integrable,} \\ 0, & \text{otherwise,} \end{cases}$$

is integrable with respect to μ.

(d) $\int_X f(x, y)\, d\mu(x)$ *is an integrable function on Y;*

(e) $\int_X \left[\int_Y f \, d\nu \right] \, d\mu = \int_{X \times Y} f d(\mu \times \nu) = \int_Y \left[\int_X f \, d\mu \right] \, d\nu.$

Proof. Because of the symmetry between x and y it suffices to prove (a),(c) and the first half of (e). If the conclusion of the theorem holds for each of two functions, it also holds for their difference, and hence it is sufficient to consider the case when f is nonnegative. Proposition 3.9 asserts that the theorem is true if f is the characteristic function of a measurable set of finite measure, and hence the theorem must be true if f is a simple function which vanishes outside a set of finite measure. Proposition 2.4 of V asserts that each nonnegative integrable function f is the limit of an increasing sequence $\langle \varphi_n \rangle$ of nonegative simple functions, and, since each φ_n is integrable and simple, it must vanish outside a set of finite measure. Thus f_x is the limit of the increasing sequence $\langle (\varphi_n)_x \rangle$ and is measurable. By the Monotone Convergence Theorem

$$\int_Y f(x, y)\, d\nu(y) = \lim \int_Y \varphi_n(x, y)\, d\nu(y),$$

and so this integral is a measurable function of x. Again by the Monotone Convergence Theorem

$$\begin{aligned} \int_X \left[\int_Y f \, d\nu \right] \, d\mu &= \lim \int_X \left[\int_Y \varphi_n \, d\nu \right] \, d\mu \\ &= \lim \int_{X \times Y} \varphi_n d(\mu \times v) \\ &= \int_{X \times Y} f \, d(\mu \times v). \end{aligned}$$

□

In order to apply the Fubini Theorem, one must first verify that f is integrable with respect to $\mu \times v$; that is, one must show that f is a measurable function on $X \times Y$ and that $\int |f| \, d(\mu \times v) < \infty$. The measurability of f on $X \times Y$ is sometimes difficult to establish, but in many cases we can establish it by topological considerations (cf. Problem 2). In the cases when μ and v are σ-finite, the integrability of f can be determined by iterated integration using the following theorem:

3.11 Theorem. *(Tonelli) Let $(X, \mathcal{A}, \mu)$ and $(Y, \mathcal{B}, \nu)$ be two σ-finite measure space, and let f be a nonnegative measurable function on $X \times Y$. Then*

i. *for almost all x the function f_x defined by $f_x(y) = f(x, y)$ is a measurable function on Y.*

i'. *for almost all y the function f_y defined by $f_y(x) = f(x, y)$ is a measurable function on X.*

ii. $\int_Y f(x, y) \, d\nu(y)$ *is a measurable function on X.*

ii'. $\int_X f(x, y) \, d\mu(x)$ *is a measurable function on Y.*

iii. $\int_X \left[\int_Y f \, d\nu\right] d\mu = \int_{X \times Y} f \, d(\mu \times \nu) = \int_Y \left[\int_X f \, d\mu\right] d\nu.$

Proof. For a nonnegative measurable function f the only point in the proof of Theorem 3.10 where the integrability of f was used was to infer the existence of an increasing sequence $\langle \varphi_n \rangle$ of simple functions each vanishing outside a set of finite measure such that $f = \lim \varphi_n$. But if μ and ν are σ-finite, then so is $\mu \times \nu$, and any nonnegative measurable function on $X \times Y$ can be to so approximated by Proposition 3.10 □

If $\mathcal{A}$ and $\mathcal{B}$ are σ-algebras on X and Y, then the smallest σ-algebra, containing the measurable rectangles is denoted by $\mathcal{A} \times \mathcal{B}$, and since $\mu \times \nu$ is obtained by the Caratheodory extension process, it is both complete and saturated. If μ and ν are both σ-finite, then the product measure on $\mathcal{A} \times \mathcal{B}$ is already saturated and the measurables sets for $\mu \times \nu$ are those which differ from sets in $\mathcal{A} \times \mathcal{B}$ by sets of measure zero.

Many authors prefer to define product measure to be the restriction of $= \mu \times \nu$ to $\mathcal{A} \times \mathcal{B}$. The advantage of taking $\mu \times \nu$ to be complete, as we have done here, is that this does what we want it to for Lebesgue measure: The product of n-dimensional Lebesgue measure with m-dimensional Lebesgue measure is $(n+m)$-dimensional Lebegue measure. Since our hypotheses for the Fubini and Tonelli theorems require only measurability with respect to the complete product measure, they are weaker than requiring measurability with respect to $\mathcal{A} \times \mathcal{B}$. The price for using these weaker hypotheses is the necessity of including the "almost all" phrases in the conclusion of the theorems. This has to be expected, since changing f arbitrarily for x in a set of measure zero doesn't change the measurability or integrability of f, but f_x can be arbitrary for those x. If however, f is measurable with respect to $\mathcal{A} \times \mathcal{B}$, then f_x is measurable for each x.

We have also used the completeness of μ to show that $\int f(x,y) d\nu$ (y) is measurable, for if μ were not complete we could only conclude that this was a function which differed on a subset of a set of measure zero from a measurable function. If, however, f is measurable with respect to $\mathcal{A} \times \mathcal{B}$, then it turns out that $\int f(x,y)\ d\nu(y)$ is measurable with respect to $\mathcal{A}$ even if μ is not complete (provided f is integrable), but the proof of this is surprisingly intricate. For a proof see Halmos [6], p.143.

The examples in the problems show that we cannot omit the hypothesis of the integrability of f from the Fubini Theorem or the hypotheses of σ-finiteness and nonnegativity from the Tonelli Theorem. Problem 7 shows the essential role played by the measurability of f in these theorems: If we omit this assumption, even for bounded functions and finite measures, we may have the iterated integrals $\int[\int f\ d\nu]\ d\mu$ and $\int[\int f\ d\mu]\ d\nu$ well defined but unequal.

PROBLEMS [6.3]

1. Let $X = Y$ be the set of positive integers, $\mathcal{A} = \mathcal{B} = \wp(x)$, and let $\nu = \mu$ be the measure defined by setting $\mu(E)$ equal to the number of points in E if E is finite and ∞ if E is an infinite set. (This measure is called the counting measure.) State the Fubini and Tonelli Theorems explicitly for this case.

2. Let $X = Y = [0,1]$, and let $\mu = \nu$ be Lebesgue measure. Show that each open set in $X \times Y$ is measurable, and hence each Borel set in $X \times Y$ is measurable.

3. Show that if $(X, \mathcal{A}, \mu)$ and $(Y, \mathcal{B}, \nu)$ are two σ-finite measure spaces, then $\mu \times \nu$ is the only measure on $\mathcal{A} \times \mathcal{B}$ which assigns the value $\mu A\ vB$ to each measurable rectangle $A \times B$. Show that a measure on $\mathcal{A} \times \mathcal{B}$ with this property need not be unique, if we do not have σ-finiteness.

4. Let $(X, \mathcal{A}, \mu)$ be any σ-finite measure space and Y the set of positive integers with ν the counting measure (Problem 1). Then Theorem 3.11 and Corollary 11.14 state the same conclusion. However, Corollary 3.5 of V is vaild even if μ is not σ-finite and hence the Tonelli Theorem is true without σ-finiteness if (Y, β, v) is this special measure space.

5. Let h and g be integrable functions on X and Y, and define $f(x,y) = h(x)g(y)$. Then f is integrable on $X \times Y$ and
$$\int_{X\times Y} f\ d(\mu \times \nu) = \int_X h\ d\mu \int_Y g\ d\nu.$$
(Note: We do not need to assume that μ and ν are σ-finite.)

6. Show that Tonelli's Theorem is still true if, instead of assuming μ and ν to be σ-finite, we merely assume that $\{<x,y>: f(x,y) \neq 0\}$ is a set of σ-finite measure.

7. The following example shows that we cannot remove the hypothesis that f be nonnegative from the Tonelli Theorem or

that f be integrable from the Fubini Theorem. Let $X = Y$ be the positive integers and $\mu = \nu$ be the counting measure. Let

$$f(x,y) = \begin{cases} 2 - 2^{-x} & if \quad x = y \\ -2 + 2^{-x} & if \quad x = y + 1 \\ 0 & otherwise. \end{cases}$$

8. The following example shows that we cannot remove the hypothesis that f be integrable from the Fubini Theorem or that μ and ν are σ-finite from the Tonelli Theorem: Let $X = Y$ be the interval $[0,1]$, with $\mathcal{A} = \mathcal{B}$ the class of Borel sets. Let μ be Lebesgue measure and ν the counting measure. Then the diagonal $\triangle = \{< x, y > \in X \times Y : x = y\}$ is measurable (is an $\mathfrak{R}_{\sigma\delta}$, in fact), but its characteristic function fails to satisfy any of the equalities in condition (c) of the Fubini and Tonelli Theorems.

9. The following example shows that the hypothesis that f be measurable with respect to the product measure cannot be omitted from the Fubini and Tonelli Theorems even if we assume the measurability of f_y and f_x and the integrability of $\int f(x,y)\, d\nu(y)$ and $\int f(x,y)\, d\mu(x)$. Let $X = Y =$ the set of ordinals less than or equal to the first uncountable ordinal Ω. Let $\mathcal{A} = \mathcal{B}$ be the σ-algebra consisting of all countable sets and their complements. Define $\mu = \nu$ by letting $\mu E = 0$ if E countable, $\mu E = 1$ otherwise. Define a subset S of $X \times Y$ by $S = \{< x, y >: x < y\}$. Then S_x and S_y are measurable for each x and y, but if is the characteristic function of S we have

$$\int \left[\int f(x,y)\, d\mu(x) \right] d\nu(y) \neq \int \left[\int f(x,y)\, d\nu(y) \right] d\mu(x).$$

Chapter 7

BANACH SPACE

Banach spaces have the algebraic structure as a vector space and topological structure as a metric space. Since they have the fine structure and wide application, Banach spaces are very important in contemporary analysis: many basic theorems can be couched in abstract terms and assert-tions about Banach spaces of one kind or another, We will begin the metric space.

7.1 Metric Space

A distance or metric function on an arbitrary set X will be defined as a function having certain properties which are possessed by the distance function in a Euclidean space.

In this section, we study the properties of metric spaces and define continututy of functions and convergence of sequences in these spaces. Considering these topics in the general setting of a metric space has two advantages. It highlights the essential properties of these concepts and shows how broad is their range of applicability. The theory can then be expanded by introducing additional properties into the space. In section 6 we study metric spaces which have the algebraic structure of a linear space.

1.1 Definition. *A metric or distance function on a set X is a real-valued function d defined on $X \times X$ which has the following properties:for all x, y, and z in X.*

D_1 $d(x,y) \geq 0$; $d(x,y) = 0$ *if and only if* $x = y$;

D_2 $d(x,y) = d(y,x)$;

D_3 $d(x,y) \leq d(x,z) + d(z,y)$.

Properties D_1 and D_2 are called, respectively, the positive definite and symmetric properties. Thus, a metric on X is a symmetric, positive definite , real-valued function on $X \times X$ which satisfies the triangle inequality D_3.

1.2 Definition. *A metric space (X, d) is a nonempty set X and a metric d defined on X.*

The elements of X are called the points of (X, d). When no confusion will arise as a result, the metric space (X, d) will be denoted simply by X.

Since the definition of a metric space was patterned on certain properties of Euclidean spaces, it is clear that these spaces are metric spaces. The complex number system C is also a metric space. As a metric space, C is the same as $\mathbb{R}^2$. Some other examples of metric spaces follow.

1.3 Example. 1) Our first example is the usual metric on $\mathbb{R}$. $d : \mathbb{R} \times \mathbb{R} \to \mathbb{R}$ defined by

$$d(x,y) = |x - y| \quad \text{for all } x, y \in \mathbb{R}$$

Properties (a) to (d) are known to hold, and so we have the metric space $(\mathbb{R}, d)$.

2) Let X be any nonempty set and let d be deined by

$$d(x,y) = \begin{cases} 0 & if \quad x = y \\ 1 & if \quad x \neq y. \end{cases}$$

Then d is a metric on X, called the **discrete metric** or **trivial metric**, and the metric space (X, d) is called a **discrete space**.

3) If X is the set of all m-tuples of real numbers and, if for $x = (\alpha^1, \cdots, \alpha^m)$ and $y = (\beta^1, \cdots, \beta^m)$,

$$d(x, y) = \max\{|\alpha^k - \beta^k| : k = 1, \cdots, m\},$$

then (X, d) is a metric space.

4) Let X be the set of all real-valued functions which are defined and continuous on the closed interval $[a, b]$ in $\mathbb{R}$ and let

$$d(f, g) = \max\{|f(t) - g(t)| : t \in [a, b]\}.$$

Then (X, d) is a metric space which is usually denoted by $C[a, b]$.

Notice that in example (3) we have defined a metric on the set of m-tuples of real numbers which is different from the Euclidean metric. The resulting metric space differs from the Euclidean space $\mathbb{R}^m$. On the other hand, there are metric spaces which for all practical purposes are the same even though they may have different sets of points.

1.4 Definition. *Let (X_1, d_1) and (X_2, d_2) be two metric spaces. A function f from X_1 into X_2 is called an* ***isometry*** *if*

$$d_2(f(x), f(y)) = d_1(x, y) \quad \text{for all } x, y \in X_1.$$

Two metric spaces (X_1, d_1) and (X_2, d_2) are said to be isometric if there exists an isometry of X_1 onto X_2.

Since an isometry preserves distance between points, it is clearly a one-to-one mapping. Considered as metric spaces, two isometric spaces are essentially the same and may be identified if we so choose.

This discussion of isometry suggests a method for definiting a metric for certain sets. Let (X_2, d_2) be a metric space, X_1, be a set, and f be a one-to-one function from X_1 into X_2. If we define

$$d_1(x, y) = d_2(f(x), f(y)) \quad \text{for all } x, y \in X_1,$$

then (X_1, d_1) is a metric space. This is easily verified.

Consider the extended real line. This is the real line with two points adjoined. These points are called (plus) infinity and minus infinity and are denoted by $(+)\infty$ and $-\infty$. The order relation in the real number system is extended by definiting $-\infty < x < \infty$ for all $x \in \mathbb{R}$. The usefulness of the extended real line is illustrated by the fact that we can say that any set of real numbers has a least upper bound in the extended real number system. In Chapter 2 we proved that a nonempty set A real numbers which is bounded above has a least upper bound. If A is empty, we take the least upper bound to be $-\infty$, and if A is not bounded above, then we take the least upper bound to be ∞.

In order to introduce a metric in the extended real line, we define

$$\begin{aligned} f(-\infty) &= -1 \\ f(x) &= \frac{x}{1+|x|}, \quad x \in \mathbb{R} \\ f(\infty) &= 1. \end{aligned}$$

Then f is a one-to-one mapping of the extended real line into $\mathbb{R}$ and letting

$$d(x, y) = |f(x) - f(y)| \quad \text{for all } x, y \in \mathbb{R} \cup \{-\infty, \infty\},$$

we obtain a metric space which is denoted by $\overline{\mathbb{R}}$. Notice that the extended real line $\overline{\mathbb{R}}$ is isometric to the metric space consisting of the set $[-1, 1]$ with the distance between points given by the Euclidean metric. This space, denoted simply by $[-1, 1]$, is called a subspace of $\mathbb{R}$.

In general, given a metric space (X, d), we can obtain new metric spaces as follows. Let E be a nonempty subset of X and let d_1 be the restriction of d to $E \times E$. Then (E, d_1) is a metric space and is said to be a **subspace** of (X, d).

In a metric space (X, d) we defie the distance between a point and a set and the distance between two sets as follows. If $x \in X$ and B is a noempty subset of X, then the distance between x and B, denoted by $d(x, B)$, is

$$d(x, B) = \inf\{d(x, y) : y \in B\}.$$

And, if A and B are nonempty subsets of X, the distance between A and B, denoted by $d(A, B)$, is

$$\begin{aligned} d(A,B) &= \inf\{d(x,B) : x \in A\} \\ &= \inf\{d(x,y) : x \in A, y \in B\}. \end{aligned}$$

A set E in a metric space (X, d) is said to be **dense** in X if $\overline{E} = X$. A metric space with a countable dense subset is said to be **separable**. For example, the set of m-tuples of rational numbers is a countable dense set in $\mathbb{R}^m$ and, hence, $\mathbb{R}^m$ is a separable space.

1.5 Proposition. *If X is a separable metric space, then X has a countable open base consisting of open balls.*

Proof. Let E be a countable dense set in X. Then the set of open balls with a point of E as center and rational radius is countable. Let G be an open set in X and take $x \in G$. There exists a rational number $r(x)$ such that $B(x; r(x)) \subset G$. Since E is dense in X there is a point $y(x) \in E$ such that $d(x, y(x)) < \frac{1}{2}r(x)$.Then

$$x \in B(y(x); \frac{1}{2}r(x)) \subset B(x; r(x)) \subset G.$$

Thus $G = \cup_{x \in G} B(y(x); \frac{1}{2}r(x))$. □

We define a Cauchy sequence in a metric space (X, d) in the expected way.

1.6 Definition. *A sequence $\{x_n\}$ of points in a metric space (X, d) is called a* ***Cauchy sequence*** *if for each $\epsilon > 0$ there exists a positive integer n_0 such that*

$$d(x_n, x_m) < \epsilon \text{ whenever } n, m \geq n_0.$$

1.7 Proposition. *A convergent sequence in a metric space is a Cauchy sequence.*

Proof. If $\lim x_n = b$, then for each $\epsilon > 0$ there exists a positive integer n_0 such that $d(x_n, b) < \epsilon/2$ whenever $n \geq n_0$. Thus, if $n, m \geq n_0$

$$d(x_n, x_m) \leq d(x_n, b) + d(b, x_m) < \epsilon. \qquad \square$$

The converse of this theorem does not hold in every metric space.

1.8 Definition. *A metric space X is* **complete** *if every Cauchy sequence in X converges to a point in X.*

We have shown that the real line $\mathbb{R}$ is a complete metric space. However, the subspace $(0,1)$ of $\mathbb{R}$ is not complete; the sequence $(1/n)$ is a Cauchy sequence in $(0,1)$ which does not converg to a point in this space. Since $\mathbb{R}$ and $(0,1)$ are homeomorphic, completeness is not a topological property.

We now prove the completeness of some inportant spaces.

1.9 Proposition. *The Euclidean space $\mathbb{R}^m$ is a complete metric space.*

Proof. Let $\{x_n\}$ be a Cauchy sequence in $\mathbb{R}^m$ where $x_n = (\alpha_n^1, \cdots, \alpha_n^m)$. Then for each $\epsilon > 0$ there exists a positive integer n_0 such that

$$d(x_r, x_s) = \left[\sum_{i=1}^{m} (\alpha_r^i - \alpha_s^i)^2\right]^{1/2} < \epsilon \quad \text{whenever } r, s \geq n_0.$$

Thus, for each $i = 1, \cdots, m$.

$$|\alpha_r^i - \alpha_s^i| < \epsilon \quad \text{whenever } r, s \geq n_0.$$

That is, each componont sequence $\{\alpha_n^i\}$ is a Cauchy sequence in $\mathbb{R}$. Since $\mathbb{R}$ is complete there exists a real number α^i such that $\alpha^i = \lim_{n\to\infty} \alpha_n^i$. We now show that the point $x = (\alpha^i, \cdots, \alpha^m)$ is the limit of the sequence $\{x_n\}$. Take $\epsilon > 0$. Then, for each $i = 1, \cdots, m$, there exists a poistive integer n_i such that

$$|\alpha_n^i - \alpha^i| < \epsilon/\sqrt{m} \quad \text{whenever } n \geq n_i.$$

Hence, if $n \geq \max\{n_i : i = 1, \cdots, m\}$, then

$$d(x_n, x) = \left[\sum_{i=1}^{m}(\alpha_n^i - \alpha^i)^2\right]^{1/2} < \left[\sum_{i=1}^{m}\frac{\epsilon^2}{m}\right]^{1/2} = \epsilon.$$

□

Consider now convergence in the metricd space $C[a, b]$. This is the space of continuous real-value functions defined on $[a, b]$ with metric

$$d(f, g) = \max\{|f(t) - g(t)| : t \in [a, b]\}.$$

In this space $\lim f_n = f$ means: for each $\epsilon > 0$ there exists a positive integer n_0 such that

$$\max\{|f_n(t) - f(t)| : t \in [a, b]\} < \epsilon \quad \text{whenever } n \geq n_0;$$

that is, for all $t \in [a, b]$

$$|f_n(t) - f(t)| < \epsilon \quad \text{whenever } n \geq n_0.$$

Thus, convergence of $\{f_n\}$ to f in $C[a, b]$ is equivalent to uniform convergence of $\{f_n\}$ to f on $[a, b]$.

1.10 Proposition. *The metric space $C[a, b]$ is complete.*

Proof. Let $\{f_n\}$ be a Cauchy sequence in $C[a, b]$. Then for $\epsilon > 0$ there exists an n_0 such that for all $t \in [a, b]$

$$|f_n(t) - f_m(t)| < \epsilon \quad \text{whenever } n, m \geq n_0. \tag{$*$}$$

Thus, for each $t \in [a, b]$, $\{f_n(t)\}$ is a Cauchy sequence in $\mathbb{R}$ and, hence, converges to a real number $f(t)$. Taking the limit with respect to m in $(*)$ we obtain, for all $t \in [a, b]$,

$$|f_n(t) - f(t)| \leq \epsilon \quad \text{whenever } n \geq n_0.$$

Thus $\{f_n\}$ converges uniformly to f on $[a, b]$. Since each of the functions f_n is continuous on $[a, b]$, f is continuous on $[a, b]$; that is, $f \in C[a, b]$ (Problem 6). Thus, we have shown that $\{f_n\}$ converges to f in $C[a, b]$. □

A set E in a metric space X is said to be complete if the subspace E of X is complete. We will estalish a connection between a set being complete and being closed. Before doing that we give a characterization of closed sets in therms of the convergence of sequences.

1.11 Proposition. *A set E in a metric space X is closed if and only if every sequence of points in E which is convergent in X converges to a point in E.*

Proof. Suppose E is closed and let $\{x_n\}$ be a sequence of points in E such that $\lim x_n = b$. Then every neighborhood of b intersects E and, hence, $b \in \overline{E} = E$.

Suppose now that E has the property that $b \in E$ whenever $\{x_n\}$ is a sequence of points in E which converges to b. Take any point $x \in \overline{E}$. Then, for each positive integer n, $B(x; 1/n) \cap E \neq \phi$. Choose a point $x_n \in B(x; 1/n) \cap E$. The sequence $\{x_n\}$ so chosen converges to x and, hence, $x \in E$. Thus, $\overline{E} \subset E$ and E is closed. □

1.12 Proposition. *If E is a complete subspace of a metric space X, then E is closed in X.*

Proof. Take $x \in \overline{E}$. Select a point $x_n \in B(x; 1/n) \cap E$. Then $\{x_n\}$ converges to x in X. Since $\{x_n\}$ is a Cauchy sequence in E, it must converge to a point $y \in E$. But the limit of a convergent sequence is unique and, hence, $y = x$. Thus, $x \in E$ and E is closed □

1.13 Proposition. *A closed set in a complete metric space is complete.*

Proof. Let E be a closed subset of X. Any Cauchy sequence of points in E is convergent in X and, hence, converges to a point in E (Proposition 1.12). Thus, E is complete. □

Proposition 1.12 and 1.13 characterize closed sets in a complete metric space: A set E in a complete metric space X is closed if and only if E is complete.

PROBLEMS [8.1]

1. Show that the functions d defined in examples (1),(2), and (3), satisfy the properties of a metric.

2. Let w, x, y, z be four points in a metric space. Establish the **quadrilateral inequality**.

$$|d(w,x) - d(x,y)| \leq d(w,z) + d(z,y).$$

   In $\mathbb{R}^2$ this inequality may interpreted as stating that the sum of the lengths of two sides of a quadrilateral is greater than or equal to the absolute value of the difference in length of the other two sides.

3. Give an example of sets A and B in a metric space such that $A \cap B = \phi$ but $d(A,B) = 0$.

4. Let X be the set of all continuous functions from $[a,b]$ into $\mathbb{R}^n$ and let d be defined by

$$d(f,g) = \max\{|f(t) - g(t)| : t \in [a,b]\}.$$

   Show that (X,d) is a complete metric space.

5. If A is a closed set in a metric space (X,d) and $x \notin A$, show that $d(x,A) > 0$.

6. Let f be a function from the metric space X into the metric space Y. Suppose whenever a sequence $\{x_n\}$ converges to x then $\{f(x_n)\}$ converges to $f(x)$. Show that f is continuous at x.

7. A sequence $\{B_n\}$ of balls is said to be **nested** if $B_n \supset B_{n+1}$ for all n. Show that a metric space X is complete if every nested sequence of closed balls in X with radii tending to zero has nonempty intersection.

8. If d is a real-valued function of $X \times X$ which for all x, y, and z in X satisfies

$$d(x,y) = 0 \text{ if and only if } x = y$$
$$d(x,y) + d(x,z) \geq d(y,z),$$

show that d is a metric on X.

7.2 Normed Linear Space

An inportant aspect of modern analysis is the recognition and exploitation of the fact that certain sets of functions form linear spaces. The basic concepts of linear or vector spaces are no doubt quite familiar to the reader. Our purpose in this chapter is to review these concepts briefly and to emphasize certain aspects that will be of particular importance for our later work. For example, many of the interesting spaces considered in analysis are of infinite dimension.

2.1 Definition. *A* ***linear (vector) space*** *X over a field F is a set of elemednts together with a function, called addition, from $X \times X$ into X and a function, called scalar multiplication, from $F \times X$ into X which satisfy the following conditions for all $x, y, z \in X$ and $\alpha, \beta \in F$:*

A_1 $(x+y)+z = x(y+z)$

A_2 $x+y = y+x$

A_3 *There is an element 0 in X such that $x+0 = x$ for all $x \in X$.*

A_4 *For each $x \in X$ there is an element $-x \in X$ such that $x+(-x) = 0$.*

S_1 $\alpha(x+y) = \alpha x + \alpha y$

S_2 $(\alpha+\beta)x = \alpha x + \beta x$

S_3 $\alpha(\beta x) = (\alpha\beta)x$

S_4 $1 \cdot x = x$

Properties $A_1 - A_4$ imply that X is an abelian group under addition and Properties $S_1 - S_4$ relate the operation of scalar multiplication to addition in X and to addition and multiplication in F.

The elements of X are called points or vectors and the elements of F are called **scalars**. In our discussions the field F will always be the real field $\mathbb{R}$ or the complex field $\mathbb{C}$. Note that a linear space over $\mathbb{C}$ can also be considered as a linear space over $\mathbb{R}$ since $\mathbb{R}$ is a subset of $\mathbb{C}$.

In general we will abide by the notational convention that Greek letters denote scalars and Roman letters denote vectors. Frequent use will be made of supersrcripts as well as subscripts.

A familiar example of a linear space is the space $V_n(\mathbb{R})$ where the vectors are n-tuples of real numbers and the scalars are real numbers with addition and scalar multiplication defined by

$$(\alpha^1, \cdots, \alpha^n) + (\beta^1, \cdots, \beta^n) = (\alpha^1 + \beta^1, \cdots, \alpha^n + \beta^)$$
$$\gamma(\alpha^1, \cdots, \alpha^n) = (\gamma\alpha^1, \cdots, \gamma\alpha^n).$$

That $V_n(\mathbb{R})$ satisfies the requirements for a linear space over $\mathbb{R}$ is easily verified. Similarly, the set of all n-tuples of complex numbers with the above definitions of addition and scalar multiplication is a linear space over $\mathbb{C}$ and is denoted by $V_n(\mathbb{C})$. In the case where $n = 1$, we understand that $V_1(F)$ is the same as F and, hence, F is a linear space over F.

Before giveing more examples of linear spaces we discuss some elementary properties which can be derived directly from the definition of a linear space. For example, by adding $-x$ to both sides of $x + y = x$, we can show that if $x + y = x$ for some $x \in X$ then $y = 0$. This implies that the zero element specified in A_3 is unique. This result can also be used to show that $0x = 0$ for all $x \in X$ since

$$0x = (0 + 0)x = 0x + 0x.$$

Similarly, we have: $\alpha 0 = 0$ for all $\alpha \in F$. Observe that we use the symbol 0 to deote both the zero vector and the zero scalar. The context indicates which meaning is intended. It is also easy to show that for each $x \in X$ the element $-x$ specified in A_4 is unique. It then follows that $-x = (-1)x$. Subtraction in X is defined by the rule: $x - y = x + (-y)$.

Many of the linear spaces of interest to us are special cases of the linear space introduced in the following proposition.

2.2 Proposition. *The set X of all functions from a nonempty set T into a field F with addition and scalar multiplication defined by*

$$[f+g](t) = f(t)+g(t) \text{ and } [\alpha f](t) = \alpha f(t); f, g \in X, \ t \in T, \ \alpha \in F$$

is a linear space over F.

Proof. Conditions A_1, A_2, and $S_1 - S_4$ of Definition 2.1 follow immediately from corresponding properties of F. The zero element in X is the function whose value is zero at all points in T. Also, for each $f \in X$ the function $-f$ is defined by $[-f](t) = -f(t)$. □

In this section we consider space which are both linear spaces and metric spaces. The Euclidean spaces $\mathbb{R}^n$ are familiar examples of such spaces. In $\mathbb{R}^n$ the metric is introduced via the length of a vector; that is, $d(x, y) = |x - y|$. In the linear spaces considered in this chapter we introduce a generalization of the length of a vector in $\mathbb{R}^n$, called a norm, and define a metric in terms of this norm.

A normed linear space is a linear space with a norm defined on it. Since such spaces are metric spaces, the theory developed for metric spaces applies to them. As one might expect, the introduction of algebraic structure allows a futher development of the theory. Let X be a linear space over the field F where F is either $\mathbb{R}$ or $\mathbb{C}$.

2.3 Definition. *A norm on X is a real-valued function. whose value at x is denoted by $\|x\|$, satisfying the following conditions for all $x, y \in X$ and $\alpha \in F$:*

(1) $\|X\| \geqslant 0$ *and* $\|x\| = 0 \iff x = 0,$

(2) $\|\alpha x\| = |\alpha| \ \|x\|$

(3) $\|x + y\| \leq \|x\| + \|y\|.$

2.4 Definition. *A linear space* X *equipped with a norm* $\|x\|$ *on it is called a* ***normed linear space*** *or simply a* ***normed space****.*

The normed space just defined is denoted by $(X, \|x\|)$*. If no confusion is likely to occur, we may denote it simply by* X*.*

From(2) we see that $\|0\| = \|0 \cdot x\| = 0\|x\| = 0$ and this along with (1) implies that $\|x\| = 0$ if and only if $x = 0$. Thus, a normed on a linear space has the basic properties of the length of a vector in a Euclidean space.

A norm $\| \cdot \|$ on X defines a metric d on X given by

$$d(x, y) = \|x - y\|.$$

then it is easy to check that d is a metric on X (Problem 2). Nothing that $d(x, 0) = \|x\|$, we now rephrase in normed linear space terminology some of the concepts considered in our discussion of metric spaces.

The set E in a normed linear space X is **bounded** if and only if there exists a positive number r such that $E \subset B(0; r)$; that is, if and only if there exists a positive number r such that $\|x\| < r$ for all $x \in E$.

The definition of continuity becomes: a function f from a normed linear space X into a normed linear space Y is **continuous at the point** $x \in X$ if for each $\epsilon > 0$ there exists a $\delta > 0$ such that

$$\|f(x) - f(a)\| < \epsilon \quad \text{whenever } \|x - a\| < \delta.$$

Notice that in this definition no notational distinction is made between the norms in the two spaces; the context makes clear which norm is intended.

In a normed linear space X the norm is a continuous function from X into $\mathbb{R}$, since for all $x, y \in X$

$$| \ \|x\| - \|y\| \ | \leq \|x - y\| \quad \text{(Problem 3).}$$

Rephrasing the definition of convergence in a normed linear space X, we have: a point b in X is the limit of the sequence $\{x_n\}$ if for each $\epsilon > 0$ there exists a positive integer n_0 such that

$$\|x_n - b\| < \epsilon \quad \text{whenever} \quad n \geq n_0.$$

If f is a function from a normed linear space X into a normed linear space Y, then we have the following result from Chapter 2 (6.2, p.73 and Problem 11, p.76):f is continuous at x if and only if $\{f(x_n)\}$ converges to $f(x)$ whenever $\{x_n\}$ converges to x.

If $(X_1, \|\cdot\|_1)$ and $(X_2, \|\cdot\|_2)$ are normed linear spaces over F, we can make $X_1 \times X_2$ a normed linear space over F as follows. First, $X_1 \times X_2$ will be a linear space over F, if for $x = (x_1, x_2)$ and $y = (y_1, y_2)$, we define

$$x + y = (x_1 + y_1, x_2 + y_2) \quad \text{and} \quad \alpha x = (\alpha x_1, \alpha x_2).$$

For the norm we can take

$$\|x\| = \|x_1\|_1 + \|x_2\|_2 \tag{A}$$

or

$$\|x\| = \max\{\|x_1\|_1, \|x_2\|_2\}. \tag{B}$$

These are the norms corresponding to the metrics introduced on $X_1 \times X_2$ in section 8.1 and the topology corresponding to either of these norms is the product topology.

Let X be a normed linear space and let f denote the addition operation on X; that is, f is the function from $X \times X$ into X such that $f(x, y) = x + y$. Suppose that the sequence $\{(x_n, y_n)\}$ converges to $\{x, y\}$ in $X \times X$. Since

$$\|f(x_n, y_n) - f(x, y)\| = \|(x_n + y_n) - (x + y)\| \leq \|x_n - x\| + \|y_n - y\|,$$

using the norm (A) in $X \times X$ we see that f is continuous. Similarly, let g denote the scalar multiplication operation on X; that is, g is the function from $F \times X$ into X such that $g(\alpha, x) = \alpha x$. Suppose that the sequence $\{(\alpha_n, x_n)\}$ converges to (α, x) in $F \times X$. Since

$$\|g(\alpha_n, x_n) - g(\alpha, x)\| = \|\alpha_n x_n - \alpha x\| \leq |\alpha_n| \; \|x_n - x\| + |\alpha_n - \alpha| \; \|x\|,$$

using the norm (B) in $F \times X$ we see that g is continuous. Thus, in a normed linear space the operations of addition and scalar multiplication are continuous.

A **linear topological space** (topological vector space) is a generalization of a normed linear space. This is a linear space with a topology such that the operations of addition and scalar multiplication are continuous are continuous. In the above discussion we have show that a normed linear space is a linear topological space. [Chap. 9]

In some linear spaces there occurs naturally a real-valued function that has the properties of a norm with the execption that a nonzero element may have value zero. A **semi-norm** on a linear space X is a real-valued function s such that

$$s(\alpha, x) = |\alpha| s(x) \text{ and } s(x+y) \leq s(x) + s(y).$$

It is easy to show that if s is a semi-norm then $s(0) = 0$ and $s(x) \geq 0$ for all $x \in X$. However, it is not true that $s(x) = 0$ implies $x = 0$.

Let $N = \{x \in X : s(x) = 0\}$. Then N is a subspace of X. If $x - y \in N$, then

$$s(x) \leq s(y) + s(x-y) = s(y)$$
$$s(y) \leq s(x) + s(y-x) = s(x)$$

and, therefore, $s(x) = s(y)$. Thus, on the quotient space X/N we can define

$$\|[x]\| = s(x)$$

and this will be a norm on the quotient space. Hence, from a linear space with a semi-norm we can obtain a normed linear space.

We will obtain some examples of normed linear spaces by putting norms on the linear spaces of n-tuples of scalars, of sequences of scalars, and of scalar-valued functions. The field of scalars is either $\mathbb{R}$ or $\mathbb{C}$ and is denoted by F when it can be either.

2.5 Example. The Euclidean n-space $\mathbb{R}^n$ i.e. the space consisting of all n-tuples of real numbers: $x = (x_1, x_2, \cdots, x_n)$ normed by $\|x\| = (\sum x_i^2)^{1/2}$

2.6 Example. **(The Space l_p^n)** On the linear space $V_n(F)$ define, for any real number $p \geq 1$. A norm called the **p-norm** by the rule

$$\|x\|_p = \left[\sum_{i=1}^{n} |\alpha^i|^p\right]^{1/p} \qquad \text{where } x = (\alpha^1, \cdots, \alpha^n).$$

That this defines a norm on $V_n(F)$ is easily verified (Problem 1). the resulting normed linear space is denoted by l_p^n. If we need to identify F, then we will call the space the real space l_p^n or the complex space l_p^n. In the real case if $p = 2$, then the p-norm is the Euclidean norm (length) and the space l_2^n is the Euclidean n-space $\mathbb{R}^n$. In the complex case if $p = 2$, the space l_2^n is called the **unitary** n-space and is denoted by $\mathbb{C}^n$.

2.7 Example. **(The Space l_∞^n)** On the linear space $V_n(F)$ define a norm, called the infinity-norm, by the rule

$$\|x\|_\infty = \max\{|\alpha^i| : i = 1, \cdots, n\} \quad \text{where } x = (\alpha^1, \cdots, \alpha^n).$$

Again, it is easy to verfy that this defines a norm on $V_n(F)$. The resulting normed linear space is denoted by l_∞^n. The notation used in this case is motivated by the fact that $\lim_{p\to\infty} \|x\|_p = \max\{|\alpha^i| : i = 1, \cdots, n\}$ (Problem 4).

2.8 Example. **(The Space l_p)** Let l_p be the subset of $V_\infty(F)$ consisting of all sequences $x = (\alpha^i)$ such that $\sum_{i=1}^{\infty} |\alpha^i|^p < \infty$ for the real number $p \geq 1$. This set is a subspace of $V_\infty(F)$. For, if $x, y \in l_p$ and $\alpha, \beta \in F$, using the Minkowski Inequality we have for any positive integer n

$$\begin{aligned}\left[\sum_{i=1}^{n} |\alpha\alpha^i + \beta\beta^i|^p\right]^{1/p} &\leq |\alpha| \left[\sum_{i=1}^{n} |\alpha^i|^p\right]^{1/p} + |\beta| \left[\sum_{i=1}^{n} |\beta^i|^p\right]^{1/p} \\ &\leq |\alpha| \left[\sum_{i=1}^{\infty} |\alpha^i|^p\right]^{1/p} + |\beta| \left[\sum_{i=1}^{\infty} |\beta^i|^p\right]^{1/p}\end{aligned}$$

and, therefore,

$$\left[\sum_{i=1}^{\infty} |\alpha\alpha^i + \beta\beta^i|^p\right]^{1/p} \leq |\alpha| \left[\sum_{i=1}^{\infty} |\alpha^i|^p\right]^{1/p} + |\beta| \left[\sum_{i=1}^{\infty} |\beta^i|^p\right]^{1/p}. \qquad \text{(C)}$$

This shows that $\alpha x+\beta y \in l_p$ and, hence, l_p is a subspace of $V_\infty(F)$. On the linear space l_p defie the p-norm by the rule

$$\|x\|_p = \left[\sum_{i=1}^{\infty} |\alpha^i|^p\right]^{1/p} \quad \text{where } x=(\alpha^i).$$

The fact that this is a norm on l_p follows easily from the definition and (C)

2.9 Example. **(The Space l_∞)** The set of all bounded sequences of scalars is also a subspace of $V_n(F)$ (Problem 5). By a bounded sequence we mean a sequencce (α^i) such that the set $\{|\alpha^i| : i \in \mathbb{N}\}$ is bounded. If on the linear space of all bounded sequences of scalars we define the infinity-norm by the rule

$$\|x\|_\infty = \sup\{|\alpha^i| : i \in \mathbb{N}\} \quad \text{where } x=(\alpha^i),$$

we obtain the normed linear space l_∞.

2.10 Example. **(The Space $L^p(E)$)** For any real number $p \geq 1$ let $\mathcal{L}_p(E)$ be the set of Lebesgue measurable scalar-valued functions f defined on a Lebesgue measurable set E in $\mathbb{R}^n$ such that $|f|^p$ is Lebesgue integrable. The Minkowski Inequality shows that if f and g are in $L^p(E)$, then $f+g \in L^p(E)$. Also, it is easy to see that if $f \in L^p(E)$ and α is a scalar, then $\alpha f \in L^p(E)$. Thus, $\mathcal{L}_p(E)$ is a linear space over F.

On $L^p(E)$ define a real-valued function s by the rule

$$s(f) = \left[\int_E |f|^p\right]^{1/p}.$$

Using the Minkowski Inequality we see that s is a semi-norm on $\mathcal{L}_p(E)$. It fails to be a norm because $s(f)=0$ does not imply $f=0$ but only that $f=0$ a.e. on E. If we let $N=\{f : f=0 \text{ a.e. on } E\}$, then the quotient space $\mathcal{L}_p(E)/N$ is a normed linear space with norm

$$\|[f]\| = \left[\int_E |f|^p\right]^{1/p}.$$

We will follow the common practice practie of not using equivalence class notation. Instead we will consider $\mathcal{L}_p(E)$ to be a normed linear space with equality meaning equality almost everywhere on E and with norm:

$$\|f\| = \left[\int_E |f|^p\right]^{1/p}$$

2.11 Example. (bf The **Space** $\mathcal{L}_\infty(E)$) Let $\mathcal{L}_\infty(E)$ denote the set of Lebesgue measurable scalar-valued function f defined on a Lebesgue measurable set E in $\mathbb{R}^n$ for which there exists a number M_f such that

$$|f(x)| \leq M_f \quad \text{for all } x \in E.$$

Such functions are said to be **essentially bounded**. Clearly, $\mathcal{L}_\infty(E)$ is a linear sapce. If we define the function s on $\mathcal{L}_\infty(E)$ by the rule

$$s(f) = \inf\{M : |f| \leq M \text{ a.e. on } E\},$$

then s is a semi-norm on $\mathcal{L}_\infty(E)$. If $N = \{f : f = 0 \text{ a.e. on } E\}$, then $\mathcal{L}_\infty(E)/N$ is a normed linear space with norm

$$\|[f]\| = \inf\{M : |f| \leq M \text{ a.e. on } E\}.$$

Again, we consider $\mathcal{L}_\infty(E)$ to be a normed linear space with norm

$$\|f\| = \inf\{M : |f| \leq M \text{ a.e. on } E\}$$

and take equality in $\mathcal{L}_\infty(E)$ to mean equality almost everywhere on E.

2.12 Example. (**The Space** $C(X, \mathbb{R})$) Let X be a compact metric space and let $C(X, \mathbb{R})$ be the linear space of all continuous real-valued functions on X with norm defined by

$$\|f\|_\infty = \max\{|f(x)| : x \in X\}.$$

It is easy to check that $C(X, \mathbb{R})$ is a normed linear space. Note that the sequence $\{f_n\}$ converges to f in $C(X, \mathbb{R})$ if and only if for each $\epsilon > 0$ there exists an n_0 such that

$$\|f_n - f\| < \epsilon \quad \text{whenever } n \geq n_0;$$

that is

$$|f_n(x) - f(x)| < \epsilon \quad \text{whenever } n \geq n_0 \text{ and } x \in X.$$

Thus, convergence of a sequence of functions in $C(X, \mathbb{R})$ is equivalent to the uniform convergence of the sequence on X.

If X is the interval $[a, b]$ in $\mathbb{R}$, then $C(X, \mathbb{R})$ is denoted by $C[a, b]$ and this corresponds to the metric space $C[a, b]$ considered in Section 1.

PROBLEMS [8.2]

1. Verify that p-norm ($p \geqslant 1$) and infinity-norm satisfy the conditions for a norm on $V_n(F)$.

2. If X is a normed linear space and $d(x, y) = \|x - y\|$ for any $x, y \in X$, show that d is a metric on X.

3. If X is a normed linear space, show that for any $x, y \in X$;

$$\Big| \|x\| - \|y\| \Big| \leq \|x - y\|.$$

4. Complete the proof of Example 3.

5. Show that the set of all bounded sequences of scalars is a subspace of $V_n(F)$ and verify that the infinity-norm satisfies the conditions for a norm on this subspace.

6. Let $\mathbb{R}_n^m$ be the space of $m \times n$ matrices with entries in $\mathbb{R}$. If for $p \geqslant 1$ we define

$$\left\| [x_j^i] \right\| = \left[\sum_{j=1}^{m} \sum_{i=1}^{n} |x_j^i|^p \right]^{\frac{1}{p}}.$$

Verify that this is a norm on $\mathbb{R}_n^m$.

7. For $p \geqslant 1$ let $C_p[a,b]$ be the set of all continuous real-valued functions on the interval $[a,b]$ with

$$\|f\|_p = \left[\int_a^b |f|^p\right]^{\frac{1}{p}}.$$

Show that $C_p[a,b]$ is a normed linear space.

7.3 Banach Space

A normed vector space becomes a metric space if we define a metric d by $d(x,y) = \|x - y\|$. When we speak about metric properties in a normed sapce we are referring to this metric.

3.1 Theorem. *Let $(X, \|\cdot\|)$ be a real normed linear space. Let d be the function on $X \times X$ defined by*

$$\begin{aligned} d:\ X \times X &\rightarrow \mathbb{R} \\ (x,y) &\rightarrow d(x,y) = \|x - y\|. \end{aligned}$$

Then d is a metric.

Proof. Trivial.

By above Theorem, we can formulate the following theorem.

3.2 Theorem. *A normed space is a metric space.*

Proof. trivial.

The notion of convergence for a sequence of real numbers generalizes to give us a notion of convergence for sequences in a normed linear space.

3.3 Definition. *A sequence $\{f_n\}$ in a normed linear space is said to converge to an element f in the space if, given $\epsilon > 0$, there is an N such that for all $n > N$ we have $\|f - f_n\| < \epsilon$. If f_n converges to f we write $f = \lim f_n$ or $f_n \rightarrow f$.*

Just as for the case of sequence of real numbers, we say that a sequence $\{f_n\}$ in a normed linear space is a Cauchy sequence if, given $\epsilon > 0$, there is an N such that for all $n \geq N$ and all $m \geq N$ we have $\|f_n - f_M\| < \epsilon$. It is easily verfied that each convergent sequence is a Cauchy sequence.

3.4 Definition. *A normed linear space is called* ***complete*** *if every Cauchy sequence in the space convergs, that is if for each Cauchy sequence $\{f_n\}$ in the space there is an element f in the space such that $f_n \to f$. A complete normed linear space is called a* ***Banach space.***

3.5 Definition. *A normed linear space X is called a* ***Banach space*** *iff X is complete in the metric defined by the norm. The simplest real Banach space is the field of real numbers normed by $\|x\| = |x|$. Other examples of Banach spaces are listed below. Verifications are left to the reader.*

3.6 Example.

1. The space l^∞ of all bounded sequences $x = (x_1, x_2, x_3, \cdots)$ of scalars is a Banach space under the norm given by

$$\|x\|_\infty = \sup_n |x_n|.$$

2. The sapce $P[a,b]$ of all polynomials defined on $[a,b]$ is a normed space under the norm given by

$$\|x\| = \max_{t \in [a,b]} |x(t)|.$$

   But $P[a,b]$ is not a Banach space.

3. The space $C[a,b]$ of all continuous functions defined on $[a,b]$ is a Banach space under the norm given by

$$\|x\| = \max_{t \in [a,b]} |x(t)|.$$

4. the space $\mathbb{Q}$ of all rational numbers with the norm $\|\cdot\|$ defined by $\|x\| = |x|$, $x \in \mathbb{Q}$ is a normed space, but not a Banach space.

3.7 Definition. *A sequence $\{x_k\}$ in a normed space X is said to be **summable** to the sum s if s is in the space and the sequence $\{s_n\}$ of the partial sums of the series $\sum_k x_k$ converges to s in X, i.e.*

$$\|s_n - s\| \to 0 \quad as \quad n \to \infty$$

or

$$\|\sum_{k=1}^{n} x_k - s\| \to 0 \quad as \quad n \to \infty.$$

If this is the case, we write $s = \sum_{k=1}^{\infty} x_k$.

Thesequence $\{x_k\}$ is said to be absolutely summable if

$$\sum_{k=1}^{\infty} \|x_k\| < \infty.$$

For a sequence of real numbers we know that absolute summability implies that the sequence is summable. While this is not true in general for sequence of elements in a normed linear space, the following proposition shows that this implication holds if the space is complete.

3.8 Theorem. *A normed space X is complete if and only if every absolutely summable sequence is summable.*

Proof. Assume that X is a complete normed space. Let $\{x_n\}$ be an absolutely summable sequence in X. Then

$$\sum_{n=1}^{\infty} \|x_n\| = M < \infty.$$

Thus, for each $\epsilon > 0$, there is an N such that

$$\sum_{n=N}^{\infty} \|x_n\| < \epsilon.$$

Let $s_n = \sum_{k=1}^{n} x_k$ be the partial sums of the series $\sum_{n=1}^{\infty} x_n$. Then, we have

$$\begin{aligned} \|s_n - s_m\| &= \left\| \sum_{k=m+1}^{n} x_k \right\| \\ &\leq \sum_{k=m+1}^{n} \|x_k\| \\ &\leq \sum_{k=N}^{\infty} \|x_k\| < \epsilon, \end{aligned}$$

for $n \geq m > N$. Thus $\{s_n\}$ is a Cauchy sequence in X and must converge to some element (say) s in X, since X is complete. Hence $\{x_n\}$ is summable in X.

Conversely, suppose each absolutely summable sequence in X is summable in X. Let $\{x_n\}$ be a Cauchy sequence in X. Then, for each k there is an integer n_k such that

$$\|x_n - x_m\| < \frac{1}{2^k}, \quad \forall\, n, m \geq n_k.$$

We may choose n_k such that $n_{k+1} > n_k$. Then $\{x_{n_k}\}$ is a subsequence of $\{x_n\}$. Set

$$\begin{aligned} y_1 &= x_{n_1} \\ y_2 &= x_{n_2} - x_{n_1} \\ &\vdots \\ y_k &= x_{n_k} - x_{n_{k-1}} \\ &\vdots \end{aligned}$$

We note that

1. $\sum_{i=1}^{k} y_i = x,$

2. $\|y_k\| < \frac{1}{2^{k-1}} \; k > 1$

and as such

$$\sum \|y_k\| \leq \|y_1\| + \sum_{k=2}^{\infty} 2^{1-k} = \|y_1\| + 1 < \infty.$$

Thus, the sequence $\{y_k\}$ is absolutely summable and hence summable to some element (say) x in X. It remains to prove that $\lim_{n\to\infty} x_n = x$. Since $\{x_n\}$ is a Cauchy sequence, given an $\epsilon > 0$, there is an N such that

$$\|x_n - x_m\| < \frac{\epsilon}{2}, \quad \forall\, n, m > N.$$

Further, since $x_{n_k} \to x$, there is a K such that

$$\|x_n - x\| < \frac{\epsilon}{2}, \quad \forall\, k \geq K.$$

Choose k so large that $k > K$ and $n_k > N$.Hence

$$\|x_n - x\| \leq \|x_n - x_{n_k}\| + \|x_{n_k} - x\| < \epsilon, \quad \forall\, n > N$$
$$\implies \quad x_n \to x.$$

□

We also showed in section 8.1 that completeness is not a topological property for metric spaces. However, for normed linear spaces we have the following useful result.

3.9 Proposition. *If X and Y are homeomorphically isomorphic normed linear spacc and X is a Banach space, then Y is a Banach space.*

Proof. Let T be a linear homeomorphism from X onto Y. Take a Cauchy sequence $\{y_n\}$ of points in Y and let $x_n = T^{-1}(y_n)$. Since there is a positive number m such that $m\|x\| \leq \|T(x)\|$ for all $x \in X$. $\{x_n\}$ is a Cauchy sequence of points in X. The completeness of X implies that $\{x_n\}$ converges to some point x in X. Since T is continuous, $\{T(x_n)\}$ converges to $T(x)$ in Y. Thus, $\{y_n\}$ converges in Y.

□

The next two Propositions follow easily.

3.10 Proposition. *Any finite-dimensional normed linear space is a Banach space.*

Proof. If X is a normed linear space over F of finite dimension n, then X is homeomorphically isomorphic to the Banach space $l_2^n(\mathbb{R}^n$ or $\mathbb{C}^h)$ □

3.11 Proposition. *A finite-dimensional subspace E of a normed linear space X is closed in X.*

Proof. Since E is a complete subspace of X it is closed in X. □

PROBLEMS [8.3]

1. Show that $\mathcal{L}_\infty(E)$ is a Banach space.

2. Let c denote the set of convergent sequences of scalars. Show that c is a closed subspace of l_∞ and hence is a Banach space.

3. Let

$$f_n(t) = \begin{cases} 0, & 0 \le t \le \frac{1}{2} - \frac{1}{n} \\ 1 + nt - \frac{n}{2}, & \frac{1}{2} - \frac{1}{n} < t < \frac{1}{2} \\ 1, & \frac{1}{2} \le t < 1 \end{cases}$$

   for $n = 2, 3, 4, \cdots$.

   (a) Show that $\{f_n\}$ is a Cauchy sequence in $C_p[0,1]$ for any $p \ge 1$.

   (b) Show that $\{f_n\}$ does not converge in $C_p[0,1]$ and, hence, $C_p[0,1]$ is not a Banach space.

4. If X is a normed linear space and $\{x_n\}$ is a Cauchy sequence in X, show that $\{\, \|x_n\| \,\}$ converges.

5. If X is a normed linear space, prove that X is a Banach space if and only if the set $\{x : \|x\| = 1\}$ is complete.

6. Let X and Y be normed linear spaces and let $X \times Y$ be the normed linear space with norm

$$\|(x, y)\| = \|x\| + \|y\|.$$

Show that $X \times Y$ is a Banach space if and only if X and Y are Banach spaces.

7. Let X be a normed linear space and let $\tilde{X}$ be the completion of X as a metric space. Prove the following:

   (a) If $\{x_n\}$ and $\{y_n\}$ are Cauchy sequences of elements of X, then $\{x_n + Y_n\}$ and $\{\alpha x_n\}$ are Cauchy sequences.

   (b) Addition and scalar multiplication can be defined in $\tilde{X}$ in the following way:

   $$\tilde{x} + \tilde{y} = [\{x_n + y_n\}] \text{ and } \alpha\tilde{x} = [\{\alpha x_n\}]$$

   where $\tilde{x} = [\{x_n\}]$ and $\tilde{y} = [\{y_n\}]$.

   (c) $\tilde{X}$ is a linear space with the operations defined in part *b*.

   (d) If we define $\|\tilde{x}\| = \tilde{d}(\tilde{x}, 0)$, then $\widetilde{X}$ is a normed linear space.

   (e) If $f(x) = [(\overline{x})]$, where $(\overline{x})$ denotes the sequence all of whose terms are x, then f is an isometric isomorphism of X onto a dense subset of $\tilde{X}$.

7.4 Baire Category

In this section we shall go more deeply into certain aspects of metric spaces. We begin by defining a set E to be **nowhere dense** if $\sim \overline{E}$ is dense. This is equivalent to stating that $\overline{E}$ contains no spheroid. For example, the integers are nowhere dense in $\mathbb{R}$, and the Cantor ternary set is nowhere dense in $[0, 1]$. Aset E is said to be of **first category** (or meager) if it is the union of a countable collection of nowhere dense sets. A set which is not of first category is said to be of **second category**, and the complement of a set of first category is called **residual**. We shall show that a complete metric space is of second category when considered as a subset of itself. We begin with the following theorem:

4.1 Theorem. *Let X be a complete metric space and $\{O_n\}$ a countable collection of dense open subsets of X. Then $\bigcap O_n$ is not empty.*

Proof. Let x_1 be a point of O_1 and S_1 a speroid (of radius r_1) which is centered at x_1 and contained in O_1. Since O_2 is dense, there must be a point x_2 in $O_2 \cap S_1$. Since O_2 is open, there is a spheroid S_2 centered at x_2 and contained in O_2, and we may take the radius r_2 of S_2 to be smaller than $\frac{1}{2}r_1$ and smaller than $r_1 - \rho(x_1, x_2)$. Then $\overline{S_2} \subset S_1$. Proceeding inductively, we obtain a sequence $\langle S_n \rangle$ of spheroids such that $\overline{S_n} \subset S_{n-1}$ and $S_n \subset O_n$ and whose radii $\langle r_n \rangle$ tend to zero. Let $\langle x_n \rangle$ be the sequence of centers of these spheres. Then for $n, m \geq N$ we have $x_n \in S_N$ and $x_m \in S_N$. Hence $\rho(x_n, x_m) \leq 2r_N$, and $\langle x_n \rangle$ is a Cauchy sequence, since $r_n \to 0$. By the completeness of X there is a point x such that $x_n \to x$. Since $x_n \in S_{N+1}$ for $n > N$, we have $x \in \overline{S}_{N+1} \subset S_N \subset O_N$. Hence $x \in \bigcap O_n$. □

4.2 Corollary. *(**Baire Category Theorem**): A complete metric space is not the union of a countable collection of nowhere dense sets.*

Proof. Let $\{E_n\}$ be a countable collection of nowhere dense sets. Then $O_n = \sim \overline{E}_n$ is a dense open set, and so there is a point $x \in \bigcap O_n$. But this means $x \notin \bigcup E_n$. □

As an application of the Baire Category Theorem we establish the following theorem, which is know as the **uniform boundedness principle**:

4.3 Theorem. *Let $\mathfrak{F}$ be a family of real-valued continuous functions on a complete metric space X, and suppose that for each*

$x \in X$ there is a number M_x such that $|f(x)| \leq M_x$ for all $f \in \mathfrak{F}$. Then there is a nonempty open set $O \subset X$ and a constant M such that $|f(x)| \leq M$ for all $f \in \mathfrak{F}$ and all $x \in O$.

Proof. For each integer m, let $E_{m,f} = \{x : |f(x)| \leq m\}$, and set $E_m = \bigcap_{\mathfrak{F}} E_{m,f}$. Since each f is continuous, $E_{m,f}$ is closed, and consequently E_m is closed. For each $x \in X$, there is an m such that $|f(x)| \leq m$ for all $f \in \mathfrak{F}$; that is, there is an m such that $x \in E_m$. Hence

$$X = \bigcup_{m=1}^{\infty} E_m.$$

Since X is a complete metric space, there is a set E_m which is not nowhere dense. Since this E_m is a closed set, it must contain some spheroid O. But for every $x \in O$ we have $|f(x)| \leq m$ for all $f \in \mathfrak{F}$. □

PROBLEMS [8.4]

1. (a) Prove that a closed set F is nowhere dense if and only if it contains no open set.

   (b) Prove that E is nowhere dense if and only if for any nonempty open set O there is a spheroid contained in $O \sim E$.

2. (a) Prove that if E is of irst category and $A \subset E$ then A is also of first category.

   (b) Prove that if $\langle E_n \rangle$ is a sequence of sets of first category then $\bigcup E_n$ is also of first category.

3. If X is a complete metric space and E is a set of first category in E, then $\sim E$ is dense in X.

4. (a) Show that on $[0, 1]$ there is a nowhere dense closed set having Lebesgue measure $1 - 1/n$.

(b) Construct a set of first category on $[0, 1]$ which has measure 1.

5. A point x in a metric space is called isolated if the set $\{x\}$ is open. Prove that a complete metric space without isolated points has an uncountable number of points.

6. (a) If $\tilde{E}$ is dense and F a closed set contained in E, then F is nowhere dense.

   (b) If E and $\tilde{E}$ are both dense, then at most one of them is an $\mathfrak{F}_\sigma$.

   (c) The set of rational numbers on $[0, 1]$ is not a $\mathcal{G}_\delta$.

   (d) Is there a real-valued function on $[0, 1]$ which is continuous on the rationals and discontinuous on the irrationals?

7. Let C be the space of continuous functions on $[0, 1]$ and set $F_n = \{f : (\exists x_0) \text{ with } |f(x) - f(x_0)| \leq n(x - x_0) \text{ for all } x, x_0 \leq x < 1\}$.

   (a) Show that F_n is a closed subset of C.

   (b) Show that F_n is nowhere dense. [Hint: Any $g \in C$ can be approximated to within $\epsilon/2$ by a polygonal function φ, and φ can be approximated to within $\epsilon/2$ by a polygonal function φ whose right-hand derivative is everywhere greater than n in absolute value.]

   (c) Show that the set D of continuous functions which have a finite derivative on the right at least one point of $[0, 1]$ is a set of first category in C.

   (d) There is a nowhere differentiable continuous function on $[0, 1]$.

7.5 The Hölder and Minkowski Inequality

Before considering a special case of normed linear spaces we will prove some inequalities which will be useful in the construction of examples of these spaces. First, we extend a familiar inequality:

$$(1+x)^n \geq 1+nx \text{ if } x \geq -1 \text{ and } n \text{ is a positive integer.}$$

5.1 Lemma. *If x and p are real numbers such that $x \geq -1$ and $p \geq 1$, then*

$$(1+x)^p \geq 1+px.$$

Proof. The result is obvious if $p = 1$. Assume $p > 1$. Let

$$f(x) = (1+x)^p - 1 - px \quad \text{for } x \geq -1.$$

Then

$$f'(x) = p(1+x)^{p-1} - p$$

and, thus, $f'(x) < 0$ for $-1 \leq x < 0$ and $f'(x) > 0$ for $x > 0$. This shows that f has a minimum value at $x = 0$ and, hence,

$$(1+x)^p - 1 - px \geq 0 \quad \text{for } x \geq -1. \qquad \square$$

If we let $1 + x = u^{1/p}/v^{1/p}$, then we can rewrite the above inequality as follows:

$$\frac{u}{v} \geq 1 + p\frac{u^{1/p}}{v^{1/p}} - p \quad \text{for } u \geq 0, v > 0$$

and this in turn gives

$$u^{1/p}v^{1-1/p} \leq \frac{1}{p}u + \left(1 - \frac{1}{p}\right)v \quad \text{for } u \geq 0, v \geq 0.$$

Then we obtain the summation form of these inequalities.

5.2 Corollary. *Let α and β be nonnegative real numbers, and suppose $0 < \lambda < 1$. Then $\alpha^\lambda\beta^{1-\lambda} \leq \lambda\alpha + (1-\lambda)\beta$ with equality only if $\alpha = \beta$.*

5.3 Hölder's Inequality. If $p > 1$ and $1/q = 1 - 1/p$, then

$$\sum_{k=1}^m |x_k y_k| \leq \left[\sum_{k=1}^m |x_k|^p\right]^{1/p} \left[\sum_{k=1}^m |y_k|^q\right]^{1/q}$$

for any complex numbers $x_1, \cdots, x_m, y_1, \cdots, y_m$.

5.4 Minkowski's Inequality. If $p \geq 1$, then

$$\left[\sum_{k=1}^{m} |x_k + y_k|^p\right]^{1/p} \leq \left[\sum_{k=1}^{m} |x_k|^p\right]^{1/p} \left[\sum_{k=1}^{m} |y_k|^q\right]^{1/p}$$

for any complex numbers $x_1, \cdots, x_m, y_1, \cdots, y_m$.

7.6 The Lebesgue L^p Space

Many of the classical spaces in analysis consist of measurable functions and most of the important norms on such spaces have been defined by integrals. The Lebesguee L^p-spaces are among such important classes. A complete understanding of these spaces needs a thorough understanding of the Lebesgue Theory of measure and integration which we have developed in the preceding chapters. We are now fully prepared to introduce the L^pspaces. These spaces have remarkable properties and are of enormous importance in analysis as well as its applications. In this section we study some spaces of functions of a real variable.

If f is a measurable function on E, then $|f|^p$ is so for each p, $0 < p < \infty$. Designate by $L^p(E)$, the class of all p-integrable functions over E, i.e.,

$$L^p(E); \left\{f : \int_E |f|^p < \infty\right\}.$$

6.1 Example.

1. Let $E = [0, 16]$ and $f : E \to \mathbb{R}$ be a function defined by $f(x) = (x)^{-1/4}$. Then $f \in L^1(E)$ but $f \notin L^4(E)$.

2. Let $E = \left[0, \dfrac{1}{2}\right]$ and the function $f : E \to \mathbb{R}$ be defined by $f(x) = \left[x \ \log^2\left(\dfrac{1}{x}\right)\right]^{-1}$. Then $f \in L^1(E)$.

3. Let $E = (0, \infty)$ and the function $f : E \to \mathbb{R}$ be defined by $f(x) = (1 + x)^{-1/2}$. Then $f \in L^p(E)$ for each $p, 2 < p < \infty$.

6.2 Theorem. *The space $L^p(E)$ is a linear space over $\mathbb{R}$.*

Proof. $f, g \in L^p(E) \Longrightarrow f + g \in L^p(E)$, since

$$\begin{aligned} |f+g|^p &\le (|f|+|g|)^p \le 2^p \max\{|f|^p, |g|^p\} \\ &\le 2^p(|f|^p + |g|^p). \end{aligned}$$

Since $|f = g|^p$ is measurable and $|f|^p + |g|^p$ is integrable, $|f + g|^p$ is integrable. Also $f \in L^p(E)$ and $\alpha \in \mathbb{R} \Longrightarrow \alpha f \in L^p(E)$. Thus theorem is proved. □

Furthermore, if $f \in L^p(E)$, then the inequalities

$$\begin{cases} 0 \le f^+ \le |f| \\ 0 \le f^- \le |f| \end{cases}$$

imply that f^+, f^- and $|f|$ are also in $L^p(E)$.

In order to define $L^\infty(E)$, let f be a real-valued and measurable function on a set E with $m(E) > 0$. A real number M is said to be an **essential bound** for the function f if

$$|f(x)| \le M \text{ a.e. on } E.$$

A function f is said to be **essentially bounded** if it has an essential bounded. In other words, a function f defined on E is essentially bounded if it is bounded except possibly on a set of measure zero. The **essential supremem** of f on E is defined by

$$\text{ess sup}\,|f(x)| = \inf\{M : |f(x) \le M \text{ a.e. on } E\},$$

or equivalently

$$\text{ess sup}\,|f(x)| = \inf\{M : m(\{x \in E : |f(x)| > M\}) = 0\}.$$

If f does not have any essential bound, then its essential supremum is defined to be ∞.

Let us designate by $L^\infty(E)$ the class of all those measurable functions defined on E which are essentially bounded on E, i.e.

$$L^\infty(E) = \{f : \text{ess sup}\,|f| < \infty\}.$$

It is not diffcult again to verify that $L^\infty(E)$ is a linear space over $\mathbb{R}$.

6.3 Example. 1. Every bounded function on E is in $L^\infty(E)$.

2. The function $f : [a,b] \to \mathbb{R}$ given by

$$f(x) = \begin{cases} 1 & \text{if } x \text{ is irrational} \\ \infty & \text{if } x \text{ is rational} \end{cases}$$

is in $L^\infty[a,b]$.

Let us now define a function $\|\cdot\|_p : L^p(E) \to \mathbb{R}$, $0 < p \le \infty$, as follows:

$$\begin{aligned} \|f\|_p &= \left(\int_E |f|^p\right)^{1/p}, \quad 0 < p < \infty \\ \|f\|_\infty &= \text{ess sup}\,|f|. \end{aligned}$$

Then $\|f\|_p$ is called the L_p-norm of f.

Note: We shall establish in Section 4 that if $1 \le p \le \infty$, then $\|\cdot\|_p$ defines a norm on $L^p(E)$, $l \le p \le \infty$. In case $1 \le p < \infty$, $\|\cdot\|_p$ is called a p-norm on $L^\infty(E)$.

We see that $\|f\| = 0$ if and only if $f = 0$ a.e. If α is a constant, then $\|\alpha f\| = |\alpha|\ \|f\|$. In the next section we shall derive two inequalities, the second of which stattes that $\|f+g\| \le \|f\| + \|g\|$, if $p \ge 1$. Henceforth, we shall always assume $p \ge 1$. A linear space is said to be a normed linear space if we have assigned a nonnegative real number $\|f\|$ to each f such that $\|\alpha f\| = |\alpha|\ \|f\|$, $\|f+g\| \le \|f\| + \|g\|$, and $\|f\| = 0 \iff f \equiv 0$. Unfortunately, the norm for the L^p spaces does not satisty the last requirement, for from $\|f\| = 0$ we can only conclude that $f = 0$ a.e. We shall, however, consider two measurable functions to be equivalent if they are equal almost everywhere; and, if we do not distinguish between equivalent functions, then the L^p spaces are normed linear spaces.

By the consequence of above we obtain the following theorem.

6.4 Theorem. *The space $L^p (1 \le p < \infty)$ is a normed space.*

A relationship between $\|\cdot\|_p (1 \le p < \infty)$ and $\|\cdot\|_\infty$ will soon be established in Theorem 6.6 which, in fact, would also justify the notation $L^\infty(E)$ for the class of essentially bounded functions on E. Before, we do so let us prove some results in the form of a lemma.

6.5 Lemma. *Let $f \in L^\infty(E)$. Then:*

(a) $|f(x)| \le \|f\|_\infty$ a.e. on E.

(b) $\|f\|_\infty = \sup\{M : m(\{x \in E : |f(x)| \ge M\}) \ne 0\}$.

Proof. (a) Let $\eta = \|f\|_\infty$. Note that

$$\{x \in E : |f(x) \ge \eta\} = \bigcup_{n=1}^{\infty} \left\{x \in E : |f(x)| > \eta + \frac{1}{\eta}\right\}.$$

Since the union of a countable collection of sets of measure zero has measure zero, we have

$$m(\{x \in E : |f(x)| \ge \|f\|_\infty\}) = 0$$

$$\Longrightarrow \qquad |f(x) \le \|f\|_\infty \text{ a.e. om } E$$

(b) It is obvious from the definition of $\|f\|_\infty$ □

6.6 Theorem. *Let E be a measurable set with $m(E) < \infty$. Then $L^\infty(E) \subset L^p(E)$ for each p with $1 \le p < \infty$. Furthermore, if $f \in L^\infty(E)$, then*

$$\|f\|_\infty = \lim_{p \to \infty} \|f\|_p.$$

Proof. Let $f \in L^\infty(E)$ and $\eta = \|f\|_\infty$. Then

$$|f(x)|^p \le \eta^p \text{ a.e. on } E$$

$$\Longrightarrow \qquad \int_E |f|^p \le \eta^p \cdot m(E).$$

Therefore, $f \in L^p(E)$ and consequently $L^\infty(E) \subset L^p(E)$. Furthermore, we note that

$$\|f\|_p \le \eta[m(E)]^{1/p}.$$

Since $[m(E)]^{1/p} \to 1$ as $p \to \infty$, we obtain

$$\lim_{p\to\infty} \sup \|f\|_p \leq \eta.$$

On the other hand, let $|f(x)| \geq \alpha$ on a set F with $m(F) > 0$. Then

$$\|f\|_p \geq \alpha[m(F)]^{1/p}$$

$$\Longrightarrow \qquad \lim_{p\to\infty} \inf \|f\|_p \geq \alpha.$$

This verifies that (cf. Lemma 6.5)

$$\sup\{\alpha : m(\{x \in E : \ |f(x)| \geq \alpha\}) \neq 0\} \leq \lim_{p\to\infty} \inf \|f\|_p.$$

Hence, the result follows from

$$\eta \leq \lim_{p\to\infty} \inf \|f\|_p \leq \lim_{p\to\infty} \sup \|f\|_p \leq \eta.$$

□

6.7 Corollary. *$L^\infty(E) \subset \cap_{p\geq 1} L^p(E)$, and the norm on $L^\infty(E)$ is equal to the limit of $\|f\|_p$ as $p \to \infty$ provided $m(E) < \infty$.*

Remark: *The result in Theorem 6.6 need not be true if $m(E) = \infty$.*

6.8 Example. Consider the function $f : \mathbb{R} \to \mathbb{R}$ given by $f(x) = C$ $(C \neq 0)$ for all x. Clearly $f \in L^\infty(\mathbb{R})$ while $f \notin L^p(\mathbb{R})$ for any p with $0 < p < \infty$.

Most of the results for the spaces $L^p(E)$ we prove in the remaining sections hold good for any measurable set E with $m(E) < \infty$. However, for the sake of simplicity, we take $E = [a, b]$. We shall denote by L^p the space $L^p[a, b]$ unless specified otherwise.

Remark: *The Holder and Minkowski inequality in previous section can be stated as follow in L^p-space.*

6.9 Theorem. ***(Riesz-Hölder Inequality)*** *Let p and q be non-negative extended real numbers such taht $\frac{1}{p}+\frac{1}{q}=1$. If $f \in L^p$ and $g \in L^p$, then $f \cdot g \in L^1$ and*

$$\int |fg| \leq \|f\|_p \|g\|_q \left[\text{or } \int |fg| \leq \left[\int |f|^p\right]^{1/p} \left[\int |g|^q\right]^{1/q} \right]$$

Equality holds if and only if, for some nonzero constants A and B, we have $A\,|f|^p = B|g|^q$ a.e.

Proof. When $p = 1$, then $q = \infty$, and the inequality is a available trivially in this case. Indeed, if $\|g\|_\infty = M$, then $|g| \leq M$ a.e., and so

$$|fg| \leq M|f| \text{ a.e.}$$

Thus $fg \in L^1$, and by integrating, we get

$$\int |fg| \leq M \int |f| \leq \|f\|_1 \|g\|_\infty.$$

Now, assume that $1 < p < \infty$ and consequently $1 < q < \infty$. The inequality is trivial if either $f = 0$ a.e. or $g = 0$ a.e. So assume $f \neq 0$ a.e. and $g \neq 0$ a.e. This gives that $\|f\|_g > 0$ and $\|g\|_q > 0$. Now, applying Lemma 5.1 with

$$\begin{cases} \lambda = \frac{1}{p} \\ \alpha = \left(\frac{|f(t)|}{\|f\|_p}\right)^p \\ \beta = \left(\frac{|g(t)|}{\|g\|_q}\right)^q, \end{cases}$$

we obtain

$$\frac{|f(t)g(t)|}{\|f\|_p\|g\|_q} \leq \frac{1}{p}\frac{|f(t)|^p}{(\|f\|_p)^p} + \frac{1}{q}\frac{|g(t)|^q}{(\|g\|_q)^q} \tag{1}$$

This gives that $fg \in L^1$, and by integrating, we find that

$$\frac{\int |fg|}{\|f\|_p\|g\|_q} \leq \frac{1}{p} + \frac{1}{q} = 1.$$

Hence

$$\int |fg| \leq \|f\|_p \|g\|_q. \tag{2}$$

Equality in (1) would occur if $\alpha = \beta$, and, consequently, in (2) if $\alpha = \beta$ holds *a.e.* In other words, if

$$\|g\|_q^q |f(t)|^p = \|f\|_p^p |g(t)|^q. \qquad \square$$

Note: The inequality (2) is homogeneous, i.e., it holds for af and bg with $a, b \in \mathbb{R}$ whenever it is so for f and g.

The Riesz-Hölder inequality implies that if $f \in L^p$ and $g \in L^p$, $(1/p) + (1/q) = 1$ then $fg \in L^1$. It is not true in general that the product of two integrable functions is also an integrable function. In fact, we have the following.

Remark: *The condition that p and q are mutually conjugate exponents cannot be dropped in Theorem 6.7 see Problem 3.*

6.10 Theorem. ***(Riesz-Minkowski Inequality)*** *Let $l \leq p \leq \infty$. Then for every pair $f, g \in L^p$, the following inequality holds:*

$$\|f + g\|_p \leq \|f\|_p + \|g\|_p.$$

Proof. The case for $p = 1$ is straightforward. If $p = \infty$, we note that

$$\begin{cases} |f| \leq \|f\|_\infty \text{ a.e.} \\ |g| \leq \|g\|_\infty \text{ a.e.} \end{cases}$$

$$\Longrightarrow \qquad |f + g| \leq \|f\|_\infty + \|g\|_\infty \text{ a.e,}$$

and hence the result follows in this case also. Thus, we now assume that $1 < p < \infty$.

Since L^p is a linear space, $f + g \in L^p$. Also, we have

$$\int |f + g|^p \leq \int |f + g|^{p-1}|f| + \int |f + g|^{p-1}|g|.$$

Let $1 < q < \infty$ be such that $\dfrac{1}{p} + \dfrac{1}{q} = 1$. Then, since $(p-1)q = p$, observe that

$$\int (|f+g|^{p-1})^q = \int |f+g|^p,$$

and therefore $|f+g|^{p-1} \in L^p$. As such, by Theorem 6.7, both

$$|f+g|^{p-1}|f| \quad \text{and} \quad |f+g|^{p-1}|g|$$

belong to L^1, and the Riesz-Hölder Inequality leads to

$$\int |f+g|^{p-1}|f| \leq \|f\|_p \|(|f+g|^{p-1})\|_q.$$

and

$$\int |f+g|^{p-1}|g| \leq \|g\|_p \|(|f+g|^{p-1})\|_q.$$

But

$$\|(|f+g|^{p-1})\|_q = \left(\int |f+g|^{(p-1)q}\right)^{1/q} = (\|f+g\|_p)^{p/q},$$

since $(p-1)q = p$. Hence

$$\int |f+g|^p \leq (\|f\|_p + \|g\|_p)(\|f+g\|_p)^{p/q}.$$

If $\|f+g\|_p$ is nonzero finite, the result follows by dividing both sides by $\|f+g\|_p^{p/q}$. Incase $\|f+g\|_p = 0$, there is nothing to prove while in case $\|f+g\|_p = \infty$, we either have $\|f\|_p = \infty$ or $\|g\|_p = \infty$ in view of the relation $|f+g| \leq |f| + |g|$, and the result is obviously true again. □

Reamrk. *Equality holds in Theorem 6.10 if and only if one of the functions f and g is a multiple of the other.*

Note: For the special case $p = q = 2$, the inequality in Theorem 6.9 is known as the Cauchy Schwarz Inequality. Cauchy(1821) first proved the inequality (Cauchy's inequality) for square summable sequences. Indeed, if $\{a_n\}$ and $\{b_n\}$ are sequences of numbers (real or complex) such that $\sum\limits_{n=1}^{\infty} |a_n|^2 < \infty$ and $\sum\limits_{n=1}^{\infty} |b_n|^2 < \infty$, then

$$\sum_{n=1}^{\infty} |a_n b_n| \leq \left(\sum_{n=1}^{\infty} |a_n|^2\right)^{1/2} \left(\sum_{n=1}^{\infty} |b_n|^2\right)^{1/2}$$

This inequality was generalized to integrals by A.Schwarz(1885).

It is convenient to denote by L^∞ the space of all bounded measurable functions on $[0,1]$(or rather all measurable functions which are bounded except possibly on a subset of measure zero). Again we identify functions which are equivalent. Then L^∞ is a linear space, and it becomes a normed linear space if we define

$$\|f\| = \|f\|_\infty = \text{ess } \sup |f(t)|,$$

where ess $\sup f(t)$ is the infimum of $\sup g(t)$ as g ranges over all functions which are equal to f almost everywhere.

We have introduced the convergence of sequence, another way of formulating the convergence of f_n to f is by noting that $f_n \to f$ if $\|f_n \to f\| \to 0$. Convergence in the space L^p, $1 \leq p < \infty$, is often referred to as **convergence in the mean of order** p. Thus a sequence of function $\{f_n\}$ is said to converge to f in the mean of order p if each f_n belongs to L^p and $\|f - f_n\|_p \to 0$. Convergence in L^∞ is nearly uniform convergence (Problem 8).

6.11 Theorem. ***(Riesz-Fischer)**: The L^p spaces are Banach space.*

Proof. Since the case $p = \infty$ is elementary, it is left to the reader, and we assume $1 \leq p < \infty$. By virtue of the preceding proposition we need only show that each absolutely summable sequence in L^p is summable in L^p to some element of L^p.

Let $\{f_n\}$ be a sequence in L^p with $\sum_{n=1}^{\infty} \|f_n\| = M < \infty$, and define functions g_n by setting $g_n(x) = \sum_{k=1}^{n} \|f_k(x)\|$. From the Minkowski inequality we have

$$\|g_n\| = \left\| \sum_{k=1}^{n} |f_k| \right\| \leq \sum_{k=1}^{n} \|f_k\| \leq M.$$

Hence

$$\int (g_n)^p \leq M^p.$$

For each x, $\{g_n(x)\}$ is an increasing sequence of (extended) real numbers and so must converge to an extended real number $g(x)$. The function g so defined is measurable, and, since $g_n \geq 0$, we have

$$\int g^p \leq M^p$$

by the Fatou Lemma. Hence g^p is integrable, and $g(x)$ is finite for almost all x.

For each x such that $g(x)$ is finite the series $\sum_{k=1}^{\infty} f_k(x)$ is an absolutely summable series of real nembers and so must be summable to a real number $s(x)$. If we set $s(x) = 0$ for those x where $g(x) = \infty$, we have deined a function s which is the limit almost everywhere of the partial sums $s_n = \sum_{k=1}^{n} f_k$. Hence s is measurable. Since $|s_n(x)| \leq g(x)$, we have $|s(x)| \leq g(x)$. Consequently, s is in L^p and we have

$$|s_n(x) - s(x)|^p \leq 2^p[g(x)]^p.$$

Since $2^p g^p$ is integrable and $|s_n(x) - s(x)|^p$ converges to 0 for almost all x, we have

$$\int |s_n - s|^p \to 0$$

by the Lebesgue convergence theorem. Thus $|s_n - s|^p \to 0$, whence $\|s_n - s\| \to 0$. Consequently, the sequence $\{f_n\}$ has in L^p the sum s. □

6.12 Theorem. *The space $L^\infty(E)$ is a Banach space.*

Proof. This will be left to the reader. □

Concluding this section some properties will be introduced. The proofs will be left to the reader (Ref [11]).

6.13 Theorem. *Let $0 < q < p \leq \infty$. Then $L^p \subset L^q$, and there exists a constant $k > 0$ such that*

$$\|f\|_q \leq k\|f\|_p \quad \forall f \in L^p.$$

6.14 Example. Take $E = (1, \infty)$ and define a function $f = E \to \mathbb{R}$ by $f(x) = x - \frac{1}{q}$ $1 \leq q < \infty$. Clearly $f \in L^p(E)$ if $p > q$ while $f \notin L^q(E)$.

6.15 Theorem. *Let $0 < q < p < \infty$. If $f \in L^p \cap L^p$, then $f \in L^r$ for all $q < r < p$.*

PROBLEMS [8.6]

1. Show that $\|f + g\|_\infty \leq \|f\|_\infty + \|g\|_\infty$.

2. Let f be a bounded measurable function on $[0, 1]$. Then $\lim\limits_{p \to \infty} \|f\|_p = \|f\|_\infty$.

3. Prove that $\|f + g\|_1 \leq \|f\|_1 + \|g\|_1$.

4. Let f be a function in L^p, $1 \leq p < \infty$. Show that given $\epsilon > 0$ there is a continuous function φ and a step function ψ such that $\|f - \varphi\| < \epsilon$ and $\|f - \psi\| < \epsilon$.

5. Let $\{f_n\}$ be a sequence of functions in L^p, $1 \leq p < \infty$, which converge almost everywhere to a function f in L^p. Show that $\{f_n\}$ converges to f in L^p if and only in $\|f_n\| \to \|f\|$. (For $p = 1$ this is just Problem 4.)

6. Prove the Minkowski inequality for sequences:

$$\|\{\xi_v + \eta_v\}\|_p \leq \|\{\xi_v\}\|_p + \|\{\eta_v\}\|_p.$$

   Here we have $1 \leq p \leq \infty$.

7. For $1 \leq p < \infty$, we denote by l^p the space of all sequences $\{\xi_i\}_{i=1}^\infty$ such that

$$\sum_{i=1}^\infty |\xi_i|^p < \infty.$$

Show that if $\{x_i\} \in l^p$ and $\{\eta_i\} \in l^q$ with $\frac{1}{p} + \frac{1}{q} = 1$, then

$$\sum_{i=1}^{\infty} |\xi_i \eta_i| \leq \|\{\xi_i\}\|_p + \|\{\eta_i\}\|_q,$$

where $(\|\{\xi_i\}\|_p)^p = \sum_{i=1}^{\infty} \|x_i\|^p$ and $\|\{\eta_i\}\|_\infty = \sup |\eta_i|$. This is the Hölder inequality for sequences.

8. Let $\{f_n\}$ be a sequence of functions in L^∞. Prove that $\{f_n\}$ converges to f in L^∞ if and only if there is a set E of measure zero such that f_n converges to f uniformly on $\widetilde{E}$.

9. Prove that L^∞ is complete.

10. Prove that l^p is complete $(1 \leq p < \infty)$ (see Problem 7).

11. Let $C = C[0,1]$ be the space of all continuous functions on $[0,1]$ and define $\|f\| = \max |f(x)|$. Show that C is a Banach space.

12. We denote by l^∞ the space of all bounded sequences of real numbers and define $\|\{\xi_v\}\|_\infty = \sup |\xi_v|$. Show that l^∞ is a Banach sapce.

13. Show that the space C of all convergent sequences of real numbers and the space C_0 of all sequences which converge to zero are Banach spaces (with the l^∞ norm).

Chapter 8

DIFFERENTIATION AND INTEGRATION

In Riemann theory of integration it is known that differentiation and integration are inverse operation of each other in the following sense:

1. If f is a Riemann integrable function over $[a,b]$, then its indefinite integral

$$F(x) = R\int_a^x f(t)\,dt$$

   defines a continuous function on $[a,b]$. Funthermore, if f is continuous at a point $x_0 \in [a,b]$, then F is differentiable there at, and $F'(x_0) = f(x_0)$.

2. If f is Riemann integrable over $[a,b]$ and if there is a differentiable function F on $[a,b]$ such that $F'(x) = f(x)$, then

$$R\int_a^x f(t)\,dt = F(x) - F(a), \quad \forall x \in [a,b].$$

   (This result is usually referred to as the Fundamental Theorem of Calculus.)

In this chapter we shall consider the sense in which differentiation is the inverse of integration. In particular we

shall be concerned with the following questions:
When does

$$\int_a^b f'(x)\,dx = f(b) - f(a)\,?$$

When does

$$\frac{d}{dx}\int_a^x f(y)\,dy = f(x)\,?$$

From the theory of Riemann integration it is known that the second relation is true if x is a point of continuity of f. We shall show that more generally this relation holds almost everywhere. Thus differentiation is the inverse of Lebesgue integration. The first question, however, is more difficult, and even using the Lebesque integral it is true only for a certain class of functions which we shall characterize.

8.1 Differention of Monotone Functions

In discussing the differentiation of an indefinite Lebesgue integral we shall have to take explicit care of the fact that a continuous function does not necessarily possess a unique derivative everywhere. By various ingenious methods, it is possible to construct nowhere differentiable continuous functions. For a function f which fails to possess a derivative in the ordinary sense, i.e., when the increment ratio

$$\frac{f(x+h)-f(x)}{h}$$

dose not tend to a unique limit as $|h| \to 0$, we make use of the following four quantities known as **Dini derivatives**:

$$\begin{aligned}
D^+ f(x) &= \overline{\lim_{h\to 0+}}\frac{f(x+h)-f(x)}{h} && \text{(upper right derivative)}\\
D_+ f(x) &= \underline{\lim_{h\to 0+}}\frac{f(x+h)-f(x)}{h} && \text{(lower right derivative)}\\
D^- f(x) &= \overline{\lim_{h\to 0-}}\frac{f(x+h)-f(x)}{h} && \text{(upper left derivative)}\\
D_- f(x) &= \underline{\lim_{h\to 0_-}}\frac{f(x+h)-f(x)}{h} && \text{(lower left derivative).}
\end{aligned}$$

We may easily prove, by using the definition, that:

1. The Dini derivatives always exist (finite or infinite) for any function f, and satisfy

$$D^+f(x) \geqslant D_+f(x), \quad D^-f(x) \geqslant D_-f(x)$$

2. If $D_-f(x) = D^-f(x)$, then the common value, written $\underline{D}f(x)$, is just the lift-hand derivative of f at x; and similarly, if $D_+f(x) = D^+f(x)$, then the common value, written $\overline{D}f(x)$, is just the right-hand dervative of f at x. In general, $\underline{D}f(x) \leqslant \overline{D}f(x)$. and if f has the derivative at x, then $\underline{D}f(x) = \overline{D}f(x)$; and conversely. Thus, the function f is **differentiable at a point** if and only if all four Dini derivatives thereat are identical and different from $\pm\infty$.

3. $$\underline{D}f(x) = -\overline{D}(-f(x)).$$

4. $$\begin{cases} \overline{D}(f+g)(x) \leqslant \overline{D}f(x) + \overline{D}g(x) \\ \underline{D}(f+g)(x) \geqslant \underline{D}f(x) + \underline{D}g(x). \end{cases}$$

Now, we give examples of functions along with their Dini derivatives.

1.1 Example. 1. Let $f : \mathbb{R} \to \mathbb{R}$ be defined by $f(x) = |x|$. Then $D^+f(0) = D_+f(0) = 1$ and $D_-f(0) = D^-f(0) = -1$. As such $\underline{D}f(0) = -1$ and $\overline{D}f(0) = 1$. Consequently, f is not differentiable at $x = 0$.

2. Let $f : \mathbb{R} \to \mathbb{R}$ be defined by

$$f(x) = \begin{cases} x \sin\left(\frac{1}{x}\right) & \text{if } x > 0 \\ 0 & \text{if } x = 0 \\ x \sin\frac{1}{x} + 4x & \text{if } x < 0 \end{cases}$$

Then $D^+f(0) = 1$, $D_+f(0) = -1$, $D^-f(0) = 5$ and $D_-f(0) = 3$.

3. Let $f : \mathbb{R} \to \mathbb{R}$ be defined by

$$f(x) = \begin{cases} x \sin\left(\frac{1}{x}\right) & \text{if } x \neq 0 \\ 0 & \text{if } x = 0. \end{cases}$$

Then $D^+ f(0) = D^- f(0) = 1$ and $D_+ f(0) = D_- f(0) = -1$.

4. Let $f : \mathbb{R} \to \mathbb{R}$ be defined by

$$f(x) = \begin{cases} x\{1 + \sin(\log x)\} & \text{if } x > 0 \\ 0 & \text{if } x = 0 \\ x + \sqrt{-x \sin^2(\log x)} & \text{if } x < 0. \end{cases}$$

Then $D^+ f(0) = 2$, $D_+ f(0) = 0$, $D^- f(0) = 1$ and $D_- f(0) = -\infty$.

5. Let $f : [0,1] \to \mathbb{R}$ be defined by

$$f(x) = \begin{cases} 0 & \text{if } x \text{ is rationals} \\ 1 & \text{if } x \text{ is irrational.} \end{cases}$$

Then $D^+ f(x) = \infty$, $D_- f(x) = -\infty$, and $D_+ f(x) = 0 = D^- f(x)$, if x is rational; $D^+ f(x) = 0 = D_- f(x)$, $D^- f(x) = \infty$ and $D_+ f(x) = -\infty$, if x is irrational.

Next, we establish the measurability of lower and upper derivatives of a function.

1.2 Theorem. *It f is any function on an interval I, then $\overline{D} f$ and $\underline{D} f$ are measurable functions.*

Proof. Ref [11–p171]

1.3 Corollary. *If f is a differentiable function on an interval I, then f' is measurable on I.*

1.4 Definition. *Let $E \subset R$. A collection $\mathfrak{I}$ of intervals is said to be a Vitali cover of the set E if for each $x \in E$ and each $\epsilon > 0$, there exists an interval $I \in \mathfrak{I}$ with $x \in I$ and $l(I) < \epsilon$.*

1.5 Example. Let $\{r_n\}$ be the enumeration of the rationals in $[a,b]$. Then the collection $\{I_{n,i}\}$ where $I_{n,i} = \left[r_n - \frac{1}{i}, r_n + \frac{1}{i}\right]$, $n, i \in \mathbb{N}$ forms a Vitali cover of $[a,b]$.

1.6 Lemma. *(**Vitali's Covering Theorem**). Let E be a set of finite outer measure and $\mathcal{J}$ a collection of intervals which cover E in the sense of Vitali. Then given $\epsilon > 0$, there is a finite disjoint collection $\{I_1, I_2, \cdots, I_N\}$ of intervals in $\mathcal{J}$ such that*

$$m^*\left(E - \bigcup_{i=1}^{N} I_i\right) < \epsilon.$$

Proof. We may assume, without any loss of generality, that each interval in $\mathcal{J}$ is closed; for otherwise we replace each interval by its closure and observe that the set of the endpoints of $I_1, I_2, \cdots, I_N$ has measure zero.

Let O be an open set such that $O \supset E$ and $m^*(O) < m^*(E) + 1 < \infty$ (cf. III-2.8). Since $\mathcal{J}$ is a Vitali cover of E, we may assume that each interval in $\mathcal{J}$ is contained in O; otherwise we could discard from $\mathcal{J}$ all those intervals which extend beyond O and the remaining intervals in $\mathcal{J}$ still form a Vitali cover of E; for if $x \in E$, then for some $\epsilon > 0$ there is a neighbourhood $N(x, 2\epsilon) \subset O$ and there is an interval $I \in \mathcal{J}$ such that $x \in I$, $l(I) < \epsilon$, and hence $I \subset O$.

We now choose, by the method of induction, a sequence $\{I_n\}$ of disjoint intervals from $\mathcal{J}$ as follows. Let I_1 be an arbitrary interval of the family $\mathcal{J}$ and let η_1 be the least upper bound of the lengths of the intervals in $\mathcal{J}$ which do not have any point in common with I_1. Note that $\eta_1 < \infty$ since $\eta_1 \leqslant m(O) < \infty$. We then choose I_2, an interval from $\mathcal{J}$ which is disjoint from I_1, such that $l(I_2) > \frac{1}{2}\eta_1$. Let η_2 be the least upper bound of the lengths of the intervals in $\mathcal{J}$ which do not have any point in common with either I_1 or I_2, and note that $\eta_2 < \infty$. Choose I_3, an interval from $\mathcal{J}$ which is disjoint from $I_1 \cup I_2$, such that $l(I_3) > \eta_2/2$. In general, having already choosen n disjoint intervals $I_1, I_2, \cdots, I_n$, we denote by $\eta_n (< \infty)$ the least upper bound of the lengths of the intervals in $\mathcal{J}$ which do not have any point in common with $\bigcup_{i=1}^{n} I_i$, and choose I_{n+1} to be an interval from $\mathcal{J}$ such that it is disjoint from the preceding intervals and $l(I_{n+1}) > \eta_n/2$. If for some n, the set $\bigcup_{i=1}^{n} I_i$ contains almost every point of the set E, the lemma is proved; otherwise we get an infinite sequence $\{I_n\}$ of intervals from $\mathcal{J}$ such that

$$\begin{cases} I_i \cap I_j = \phi, & i \neq j \\ l(I_{n+1} > \frac{\eta_n}{2}, & \eta_n < \infty,\ n = 1, 2, 3, \cdots \end{cases}$$

Observe that $\{\eta_n\}$ is a monotonic decreasing sequence of nonnergative real numbers.

Since $\bigcup_{n=1}^{\infty} I_n \subset O$, we have

$$\sum_{n=1}^{\infty} l(I_n) \leqslant m(O) < \infty.$$

Thus, for a given $\epsilon > 0$, there is an integer n such that

$$\sum_{n=N+1}^{\infty} l(I_n) < \frac{\epsilon}{5}.$$

Write

$$J = E - \bigcup_{n=1}^{N} I_n.$$

The lemma will be proved if we establish that $m^*(J) < \epsilon$. Let $x \in J$. Then $x \notin \bigcup_{n=1}^{N} I_n$. Since $\bigcup_{n=1}^{N} I_n$ is a closed set not containing x, we can find an interval I in $\mathcal{J}$ such that $x \in I$ and $l(I)$ so small that I does not meet any of the intervals $I_1, I_2, \cdots, I_N$, i.e.

$$I \cap I_i = \phi, \quad i = 1, 2, \cdots, N.$$

Thus, we may have

$$l(I) \leqslant \eta_N < 2l(I_{N+1}).$$

If the interval I does not meet with I_{N+1} also, we should have

$$l(I) \leqslant \eta_{N+1}.$$

If I did not have any point in common with any of the intervals in the sequence $\{I_n\}$, then we should have

$$l(I) \leqslant \eta_n \, , \quad \forall n$$

which is impossible since

$$\eta_n < 2l(I_{n+1}) \to 0 \quad \text{as } n \to \infty.$$

As such, I must meet at least one of the intervals in the sequence $\{I_n\}$. Let η_0 be the smallest integer such that I meets I_{n0}. Clearly, $n_0 > N$ and $l(I) \leqslant \eta_{n_0 - 1} < 2l(I_{n_0})$. Since $x \in I$ and I has a point in common with I_{n0}, it follows that the distance of x from the midpoint of I_{n0} is at most $l(I) + \frac{1}{2}l(I_{n0}) < \frac{5}{2}l(I_{n0})$. Thus, if J_{n_0} is an interval concentric with I_{n_0} such that $l(J_{n_0}) = 5l(I_{n_0})$, we find that $x \in J_{n_0}$. In other words, for every $x \in j$, $\exists$ an $n \geqslant N+1$ such that $x \in J_n$ and $l(J_n) = 5l(I_n)$, and as such

$$\begin{aligned} J &\subset \bigcup_{n=N+1}^{\infty} J_n \\ m^*(J) &\leqslant \sum_{n=N+1}^{\infty} l(J_n) \\ &= 5 \sum_{n=N+1}^{\infty} l(I_n) < \epsilon. \end{aligned}$$

This completes the proof of the lemma. □

As an application of Theorem 1.6 we have the following significant result which generalizes the result: 'The union of a countable collection of intervals is a measurable set'.

1.7 Corollary. *The union of any collection of intervals is a measurable set.*

Proof. Let $\{I_\lambda : \lambda \in \Lambda\}$ be a collection of intervals and let

$$E = \bigcup_{\lambda \in \Lambda} I_\lambda .$$

Denote by $\mathfrak{J}$ the collection of all closed intervals J such that $J \subset I_\lambda$ for at least one $\lambda \in \Lambda$. Then $\mathfrak{J}$ is a Vitali cover of E, and it follows that for each $\epsilon > 0$, there exists a finite collection $\{J_1, J_2, \cdots, J_n\}$ of disjoint intervals in $\mathfrak{J}$ such that

$$m^*\left(E - \bigcup_{i=1}^{n} J_i\right) < \epsilon$$

$\Longrightarrow$ $$m^*\left(E - \cup_{i=1}^{n} J_i\right) = 0.$$
But $J_i \subset E$, for all i. Therefore

$$E = \left[\bigcup_{I=1}^{\infty} J_i\right] \cup \left[E - \bigcup_{I=1}^{\infty} J_i\right].$$

Hence the measurability of E follows from that of $\bigcup_{i=1}^{\infty} J_i$, and $E - \bigcup_{i=1}^{\infty} J_i$. □

1.8 Theorem. ***(Lebesgue's Theorem)*** *Let f be an increasing real-valued function defined on $[a, b]$. Then f is differentiable a.e. and the derivative f' is mceasurable. Furthermore,*

$$\int_a^b f'(x)\,dx \leqslant f(b) - f(a). \tag{1}$$

Proof. Let us first prove that the points of the open interval (a, b) at which not all the four Dini derivatives are equal form a set of measure zero. Since

$$D_+f(x) \leqslant D^+f(x) \quad \text{and} \quad D_-f(x) \leqslant D^-f(x)$$

it is sufficient to show that each of the following four sets of points of (a, b) is of measure zero:

$$\begin{aligned} A &= \{x \in (a,b),\ D_-f(x) < D^+f(x)\} \\ B &= \{x \in (a,b),\ D_+f(x) < D^-f(x)\} \\ C &= \{x \in (a,b),\ D_-f(x) < D^-f(x)\} \\ D &= \{x \in (a,b),\ D_+f(x) < D^+f(x)\}. \end{aligned}$$

We establish the result only for the set A, and the result in respect of the sets B, C and D can be proved similarly.

Define, for given rationals, u and v with $u < v$, the set

$$A_{u,v} = \{x \in (a,b) : \; D_-f(x) < u < v < D^+f(x)\}.$$

Clearly, A is the union of a countable family $\{A_{u,v}\}$ of sets, where u and v vary over all rationals. Thus, in order to show $m^*(A) = 0$, it is sufficient to establish $m^*(A_{u,v}) = 0$. Let $m^*(A_{u,v}) = \alpha$. Then given an $\epsilon > 0$, $\exists$ an open set $O \supset A_{u,v}$ such that

$$m(O) < \alpha + \epsilon.$$

Let x be any arbitrary point of $A_{u,v}$, since $D_-f(x) < u$, there is an arbitraty small interval (intervals) $[x-h, x] \subset O$ such that

$$f(x) - f(x-h) < uh.$$

This way we obtain a Vitali cover $\mathcal{J}$ of $A_{u,v}$. By Lemma 1.6, there is a finite subcollection $\{I_1, I_2, \cdots, I_N\}$ of disjoint intervals in $\mathcal{J}$ such that

$$m^*\left(A_{u,v} - \bigcup_{i=1}^{N} I_i\right) < \epsilon.$$

Write $I_i = [x_i - h_i, x_i]$, $i = 1, 2, \cdots, N$. Then summing over these N intervals, we have

$$\begin{aligned} \sum_{i=1}^{N}[f(x_i) - f(x_i - h_i)] &< u\sum_{i=1}^{N} h_i \\ &< m(O) \\ &< u(\alpha + \epsilon). \end{aligned}$$

Let y be an arbitrary point of the set

$$E = A_{u,v} \cap \left[\bigcup_{i=1}^{N} \text{Int } (I_i)\right].$$

Since $D^+f(y) > v$, there is a small interval, and hence intervals of the form $[y, y+k] \subset I_i$, for some $i = 1, 2, \cdots, N$ such that

$$f(y+k) - f(y) > vk.$$

This way we obtain a Vitali cover $\mathcal{J}^*$ of E. Again, by Lemma 1.6 there is a finite subcollection $\{J_1, J_2, \cdots, J_M\}$ of disjoint intervals in $\mathcal{J}^*$ such that

$$m^*\left(E - \bigcup_{j=1}^{M} J_j\right) < \epsilon.$$

Since one can easily verify that $m^*(E) > \alpha - \epsilon$, it follows that

$$m^*\left(E \cap \left[\bigcup_{j=1}^{M} J_j\right]\right) > \alpha - 2\epsilon.$$

If we write $J_j = [y_j, y_j + k_j]$, $j = 1, 2, \cdots, M$, then summing over these M intervals, we obtain

$$\begin{aligned}\sum_{j=1}^{M}[f(y_j + k_j) - f(y_i)] &> v\sum_{j=1}^{M} k_j \\ &> v(\alpha - 2\epsilon).\end{aligned}$$

But each J_j is contained in some I_i. Thus, for a fixed i, if we sum over those j for which $J_j \subset I_i$, we get

$$\sum[f(y_j + k_j) - f(y_j)] \leqslant f(x_i) - f(x_i - h_i),$$

since f is an increasing function. Therefore, summing the right-hand side of the above over $i = 1, 2, \cdots, N$, we obtain

$$\sum_{i=1}^{N}[f(x_i) - f(x_i - h_i)] \geqslant \sum_{j=1}^{M}[f(y_j + k_j) - f(y_i)]$$

$$\Longrightarrow \qquad u(\alpha + \epsilon) > v(\alpha - 2\epsilon).$$

Since this is true for each $\epsilon > 0$, we must have $u\alpha \geqslant v\alpha$. But $u < v$. This implies that $\alpha = 0$, and so $m^*(A) = 0$.

We have thus established that

$$g(x) = \lim_{h\to 0} \frac{f(x+h) - f(x)}{h}$$

is defined a.e. in $[a, b]$, and that f is differentiable wherever g is finite. Write

$$g_n(x) = \frac{f\left(x + \frac{1}{n}\right) - f(x)}{\frac{1}{n}},$$

and set $f(x) = f(b)$ for $x \geqslant b$. Then we note that $\{g_n\}$ is a sequence of nonnegative functions, since f is increasing, such that $g_n \to g$ a.e. Moreover, g is measurable. Hence, by Fatou's Lemma, we have

$$\begin{aligned}
\int_a^b g &\leqslant \underline{\lim}_{n\to\infty} \int_a^b g_n \\
&= \underline{\lim}_{n\to\infty} n \int_a^b \left[f\left(x + \frac{1}{n}\right) - f(x)\right] dx \\
&= \underline{\lim}_{n\to\infty} \left[n \int_b^{b+1/n} f - n \int_a^{a+1/n} f\right] \\
&= \underline{\lim}_{n\to\infty} \left[f(b) - n \int_a^{a+1/n} f\right] \\
&\leqslant f(b) - f(a) \\
&< \infty.
\end{aligned}$$

This verifies that g is integrable and hence finite a.e. in $[a, b]$. Consequently f is differentiable a.e. and $g = f'$ a.e. □

Remark: *Strict inequality holds in Theorem 1.8*

1.9 Example.

1. Consider the function $f : [0, 1] \to \mathbb{R}$ given by

$$f(x) = \begin{cases} 0 & \text{if } 0 \leqslant x \leqslant \frac{1}{2} \\ 1 & \text{if } \frac{1}{2} < x \leqslant 1. \end{cases}$$

Then $f'(x) = 0$ a.e. on $[0, 1]$; and

$$\int_0^1 f' = 0 < 1 = f(1) - f(0).$$

2. The Cantor function ψ satisfies the hypothesis on $[0, 1]$ and in addition $\psi'(0) = 0$ a.e. on $[0, 1]$. Thus

$$\int_0^1 \psi' = 0 < 1 = \psi(1) - \psi(0).$$

We know that equality, in the case of the Riemann integral, holds in 1 if f' is continuous in $[a, b]$. However, we shall obtain equality in the case of the Lebesgue integral under conditions less restrictive than that a continuity of f', but of course stronger than the hypothesis of this theorem, cf. Theorem 3.6

PROBLEMS [10.1]

1. Let f be the function defined by $f(0) = 0$ and $f(x) = x \sin (1/x)$ for $x \neq 0$. Find $D^+f(0)$, $D_+f(0)$, $D^-f(0)$, and $D_-f(0)$.

2. If the function f assumes its maximum at c, then $D^+f(c) \leq 0$ and $D_-f(c) \geq 0$.

3. If f is continuous on $[a, b]$ and one of its derivates (say D^+) is everywhere nonnegative on $[a, b]$, then $f(b) \geq f(a)$. [Hint: First show this result for a function g for which $D^+g \geq \epsilon > 0$. Apply this to $g = f + \epsilon x$.]

4. Show by an example that the condition f is an increasing function in Theorem 1.8 is necessary.

8.2 Functions of Bounded Variation

2.1 Definition. *A function f defined on an interval [a,b] is said to be of* ***bounded variation*** *if there is a constant $K > 0$ such that for every partition*

$$P = \{a = x_0 < x_1 < \cdots < x_i < \cdots < x_n = b\}$$

of $[a,b]$, we have

$$\sum_{i=1}^{n} |f(x_i) - f(x_{i-1})| \leqslant K.$$

In fact, the variation of a function f defined on $[a,b]$ with respect to a partition P of $[a,b]$ is given by

$$V_a^b(f,P) = \sum_{i=1}^{n} |f(x_i) - f(x_{i-1})|,$$

and the total variation of f on $[a,b]$ by

$$T_a^b(f) = \sup_P V_a^b(f,P).$$

The total varation $T_a^b(f)$ may be finite or infinite. It follows from Definition 2.1 that f is of bounded variation on $[a,b]$ if and only if $T_a^b(f) < \infty$.

Futher, let f be defined on $\mathbb{R}$. Define

$$T_{-\infty}^{\infty}(f) = \lim_{a\to\infty} T_{-a}^{a}(f).$$

If $T_{-\infty}^{\infty}(f)$, then f is said to be of bouneded variation on $\mathbb{R}$.

We now give a few simple observations with regard to the functions of bounded variation.

(a) *A bounded monotone function is a function of bounded variation and*

$$T_a^b(f) = |f(b) - f(a)|.$$

(b) *A function of bounded variation is necessarily bouneded, but not conversely.*

Let $x \in [a,b]$ be arbitrary. Then

$$|f(x) - f(a)| + |f(b) - f(x)| \leqslant T_a^b(f)$$

$$\Longrightarrow \qquad |f(x)| \leqslant f(a) + T_a^b(f).$$

For the converse, consiedr a function $f : [0,1] \to \mathbb{R}$ given by

$$f(x) = \begin{cases} x \sin\left(\frac{\pi}{x}\right) & \text{if } x \neq 0 \\ 0 & \text{if } x = 0. \end{cases} \tag{2}$$

The function f is clearly bouneded. To show that f is not of bounded variation, consider a partition $\left\{1, \frac{2}{3}, \frac{2}{5}, \frac{2}{7}, \cdots \frac{2}{2n+1}, 0\right\}$ of $[0,1]$, n being a positive integer. For this partition, we have

$$\begin{aligned} \left|f(1) - f\left(\frac{2}{3}\right)\right| &+ \left|f\left(\frac{2}{3}\right) - f\left(\frac{2}{5}\right)\right| \\ &+ \cdots + \left|f\left(\frac{2}{2n+1}\right) - f(0)\right| \\ &= \frac{2}{3} + \left(\frac{2}{3} + \frac{2}{5}\right) + \cdots + \left(\frac{2}{2n+1}\right) \\ &= 4\left(\frac{1}{3} + \frac{1}{5} + \frac{1}{7} + \cdots + \frac{1}{2n+1}\right) \end{aligned}$$

Since $\sum\limits_n \dfrac{1}{2n+1}$ is divergent, its partial sum

$$\left(\frac{1}{3} + \frac{1}{5} + \frac{1}{7} + \cdots + \frac{1}{2n+1}\right)$$

is not bounded above. Hence $T_0^1(f) = \infty$.

(c) *A function of bounded variation need not be continuous.*

Consider the function $f : [0,2] \to \mathbb{R}$ defined by $f(x) = [x]$, where $[x]$ denotes the greatest integer not greater than x. Here f is a function of bounded variation but not continuous in $[0,2]$.

(d) *A continuous function need not be of bounded variation.*

The function f defined in (2) is continuous on $[0,1]$ but not of bounded varition.

(e) If f and g are two functions of bounded variation on $[a,b]$ and λ is a constant, then λf and $f+g$ are functions of bounded variation on $[a,b]$. Furthermore,

$$T_a^b(f+g) \leqslant T_a^b(f) + T_a^b(g)$$

and

$$T_a^b(\lambda f) = |\lambda|\, T_a^b(f).$$

Let $P = \{a = x_0 < x_1 < \cdots < x_n = b\}$ be a partition of $[a, b]$. Then

$$\begin{aligned} &\sum_{i=1}^{n} |f(x_i) + g(x_i) - f(x_{i-1}) - g(x_{i-1})| \\ \leqslant &\sum_{i=1}^{n} |f(x_i) - f(x_{i-1})| + \sum_{i=1}^{n} |g(x_i) - g(x_{i-1})|. \end{aligned}$$

Taking suprema on both the sides over all partitions P of $[a, b]$, and noting that

$$\begin{aligned} \sup\{x + y \quad &: \quad x \in A \text{ and } y \in B\} \\ &\leqslant \quad \sup\{x : x \in A\} + \sup\{y : y \in B\}, \end{aligned}$$

we immediately get $T_a^b(f + g) \leqslant T_a^b(f) + T_a^b(g)$. This also verifies that $f + g$ is a function of bounded variation.

The other part is an immediate consequence of the definition of $T_a^b(f)$.

Remark: *The set of all functions of bounded variation on a given interval $[a, b]$, denoted by $BV[a, b]$, forms a linear space unlike the set of all monotone functions. In fact, the sum of two monotone functions need not be monotone, e.g. $f(x) = \sin x + 2x$ and $g(x) = \sin x - 2x$ defined on $[-\pi, \pi]$.*

(f) *If f is a function of bounded variation on $[a, b]$ and c is a point of $[a, b]$, then f is of bounded variation on $[a, c]$ and $[c, b]$ both, and conversely. Furthermore,*

$$T_a^b(f) = f_a^c(f) + T_c^b(f).$$

(g) *f is a function of bounded variation on $[a, b]$, then the function $vf : [a, b] \to \mathbb{R}$ defined by $vf(x) = T_a^x(f)$ is known as the* ***variation function****. It may easily be verified that vf is an increasing function.*

2.2 Theorem. ***(Jordan Decomposition Theorem)***

A function f defined on $[a, b]$ is of bounded variation if and only if it can be expressed as a difference of two monotone increasing functions (real-valued) on $[a, b]$.

Proof. Let f be of bounded variation, on $[a, b]$. Define g and h by

$$\begin{cases} g = \frac{1}{2}(vf + f) \\ h = \frac{1}{2}(vf - f). \end{cases}$$

so that $f = g - h$.

Now, if x_1, x_2 is any pair of points in $[a, b]$ with $x_2 > x_1$, then

$$\begin{cases} g(x_2) - g(x_1) = \frac{1}{2}[\{vf(x_2) - vf(x_1)\} + \{f(x_2) - f(x_1)\}] \\ h(x_2) - h(x_1) = \frac{1}{2}[\{vf(x_2) - vf(x_1)\} + \{f(x_2) - f(x_1)\}]. \end{cases}$$

but, f being of bounded variation on $[a, b]$, in particular on $[x_1, x_2]$, we have

$$|f(x_2) - f(x_1)| \leqslant T_{x_1}^{x_2}(f) = vf(x_2) - vf(x_1).$$

Hence $g(x_2) > g(x_1)$ and $h(x_2) > h(x_1)$ verifying that g and h both are monotone increasing (real-valued) functions.

Conversely, if $f = g - h$, where g and h are monotone increasing functions, then for any partiton P of $[a, b]$. we have

$$\begin{aligned} \sum_{i=1}^{n} |f(x_i) - f(x_{i-1})| &\leqslant \sum_{i=1}^{n}[g(x_i) - g(x_{i-1})] + \sum_{i=1}^{n}[h(x_i) - h(x_{i-1})] \\ &= g(b) - g(a) + h(b) - h(a) \end{aligned}$$

$$\Longrightarrow \qquad T_a^b(f) < \infty.$$

Hence f is a function of bounded variation. □

2.3 Corollary. *If f is a function of bounded variation on $[a, b]$, then f' exists a.e. on $[a, b]$.*

Proof. It follows from Theorems 1.6 and 2.2. □

2.4 Corollary. *If f is a function of bounded variation, then it is measurable.*

Proof. By Corollary 2.3, f is continuous a.e. in $[a, b]$ and the result follows by Theorem IV-1.23. □

Note: The functions g and h in the decomposition of f are called, respectively, the positive and the negative variation functions of f on $[a, b]$. Also, observe that

$$vf(x) = g(x) + h(x), \quad x \in [a, b].$$

The positive and the negative variation functions of f are, respectively, denoted by $P_a^x(f)$ and $N_a^x(f)$. Thus, the positive and the negative variations of f on $[a, b]$ are, respectively, given by $P_a^b(f)$ and $N_a^b(f)$. Also

$$T_a^b(f) = P_a^b(f) + N_a^b(f).$$

2.5 Theorem. *Let f be a function of bounded variation on $[a, b]$. Then f is continuous at a point in $[a, b]$ if and only if its variation funtion vf is so.*

Proof. Let f be continuous at a point x^* in $[a, b]$. Then for each $\epsilon > 0$, there exists a $\delta > 0$ such that

$$|f(x) - f(x^*)| < \frac{\epsilon}{2}, \quad \text{for } |x - x^*| < \delta.$$

By the definition of $T_{x^*}^b(f)$, $\exists$ a partition

$$P = \{x^* = x_0 < x_1 < x_2 \cdots < x_n = b\}$$

of $[x^*, b]$ such that

$$\sum_{i=1}^{n} |f(x) - f(x_{i-1})| > T_{x^*}^b(f) - \frac{\epsilon}{2}.$$

Assume that the first interval, i.e. $[x_0, x_1]$ in P is of length less than δ, otherwise it could be made so by introducing an additional point between x_0 and x_1, which of course would not decrease the sum on the right of the above and, as such, the above inequality would still remain valid. Therefore,

$$\begin{aligned} T_{x^*}^b(f) - \frac{\epsilon}{2} &< |f(x_1) - f(x_0)| + \sum_{i=2}^{n} |f(x_i) - f(x_{i-1})| \\ &< \frac{\epsilon}{2} + T_{x_1}^b(f) \\ &\Longrightarrow T_{x^*}^b(f) - T_{x_1}^b(f) < \epsilon \\ &\Longrightarrow T_{x^*}^{x_1}(f) < \epsilon. \end{aligned}$$

In other words, $vf(x_1) - vf(x^*) < \epsilon, \quad 0 < x_i - x^* < \delta$

$$\Longrightarrow \qquad \lim_{x \to x^*+0} vf(x^*).$$

Similarly

$$\lim_{x \to x^*-0} vf(x) = vf(x^*).$$

Hence, vf is continuous at x^*.

Conversely, if the variation function vf is continuous at a point x^* in $[a, b]$, to each $\epsilon > 0$, $\exists$ a $\delta > 0$ such that

$$|vf(x) - vf(x^*)| < \epsilon, \quad \text{for } |x - x^*| < \delta.$$

The result now follows from the fact that

$$|f(x) - f(x^*)| \leqslant |vf(x) - vf(x^*)|. \qquad \square$$

2.6 Corollary. *A continuous function is of bounded variation if and only if it can vbe expressed as a difference of two continuous monotone increasing functions.*

Proof. It follows from Theorems 2.2 and 2.5 $\square$

Remark: *If f is a continuous function of bounded variation, it may also be expressed as the difference of two monotone increasing functions which have discontinuities (even countably many), e.g.*

$$f = (g + k) - (h + k),$$

where k is an increasing function having discontinuities.

PROBLEMS [10.2]

1. Show that the function $f : [0,1] \to \mathbb{R}$ defined by

$$f(x) = \begin{cases} x^\alpha \sin\left(\frac{1}{x^\beta}\right) & \text{if } 0 < x \leqslant 1 \\ 0 & \text{if } x = 0. \end{cases}$$

is of bounded variation if $\alpha > \beta$ but not if $\alpha \leqslant \beta$.

2. Let f be defined by $f(0) = 0$ and $f(x) = x^2 \sin\left(\frac{1}{x^2}\right)$ for $x \neq 0$. Is f of bounded variation on $[-1,1]$?

3. Show that the function $f : [0,1] \to \mathbb{R}$ defined by

$$f(x) = \begin{cases} x \cos\left(\frac{\pi x}{2}\right) & \text{if } 0 < x < 1 \\ 0 & \text{if } x = 0. \end{cases}$$

is continuous but not of bounded variation on $[0,1]$

4. Show that a step function defined on $[a,b]$ is of bounded variation on $[a,b]$.

5. Let $\{f_n\}$ be a sequence of functions defined on $[a,b]$. If $f_n \to f$ on $[a,b]$, then $T_a^b(f) \leqslant \underline{\lim}_{n\to\infty} T_a^b(f_n)$.

6. Let f be a function of bounded variation, Prve that the set of points of descontinuity of f is countable.

7. For any function $f : [a,b] \to \mathbb{R}$ and every $x \in [a,b]$, prove

$$T_a^b(f) + T_a^b(f) = T_a^b(f).$$

Then show that $x \to T_a^b(f)$ defines a nondecreasing function $g : [a,b] \to \mathbb{R}^*$.

8. For arbitraty functions $f,\ g : [a,b] \to \mathbb{R}$ and any real number prove:

$$T_a^b(f+g) \leq T_a^b(f) + T_a^b(g), \text{ and } T_a^b(cf) = |c| T_a^b(f).$$

9. Let $f : [a, b] \to \mathbb{R}$ be any continuous function such that one of it fout Dini derivates is nonnegative everywhere in the open interve (a, b). Prove that f is nondecreasing.

10. Prove that the points of discontinuity of any function $f : [a, b] \to I$ of bounded varation is countable.

8.3 Differention of an Integral

In this section we shall show that the derivative of the indefinite integral of an integrable function is equal to the integrand almost everywhere.

If f is an integrable function on an interval $[a, b]$, then f is integrable on any interval $[a, x] \subset [a, b]$. The function F given by

$$F(x) = \int_a^x f(t)\, dt + c,$$

where c is a constant, called the **indefinite integral** of f.

3.1 Theorem. *Let f be an integrable function on $[a, b]$. Then the indefinite integral F defined by*

$$F(x) = \int_a^x f(t)\, dt$$

is a continuous function of bounded variation on $[a, b]$.

Proof. Let x_0 be any point in $[a, b]$. Then

$$\begin{aligned} |F(x) - F(x_0) &= \left|\int_{x_0}^x f(t)\, dt\right| \\ &\leqslant \left|\int_{x_0}^x |f(t)|\, dt\right| \end{aligned}$$

But f being integrable, the function $|f|$ is integrable over $[a, b]$, (cf. IV-5). Therefore, given $\epsilon > 0$, there is a $\delta > 0$ such that for every measurable set $A \subset [a, b]$ with $m(A) < \delta$, we have (cf. IV-4)

$$\int_A |f| < \epsilon.$$

In particular

$$\left|\int_{x_0}^{x} |f(t)|\, dt\right| < \epsilon, \quad \text{for } |x - x_0| < \delta.$$

Hence $|F(x) - F(x_0)| < \epsilon$, whenever $|x - x_0| < \delta$. This proves the continuity of F at x_0, and hence in $[a, b]$.

In order to establish that F is a function of bounded variation, let

$$P = \{a = x_0 < x_1 < x_2 \cdots < x_n = b\}$$

be a partition of $[a, b]$. Then

$$\begin{aligned}\sum_{i=1}^{n} |F(x_i) - F(x_{i-1})| &= \sum_{i=1}^{n} \left|\int_{x_{i-1}}^{x_i} f(t)\, dt\right| \\ &\leqslant \sum_{i=1}^{n} \int_{x_{i-1}}^{x_i} |f(t)|\, dt \\ &= \int_a^b |f(t)|\, dt\end{aligned}$$

$$\Longrightarrow \qquad T_a^b(F) = \int_a^b |f(t)|\, dt < \infty.$$

Hence the result follows. □

3.2 Theorem. *Let f be an integrable function on $[a, b]$. if*

$$\int_a^x f(t)\, dt = 0,$$

for all $x \in [a, b]$, then $f = 0$ a.e. in $[a, b]$.

Proof. Let if possible, $f \neq 0$ a.e. in $[a, b]$. Suppose $f(t) > 0$ on a set E of positive measure. Then there exists a closed set $F \subset E$ with $m(F) > 0$. Put $A = [a, b] - F$. Then A is an open set and we have

$$0 = \int_a^b f(t)\, dt = \int_{A \cup F} f(t)\, dt = \int_A f(t)\, dt + \int_F f(t)\, dt$$

$$\Longrightarrow \qquad \int_A f(t)\,dt = -\int_F f(t)\,dt.$$

But $f(t) > 0$ on F with $m(F) > 0$ implies

$$\int_F f(t)\,dt \neq 0.$$

Therefore

$$\int_A f(t)\,dt \neq 0.$$

Now, A being an open wet, it can be expressed as a union of countable collection $\{(a_n, b_n)\}$ of disjoint open intervals. Thus

$$\begin{aligned} 0 \quad &\neq \quad \int_A f(t)\,dt = \sum_n \int_{a_n}^{b_n} f(t)dt \\ &\Longrightarrow \quad \int_{a_n}^{b_n} f(t)\,dt \neq 0 \quad \text{for some } n \\ &\Longrightarrow \quad \text{either} \int_a^{a_n} f(t)\,dt \neq 0 \quad \text{or} \quad \int_a^{b_n} f(t)\,dt \neq 0. \end{aligned}$$

In either case, we see that if f is positive on a set of positive measure, then for some $x \in [a, b]$ we have

$$\int_a^x f(t)\,dt \neq 0.$$

Similar assertion is pbtained if f is negative on a set of positive measure. Hence the result follows by contradiction. □

3.3 Theorem. *Let f be a bounded and measurable function defined on $[a, b]$. If*

$$F(x) = \int_a^x f(t)\,dt + F(a),$$

then $F'(x) = f(x)$ a.e. in $[a, b]$.

Proof. Since f is bounded and measutable, it is integrable (cf. IV-2). Therfore, in view of Theorem 3.1, F is a continuous function of bounded variation on $[a, b]$ and hence F' exists a.e. in $[a, b]$, cf. Corollary 2.3 Let $|ff| \leqslant K$. Set

$$f_n(x) = \frac{F(x+h) - F(x)}{h},$$

with $h = \frac{1}{n}$. Then

$$f_n(x) = \frac{1}{h}\int_x^{x+h} f(t)\, dt$$

$\Longrightarrow$

$$|f_n| \leqslant K.$$

Also, $f_n \to F'$ a.e. If $c \in [a, b]$ is arbitrary, then the Bounded Convergence Theorem (cf.IV-3.3) implies that

$$\begin{aligned}
\int_a^c F'(x)\, dx &= \lim_{n\to\infty} \int_a^c f_n(x)\, dx \\
&= \lim_{h\to 0} \frac{1}{h}\int_a^c [F(x+h) - F(x)]\, dx \\
&= \lim_{h\to 0} \left[\frac{1}{h}\int_c^{c+h} F(x)\, dx - \frac{1}{h}\int_a^{a+h} F(x)\, dx\right].
\end{aligned}$$

But F being a continuous function, we note that

$$\begin{aligned}
\lim_{h\to 0} \frac{1}{h}\int_c^{c+h} F(x)\, dx &= \lim_{h\to 0} \frac{1}{h} R\int_c^{c+h} F(x)\, dx \\
&= \lim_{h\to 0} \frac{1}{h} F(c+\theta h), \quad 0 < \theta < 1 \\
&= F(c);
\end{aligned}$$

and similatly

$$\lim_{h\to 0} \frac{1}{h}\int_a^{a+h} F(x)\, dx = F(a).$$

Thus

$$\int_a^c F'(x)\, dx = F(c) - F(a) = \int_a^c f(x)\, dx$$

$\Longrightarrow$

$$\int_a^c [F'(x) - f(x)] = 0.$$

This is true for all c in [a,b]. Hence, by Theorem 3.2. we have $F' = f$ a.e. □

Note that in Theorem, 3.3. the function considered is bounded and measurable, We extend it, in the following theorem, to any measurable function which of course is integrable.

3.4 Theorem. *If f is bounded and measurable on $[a, b]$ and*

$$F(x) = \int_a^x f(t)\,dt + F(a),$$

then $F'(x) = f(x)$ for almost all x in $[a, b]$.

Proof. By 3.1, F is of bounded variation over $[a, b]$, and so $F'(x)$ exists for almost all x in $[a, b]$. Let $|f| \leq K$. Then setting

$$f_n(x) = \frac{F(x+h) - F(x)}{h},$$

with $h = 1/n$, we have

$$f_n = \frac{1}{h}\int_x^{x+h} f(t)\,dt,$$

and so

$$|f_n| \leq K$$

Since

$$f_n(x) \to F'(x) \text{ a.e.},$$

the bounded convergence theorem implies that

$$\begin{aligned} \int_a^c F'(x)\,dx &= \lim \int_a^c f_n(x)\,dx = \lim_{h\to 0} \frac{1}{h}\int_a^c (F(x+h) - F(x))\,dx \\ &= \lim \left[\frac{1}{h}\int_c^{c+h} F(x)\,dx - \frac{1}{h}\int_a^{a+h} F(x)\,dx\right] \\ &= F(c) - F(a) = \int_a^c f(x)\,dx, \end{aligned}$$

since F is continuous, Hence

$$\int_a^c \{F'(x) - f(x)\}\,dx = 0$$

for all $c \in [a, b]$, and so

$$F'(x) = f(x) \text{ a.e.}$$

by Theorem 3.2 □

Remark: *1. An indefinite infegral need not be differentiable everywhere, and even if it is defferentiable, it neek not follow that $F' = f$ everywhere, cf.Examples 1 and 2 in 3.5.*

2. The condition $F' = f$ a.e.is not sufficient to ensure that F is an indefinite integral of f. Consider for instance examples in Theorem 1.9.

3.5 Example.

1. Consider the function $f : [0,2] \to \mathbb{R}$ given by

$$f(x) = \begin{cases} 1 & \text{if } 0 \leqslant x \leqslant 1 \\ 2 & \text{if } 1 < x \leqslant 2. \end{cases}$$

Then

$$F(x) = \int_0^x f(t)\,dt = \begin{cases} x & \text{if } 0 \leqslant x \leqslant 1 \\ 2x - 1 & \text{if } 1 < x \leqslant 2. \end{cases}$$

Here F defines a function which is continuous but not differentiable in $[0,2]$. In fact, F is not defferentiable at $x = 1$.

2. Consider the function $f : [0,1] \to \mathbb{R}$ given by

$$f(x) = \begin{cases} \frac{1}{q} & \text{if } x = p/q \\ 0 & \text{if otherwise.} \end{cases}$$

Then

$$F(x) = \int_0^x f(t)\,dt = 0.$$

Here F defines a function differentiable in $[0,1]$. However, $F(x) \neq f(x)$ for $x = p/q$ in [0,1].

3.6 Theorem. *Let f be an integrable function on $[a,b]$, and suppose*

$$F(x) = F(a) + \int_a^x f(t)\,dt.$$

Then $F'(x) = f(x)$ for almost all x in $[a,b]$.

Proof. Without loss of generality we may assume $f \geq 0$. Let f_n be defined by $f_n(x) = f(x)$, if $f(x) \leq n$, and $f_n(x) = n$ if $f(x) > n$. Then $f - f_n \geq 0$, and so

$$G_n(x) = \int_a^x f - f_n$$

is an increasing function of x, which must have a derivative almost everywhere, and this derivative will be nonnegative. Now by Theorem 3.4.

$$\frac{d}{dx}\int_a^x f_n = f_n(x) \quad \text{a.e.},$$

and so

$$\begin{aligned} F'(x) &= \frac{d}{dx}G_n + \frac{d}{dx}\int_a^x f_n \\ &\geq f_n(x) \quad \text{a.e.} \end{aligned}$$

Since n is arbitrary,

$$F'(x) \geq f(x) \quad \text{a.e.}$$

Consequently,

$$\int_a^b F'(x)\,dx \geq \int_a^b f(x)\,dx = F(b) - F(a).$$

Thus by Theorem 1.8 we have

$$\int_a^b F'(x)\,dx = F(b) - F(a) = \int_a^b f(x)\,dx,$$

and

$$\int_a^b (F'(x) - f(x))\,dx = 0.$$

Since $F'(x) - f(x) \geq 0$, this implies that $F'(x) - f(x) = 0$ a.e., and so $F'(x) = f(x)$ a.e. □

PROBLEMS [10.3]

1. If f is the greatest integer function and $F(x) = \int_0^x f(t)\,dt$, detemine F on $[0,5]$ and verify that $F'(x) = f(x)$ a.e. in $[0,5]$.

2. Let
$$f(x) = \begin{cases} 1 & \text{if } x \text{ is rational} \\ 0 & \text{if } x \text{ is irrational.} \end{cases}$$
If F is as in Problem 1 then verify that $F'(x) = f(x)$ a.e. in $[0,1]$.

8.4 Absolute Continuity

We now introduce a special class of functions of bounded variation, namely, the class of absolutely continuous functions. This class is important for a number of applications and helps to obtain results already proved under less stringent conditions In particular, as we shall see in the next section, it charaterises the class of indefinite integrals.

4.1 Definition. *A real-valued function f defined on an interval $[a,b]$ is* ***absolutely continuous*** *on $[a,b]$, if for each $\epsilon > 0$ ther exists a $\delta > 0$ such that*

$$\sum_{i=1}^{n} |f(x'_i) - f(x_i)| < \epsilon,$$

for every finite pairwise disjoint collection $\{(x_i, x'_i) : i = 1, 2, \cdots, n\}$ of open intervals in $[a,b]$ with $\sum_i^n |x' - x_i| < \epsilon$.

Remark: *1. Every absolutely continuous function is uniformly continuous, as may be seen by choosing a single interval $(x, x') \subset [a,b]$. However, a uniformly continuous function need not be absolutely continuous; for example, the Cantor function is uniformly continuous in $[0,1]$ but not absolutely* *continuous. The function f in Problem 1. (for $\alpha = \beta = 1$) is also an example of a continuous function which is absolutely continuous.*

2. *In Definition 4.1. we may replace 'finite' with 'countably infinite'. In fact, suppose that given any $\epsilon > 0$, there is a $\delta > 0$ such that*

$$\sum_{i=1}^{n} |f(x_i') - f(x_i)| < \epsilon' < \epsilon,$$

for any finite collection $\{(x_i, x_i')\}$ of pairwise disjoint intervals of total length less than δ. Consider any countably infinite collection of pairwise disjoint intervals $(y_i, y_i') \subset [a, b]$ of total length less than δ. Then, obviously

$$\sum_{i=1}^{n} |f(y_i') - f(y_i)| < \epsilon',$$

for every n. Hence, length $n \to \infty$, we get

$$\sum_{i=1}^{\infty} |f(y_i') - f(y_i)| \leqslant \epsilon' < \epsilon.$$

4.2 Theorem. *If f is absolutely continuous on $[a, b]$, then f is of bounded variation on $[a, b]$.*

Proof. For $\epsilon = 1$, there is a $\delta > 0$ such that for every finite collection $\{[x_i, x_i'], i = 1, 2, \cdots, n\}$ of pairwise disjoint intervals in $[a, b]$ with $\sum_{i=1}^{n} |x_i' - x_i| < \delta$, we have

$$\sum_{i=1}^{n} |f(x_i') - f(x_i)| < i.$$

Select a natural number

$$N > \frac{b-a}{\delta}.$$

Divide $[a, b]$ by means of points

$$c = c_0 < c_1 < c_2 \cdots < c_N = b$$

such that $c_j - c_{j-1} < \delta$, for $j = 1, 2, \cdots, N$. Therefore, for every finite collection $\{[x_i, x_i']\}$ of pairwise disjoint subintervals in $[c_{j-1}, c_j]$, we have

$$\sum_i |f(x_i') - f(x_i)| < 1$$

$$\Longrightarrow \qquad T_{c_{j-1}}^{c_j}(f) \leqslant 1, \; j = 1, 2, \cdots, N.$$

Hence

$$T_a^b(f) = \sum_{j=1}^N T_{c_{j-1}}^{c_j}(f) \leqslant N < \infty,$$

and this proves that f is function of bounded variation. □

Note: The result in Theorem 4.2 does not hold good when a or b is allowed to be infinite.

4.3 Corollary. *If f is an absolutely continuous function, it has a derivative almost everywhere.*

Proof. It follows from Theorem 4.2 and Corollary 2.3.

Remark: *The converse of Theorem 4.2 need not be true. There are fuctions of bounded variation which are not absolutely continuous; one such function is the Cantor function.*

4.4 Theorem. *Let f be a function of bounded variation on $[a, b]$. Then f is absolutely continuous on $[a, b]$ if and only if the variation function vf given by $vf(x) = T_a^x(f)$, is absolutely continuous on $[a, b]$.*

Proof. The absolute continuity of vf clearly implies that of f since $|f(x_i) - f(x_{i-1})| \leqslant vf(x_i) - vf(x_{i-1})$ for $a \leqslant x_{i-1} < x_i < b$.

On the other hand, assume f to be absolutely continuous, Then-given an $\epsilon > 0$, there exists a $\delta > 0$ such that, for any finite collection $\mathcal{C} = \{(x_i, x_i'), \; i = 1, 2, \cdots, n\}$ of pairwise disjoint intervals in [a,b] with $\sum_{i=1}^n |x_i' - x_i| < \delta$, we have

$$\sum_{i=1}^n |f(x_i') - f(x_i)| < \epsilon.$$

For each i, let

$$P_i = \{x_i = a_0^i < a_1^i < a_2^i < \cdots < a_{mi}^i = x_i'\}$$

be a partition of $[x_i, x_i']$. Since

$$\sum_{i=1}^{n}\sum_{j=1}^{m_i} |a_j^i - a_{j-1}^i| = \sum_{i=1}^{n} |x_i' - x_i| < \delta,$$

we have

$$\sum_{i=1}^{n}\sum_{j=1}^{m_i} |f(a_j^i) - f(a_{j-1}^i)| < \epsilon.$$

Fixing the collection $\mathcal{C}$ but varying the partitions P_i of each $[x_i, x_i']$, we have, upon taking the supremum over all such partitions P_i,

$$\sum_{i=1}^{n} T_{x_i}^{x_i'}(f) < \epsilon$$

$$\Longrightarrow \qquad \sum_{i=1}^{n} vf(x_i') - vf(x_i) < \epsilon.$$

This proves vf is absolutely continuous. □

4.5 Theorem. *If f is absolutely continuous on $[a, b]$ and $f' = 0$ a.e. then f is a constant function.*

Proof. Ref [18]–P105.

4.6 Theorem. *A function F is an indefinite integral if and only if it is absolutely continuous.*

Proof. If F is an indefinite integral, then F is absolutely continuous. Suppose on the other hand that F is absolutely continuous on $[a, b]$. Then F is of bounded variation, and we may write

$$F(x) = F_1(x) - F_2(x),$$

where the functions F_i are monotone increasing. Hence $F'(x)$ exists almost everywhere and

$$|F'(x)| \leq F_1'(x) + F_2'(x).$$

Thus

$$\int |F'(x)|\, dx \le F_1(b) + F_2(b) - F_1(a) - F_2(a)$$

by Theorem 1.8 and $F'(x)$ is integrable. Let

$$G(x) = \int_a^x F'(t)\, dt.$$

Then G is absolutely continuous and so is the function $f = F - G.$. It follos from Theorem 3.7 that $f'(x) = F'(x) - G'(x) = 0$ a.e. and so f is constant by Theorem 4.5. Thus

$$F(x) = \int_a^x F'(t)\, dt + F(a). \qquad \square$$

4.7 Corollary. *Every absolutely continuous function is the indefinite integral of its derivative.*

4.8 Example. Let $F : [0,1] \to \mathbb{R}$ be a function defined by

$$F(x) = \begin{cases} x^2 \sin\left(\frac{\pi}{x^2}\right) & \text{if } x \neq 0 \\ 0 & \text{if } x = 0. \end{cases}$$

The derivative F' exists on $[0,1]$ while F' is not integrable on $[0,1]$. In fact

$$\int_0^1 \frac{1}{x} \left|\cos\left(\frac{\pi}{x^2}\right)\right| dx = \infty.$$

Note: The class of absolutely continuous functions over $[a, b]$ is identical with the class of functions obtained by integrating Lebesgue integrable functions over $[a, b]$ except that the corresponding functions in two classes differ at the most by a constant.

PROBLEMS [10.4]

1. Let f be absolutely continuous in the interval $[\epsilon, 1]$ for each $\epsilon > 0$. Does the continuity of f at 0 imply that f is absolutely continuous on [0,1]? What if f is also of bounded variation on $[0,1]$?

2. Let f be absolutely continuous on [a,b]. Show that

$$T_a^b(f) = \int_a^b |f'| \text{ and } P_a^b(f) = \int_a^b [f']^+.$$

3. The Cantor ternary function is continuous and monotone but not absolutely continuous.

4. A monotone function f on $[a, b]$ is called singular if $f' = 0$ a.e.

   (a) Show that any monotone increasing function is the sum of an absolutely continuous function and a singular function.

   (b) Construct a function f which is monotone and continuous on $[0, 1]$ but which is not absolutely continuous on any subinterval of $[0, 1]$.

   (c) Show that there is a strictly increasing singular function on $[0, 1]$.

5. **Change of variable.** Let g be a monotone increasing absolutely continuous function on $[a, b]$ with $g(a) = c$, $g(b) = d$.

   (a) Show that for any open set $O \subset [c, d]$

   $$mO = \int_{g^{-1}[O]} g'(x)\,dx.$$

   (b) Let $H = \{x : g'(x) \neq 0\}$. If E is a subset of $[c, d]$ with $mE = 0$, then $g^{-1}(E) \cap H$ has measure zero.

   (c) If E is a measurable subset of $[c, d]$, then $F = g^{-1}[E] \cap H$ is measurable and

   $$mE = \int_F g' = \int_a^b \chi_E(g(x))g'(x)\,dx.$$

   (d) If f is a nonnegative measurable function on $[c, d]$, then $(f \circ g)g'$ is measurable on [a,b] and

   $$\int_c^d f(y)\,dy = \int_a^b f(g(x))g'(x)\,dx.$$

8.5 The Daniell Approach

The preceding classical approach to integration is based on the theory of measure. It is sometimes convenient to introduce integration directly without using the concept of measure. In some parts of analysis, the attention is focused much more on integrals than on measures. Therefore, a theory of integration was developed by Daniell which does not require outer measures and measures as preliminaries. In this section, we shall sketch Stone's much-improved formulation of the Daniell integration in a form which is as brief and uncomplicated as possible.

Consider an arbitrarily given set X and let $\mathcal{L}$ denote any linear lattice of real-valued functions on X. By this, we mean that $\mathcal{L}$ is a linear space in the sense of (9.4) such that it contains the functions $f \vee g$ and $f \wedge g$ for any two members f and g of $\mathcal{L}$.

A linear functional $I : \mathcal{L} \to \mathbb{R}$ on the linear space $\mathcal{L}$ is said to be positive iff $I(f) \geq 0$ holds for every nonnegative function $f : X \to \mathbb{R}$ in $\mathcal{L}$. If I is positive and $f \leq g$, then $I(f) \leq I(g)$.

By a **Daniell integral** on $\mathcal{L}$, we mean a positive functional

$$I : \mathcal{L} \to \mathbb{R}$$

such that, for any sequence $\{f_n | n \in \mathbb{N}\}$ of nonnegative functions in $\mathcal{L}$,

$$I(f) \leq \sum_{n=1}^{\infty} I(f_n)$$

holds for every function $f : X \to \mathbb{R}$ in $\mathcal{L}$ satisfying

$$f(x) \leq \sum_{n=1}^{\infty} f_n(x), \quad (x \in X).$$

This condition is clearly equivalent to each of the following conditions;

D′. If $\langle f_n \rangle$ is a sequence of functions in $\mathcal{L}$ which decrease to zero at each point, then $\lim I(f_n) = 0$

D$''$. If $\langle f_n \rangle$ is an increasing sequence of functions in $\mathfrak{L}$, and if f is a function in $\mathfrak{L}$ such that $f \leq \lim f_n$, then $I(f) \leq \lim I(f_n)$

5.1 Examples of Daniell integrals. (1) Let X be a closed interval [a,b] of the real line $\mathbb{R}$ and let $\mathfrak{L}$ denote the linear lattice $C(X)$ of all continuous functions $f : X \to \mathbb{R}$. Then the functional $I : \mathfrak{L} \to \mathbb{R}$ defined by the Riemann integral

$$I(f) = \int_a^b f(x)\,dx$$

for every $f \in \mathfrak{L} = C(X)$ is a Daniell integral.

(2) Let $(X, \mathfrak{M}, \mu)$ be any complete measure space. Let $\mathfrak{L}$ denote the linear lattice of all measurable functions $f : X \to \mathbb{R}$ which are integrable with respect to μ. Then the functional $I : \mathfrak{L} \to \mathbb{R}$ defined by

$$I(f) = \int f\,d\mu$$

for every $f \in \mathfrak{L}$ is a Daniell integral.

Now let $I : \mathfrak{L} \to \mathbb{R}$ denote an arbitrarily given Daniell integral on a linear lattice of real-valued functions on a set X. Consider the set $\mathfrak{L}^*$ of all functions $f : X \to \mathbb{R}^*$ such that f is the pointwise limit of some nondecreasing sequence in $\mathfrak{L}$.

Let $f \in \mathfrak{L}^*$ be arbitrarily given. Then there exists a nondecreasing sequence $\sigma = \{\phi_n | n \in \mathbb{N}\}$ in $\mathfrak{L}$ such that $\phi_n(x) \to f(x)$ for every $x \in X$.

Since the functional is positive, the sequence $\{I(\phi_n) | n \in \mathbb{N}\}$ is a nonndecreasing sequence of real numbers. Hence we obtain a well-defined member

$$I(\sigma) = \lim_{n \to \infty} I(\phi_n)$$

of the extended real line $\mathbb{R}^*$.

5.2 Lemma. *The extended real number $I(\sigma)$ does not depend on the choice of the sequence σ.*

Proof. Let $\tau = \{\psi_n | n \in \mathbb{N}\}$ be another nondecreasing sequence in $\mathcal{L}$ such that f is the pointwise limit of τ. It suffices to show that $I(\sigma) \leq I(\tau)$.

For this purpose, consider the $\{\xi_n | n \in \mathbb{N}\}$ sequence of nonnegative functions in $\mathcal{L}$ defined by

$$\xi_n = \psi_n - \psi_{n-1}, \quad \psi_0 = 0.$$

Then, for every $x \in X$, we have

$$\phi_n(x) \leq f(x) = \sum_{i=1}^{\infty} \xi_i(x).$$

It follows from the definition of a Damiell integral that

$$I(\phi_n) \leq \sum_{i=1}^{\infty} I(\xi_i) = I(\tau).$$

Since this holds for all $n \in \mathbb{N}$, we obtain

$$I(\sigma) = \lim_{n \to \infty} I(\phi_n) \leq I(\tau).$$

This completes the proof of(5.2). □

Because of (5.2), we may extend the function $I : \mathcal{L} \to \mathbb{R}$ to a function

$$I^* : \mathcal{L} \to \mathbb{R}^*$$

by taking $I^*(f) = I(\sigma_f)$ for every $f \in \mathcal{L}^*$, where σ_f denotes any nondecreasing sequence in $\mathcal{L}$ such that f is the pointwise limit of σ_f.

Now let us consider an arbitrarily given function

$$f : X \to \mathbb{R}^*.$$

Let $U(f)$ denote the subset of $\mathcal{L}^*$ defined by

$$U(f) = \{\xi \in \mathcal{L}^* | \xi \geq f\}.$$

Then we can define the **upper Daniell integral** of f (with respect to I) as the extended real number

$$\tilde{I}(f) = \inf\{I^*(\xi),\ \xi \in U(f)\}$$

with the understanding that infimum of an empty subset of $\mathbb{R}^*$ is ∞. On the other hand, we define the lower Daniell integral of f by

$$\underline{I}(f) = -\overline{I}(-f).$$

The properties of the upper Daniell integral $\overline{I}$ and the lower Daniell integral $\underline{I}$ are given in the following proposition.

5.3 Proposition. *For an arbitrary given daniel integral $I : \mathfrak{L} \to \mathbb{R}$. on a linear lattice $\mathfrak{L}$ of real-valued functions on a set X, the upper and lower Daniel integrals $\overline{I}$, $\underline{I}$ have the following properties.*

(a) *For any two functions $f, g : X \to \mathbb{R}^*$, we have*

$$\overline{I}(f+g) \leq \overline{I}(f) + \overline{I}(g).$$

(b) *For any function $f : X \to \mathbb{R}^*$ and any nonnegative real number α, we have*

$$\overline{I}(\alpha f) = \alpha\overline{I}(f).$$

(c) *For any two functions $f, g : X \to \mathbb{R}^*$ with $f \leq g$, we have*

$$\overline{I}(f) \leq \overline{I}(g), \quad \underline{I}(f) \leq \underline{I}(g).$$

(d) *For every function $f : X \to \mathbb{R}^*$, we have*

$$\underline{I}(f) \leq \overline{I}(f).$$

(e) *If a function $f : X \to \mathbb{R}^*$ is in L^*, then we have*

$$I^*(f) = \underline{I}(f) = \overline{I}(f).$$

(f) *If a function $f : X \to \mathbb{R}^*$ is positive sum of a sequence*

$$\{f_n : X \to \mathbb{R}^* | n \in \mathbb{N}\}$$

of nonnegative functions, then we have

$$\overline{I}(f) \leq \sum_{n=1}^{\infty} \overline{I}(f_n).$$

Proofs of these properties are reasonably easy and can be found in [18, pp. 240 - 243].

A function $f : X \to \mathbb{R}^*$ is said to be **integrable** with respect to I iff $\overline{I}(f)$ and $\underline{I}(f)$ are finite and equal. In this case, we shall denote their common value by $I_1(f)$.

Let $\mathcal{L}_1$ denote the set of all functions $f : X \to \mathbb{R}$ which are integrable with respect to I. Then one can easily verify that $\mathcal{L}_1$ is a linear lattice containing the given linear lattice $\mathcal{L}$. Furthermore, the given Daniell integral $I : \mathcal{L} \to \mathbb{R}$ extends to a positive functional

$$I_1 : \mathcal{L}_1 \to \mathbb{R}$$

by taking $I_1(f) = \underline{I}(f) = \overline{I}(f)$ for every $f \in \mathcal{L}_1$.

To prove that I_1 is a Daniell integral, let us establish the following theorem which is an analogue of (4.6 of Chapter IV).

5.4 Theorem. *For any non-decreasing sequence $\{f_n | n \in \mathbb{N}\}$ of integrable functions $f_n : X \to \mathbb{R}^*$ with respect to I, the function $f : X \to \mathbb{R}^*$ defined by*

$$(x) = \lim_{n \to \infty} f_n(x)$$

for every $x \in X$ is integrable with respect to I, iff

$$l = \lim_{n \to \infty} I_1(f_n) < \infty.$$

In case f is integrable with respect to I, we have $I_1(f) = l$.

Proof. Necessity: Assume $l = \infty$. Because of $f_n \le f$, we have

$$I_1(f_n) = \overline{I}(f_n) \le \overline{I}(f)$$

according to (c) of 5.3. Since this is true for every $n \in \mathbb{N}$, we obtain $\overline{I}(f) = \infty$ and hence f is not integrable.

Sufficiency. Assume $l < \infty$. Since the function $f - f_1$ is pointwise sum of the sequence $\{f_{n+1} - f_n | n \in \mathbb{N}\}$ of nonnegative functions, it follows from (f) of 5.3 that

$$\overline{I}(f - f_1) \le \sum_{n=1}^{\infty} \overline{I}(f_{n+1} - f_n) = l - I_1(f_1).$$

Hence, by (a) of (5.3), we have

$$\overline{I}(f) \le \overline{I}(f - f_1) + \overline{I}(f_1) \le l.$$

On the other hand, $f_n \le f$ implies

$$I_1(f_n) = I(f_n) \ge I(f)$$

according to (c) of (5.3). Since this is true for every $n \in \mathbb{N}$, we have

$$l \le I(f) \le \overline{I}(f) \le l.$$

This implies that f is integrable with respect to I and $I_1(f) = l$. □

5.5 Corollary. *The positive functional $f_1 : \mathfrak{L}_1 \to \mathbb{R}$ is a Daniell integral on $\mathfrak{L}_1$.*

Proof. Let $\{f_n | n \in \mathbb{N}\}$ be any sequence of nonnegative functions in $\mathfrak{L}_1$ and let $f : X \to \mathbb{R}$ denote any function in $\mathfrak{L}_1$ satisfying

$$f(x) \le \sum_{n=1}^{\infty} f_n(x), \quad (x \in X).$$

It suffices to establish

$$I_1(f) \le l = \sum_{n=1}^{\infty} I_1(f_n).$$

If $l = \infty$, there is nothing to prove. Hence we assume $l < \infty$.

Consider the nondecreasing sequence $\{g_n | n \in \mathbb{N}\}$ of partial sums

$$g_n = \sum_{i=1}^{n} f_i.$$

Let $g : X \to \mathbb{R}^*$ denote the function defined by

$$g(x) = \lim_{n\to\infty} g_n(x) = \sum_{n=1}^{\infty} f_n(x).$$

Since

$$\lim_{n\to\infty} I_1(g_n) = \sum_{n=1}^{\infty} I_1(f_n) = l < \infty,$$

it follows from 5.4 that g is integrable with respect to I and $I_1(g) = l$.

Since $f \le g$, we obtain

$$I_1(f) = \overline{I}(f) \le \overline{I}(g) = I_1(g) = l$$

according to (c) of 5.3. This proves 5.5. □

A nonnegative function $f : X \to \mathbb{R}$ is said to be **measurable** with respect to I iff $f \wedge \xi \in \mathfrak{L}_1$ for every $\xi \in \mathfrak{L}_1$. A subset E of X is said to be **measurable** with respect to I iff its characteristic function $\chi_E : X \to \mathbb{R}$ is measurable with respect to I.

The given Daniell integral $I : \mathfrak{L} \to \mathbb{R}$ is said to be **regular** iff X is measurable with respect to I. Hereafter, we assume that $I : \mathfrak{L} \to \mathbb{R}$ is a regular Daniell integral. In this case, one can easily verify that the collection $\mathfrak{M}$ of all subsets of X which are measurable with respect to I is a Boolean σ-algebra.

Define a function $\mu : \mathfrak{M} \to \mathbb{R}^*$ by taking

$$\mu(E) = \begin{cases} I_1(\chi_B), & \text{if } \chi_B \in \mathfrak{L}_1, \\ \infty, & \text{if } \chi_B \notin \mathfrak{L}_1. \end{cases}$$

It is not difficult to verify the μ is a measure on $\mathfrak{M}$. Thus we obtain a measure space $(X, \mathfrak{M}, \mu)$. The relation Between I_1 and μ is given by the following theorem.

5.6 Theorem. ***(Stone's Theorem).*** *If $I : \mathcal{L} \to \mathbb{R}$ is a regular Daniell integral, then a function $f : X \to \mathbb{R}^*$ is integrable with respect to μ iff it is integrable with respect to I. Furthermore, we have*

$$\int f \, d\mu = I_1(f)$$

in case f is integrable.

A proof of (5.6) can be found in [18,pp.246-247].

This identifies the Daniell approach to integration with the classical approach by measure theory. Consequently, statements analogous to (3.5) and (3.10) are also valid to Daniell integrals.

PROBLEMS [10.5]

1. For an arbitrarily given Daniell integral $I : \mathcal{L} \to \mathbb{R}$, prove:

   (a) A nonnegative function $f : X \to \mathbb{R}^*$ is in $\mathcal{L}^*$ iff there exists a sequence $\{f_n | n \in \mathbb{N}\}$ of nonnegative functions in $\mathcal{L}$ satisgying
   $$f(x) = \sum_{n=1}^{\infty} f_n(x)$$
   for every $x \in X$.

   (b) For every sequence $\{f_n | n \in \mathbb{N}\}$ of nonnegative functions in $\mathcal{L}^*$, the function $f : X \to \mathbb{R}^*$ defined by
   $$f(x) = \sum_{n=1}^{\infty} f_n(x)$$
   for every $x \in X$ is in $\mathcal{L}^*$ and
   $$I^*(f) = \sum_{n=1}^{\infty} I^*(f_n).$$

2. Let $I : \mathcal{L} \to \mathbb{R}$ be an arbitrary regular Daniell integral and let $\mathcal{B}$ denote the smallest Boolean σ-algebra of subsets of X such that every function in $\mathcal{L}$ is measurable with respect to

$\mathcal{B}$. Prove the following **Uniqueness Theorem**: There exists a unique measure $v : \mathcal{B} \to \mathbb{R}^*$ such that every function $f \in \mathcal{L}$ is integrable with respect to v and

$$\int f \, dv = I(f).$$

3. Let X denote an arbitrary compact Hausdorff space and consider the smallest Boolean σ-algebra $\mathcal{B}$ of subsets of X such that every continuous function $f : X \to \mathbb{R}$ is measutable with respect to $\mathcal{B}$. By a **Baire measure** in X, we mean a finite measure $\mu : \mathcal{B} \to \mathbb{R}$. By **signed Baire measure** we mean a finite signed measure $\mu : \mathcal{B} \to \mathbb{R}$. Prove the following two assertions for the Banach space $C(X)$ of all continuous functions:

   (a) For every positive functional $F : C(X) \to \mathbb{R}$, there exists a unique Baire measure $\mu : \mathcal{B} \to \mathbb{R}$ such that

   $$\int f \, d\mu = F(f)$$

   holds for every $f \in C(X)$.

   (b) For every continuous functional $F : C(X) \to \mathbb{R}$, there exists a unique signed measure $\mu : \mathcal{B} \to \mathbb{R}$ such that $|\mu|(X) = \|F\|$ and

   $$\int f \, d\mu = F(f)$$

   holds for every $f \in C(X)$. Here, the integration is defined by

   $$\int f \, d\mu = \int f \, d\mu^+ - \int f \, d\mu^-.$$

8.6 Haar Measure

Thoroghout the present section, let X denote any locally compact topological group as defined in chapter 7.

By a **Haar measure** in X, we mean a measure $\mu : \mathfrak{M} \to \mathbb{R}^*$ on a Boolean σ-algebra $\mathfrak{M}$ of subsets of X satisfying the following three conditions;

(HM1) μ is **compact-finite**, that is, every compact subset K of X is in $\mathfrak{M}$ and with $\mu(K) < \infty$.

(HM2) μ is **nontrivial**, that is, there exists a compact subset A of X with $\mu(A) > 0$.

(HM3) μ is **left invariant**, that is, for every $x \in X$ and every $E \in \mathfrak{M}$, the subset xE is in $\mathfrak{M}$ and

$$\mu(xE) = \mu(E).$$

The remainder of the section is devoted to the construction of a **Haar measure** in X.

Let K be any compact subset of X and let U denote any neighborhood of the neutral element (identity, unit, or zere) e of X. By the ratio $K : U$, we mean the least nonnegative integer n such that there are n points $x_1, \cdots, x_n$ of X satisfying

$$K \subset x_1 U \cup \cdots \cup x_n U.$$

Obviously, $K : U = 0$ iff $K = \phi$.

Since X is locally compact, there exists a compact neighborhood A of the neutral element e. Let us emphasize that A is arbitrarily selected and will be used throughout the construction.

Let $\mathcal{C}$ denote the collection of all compact subsets of X and let $\mathcal{F}$ denote the family of all open neighborhoods of the neutral element e in X. For each U in $\mathcal{F}$, define a function

$$\lambda_U : \mathcal{C} \to \mathbb{R}$$

by taking

$$\lambda_U(K) = \frac{K : U}{A : U}$$

for every $K \in \mathcal{C}$.

6.1 Lemma. *for every $U \in \mathcal{F}$, the function $\lambda_U : \mathcal{C} \to \mathbb{R}$ has the following properties;*

(a) $\lambda_u(\phi) = 0$ *and* $\lambda_u(A) = 1$.

(b) $0 \leq \lambda_u(K) \leq K : A$ *for every* $K \in \mathcal{C}$

(c) $\lambda_u(xK) = \lambda_u(K)$ *for every* $x \in X$ *and every* $K \in \mathcal{C}$

(d) For any two members K and L of $\mathcal{C}$, $K \subset L$ implies

$$\lambda_u(K) \leq \lambda_u(L).$$

(e) For any two members K and L of $\mathcal{C}$, we have

$$\lambda_u(K \cup L) \leq \lambda_u(K) + \lambda_u(L).$$

(f) For any two members K and L of $\mathcal{C}$ satisfying $KU^{-1} \cap LU^{-1} = \phi$, we have $\lambda_u(K \cup L) = \lambda_u(K) + \lambda_u(L)$.

Proof. (a) is obvious.

To establish (b), let $K : A = p$ and $A : U = q$. Then there are points $y_1, \cdots, y_p$ and $z_1, \cdots, z_q$ of X such that

$$K \subset y_1 A \cup \cdots \cup y_p A, \quad A \subset z_1 U \cup \cdots \cup z_q U.$$

Hence we obtain pq points $x_{ij} = y_i z_j$ of X such that K is contained in the union of the pq sets $x_{ij}U$. This implies $0 \leq K : U \leq pq$ and therefore $0 \leq \lambda_U(K) \leq q = K : A$.

To establish (c), let $m = xK : U$ and $n = K : U$. Then there are points $y_1, \cdots, y_m$ and $z_1, \cdots, z_n$ of X such that

$$xK \subset y_1 U \cup \cdots \cup y_m U, \quad K \subset z_1 U \cup \cdots \cup z_n U.$$

It follows that we also have

$$K \subset x^{-1} y_1 U \cup \cdots \cup x^{-1} y_m U, \quad xK \subset x z_1 U \cup \cdots \cup x z_n U.$$

These imply $m \geq n$ and $n \geq m$ respectively. Hence we obtain $m = n$. This proves (c).

(d) and (e) are obvious.

To establish (f), we observe that it suffices to prove

$$\lambda_U(K \cup L) \geq \lambda_U(K) + \lambda_U(L)$$

because of (e). Let $n = (K \cup L) : U$. Then there are n points $x_1, \cdots, x_n$ of X such that

$$K \cup L \subset x_1 U \cup \cdots \cup x_n U.$$

Let i denote any integer satisfying $1 \leq i \leq n$. If $K \cap x_i U \neq \phi$, we have $x_i \in KU^{-1}$. similarly, if $L \cap x_i U \neq \phi$, we have $x_i \in LU^{-1}$. Since

$$KU^{-1} \cap LU^{-1} = \phi,$$

it follows that $n \geq (K : U) + (L : U)$. This implies

$$\lambda_U(K \cup L) = \frac{n}{A : U} \geq \lambda_U(K) + \lambda_U(L). \qquad \square$$

For every member K of $\mathcal{C}$, let I_K denote the closed interval $[0, K : A]$. Thus we obtain an indexed family $\{I_K | K \in \mathcal{C}\}$ of compact spaces I_K. According to Tychonoff's theorem (5.2 of Chapter 7), the topological product

$$\Phi = \Pi_{K \in \mathcal{C}} I_K$$

as defined in Chapter 7 is compact. By the definition of topological product, the members of Φ are the functions $\phi : \mathcal{C} \to \mathbb{R}$ such that $\phi(K) \in I_K$ holds for every $K \in \mathcal{C}$. Because of (b) in (6.1), we have $\lambda_U \in \Phi$ for every $U \in \mathcal{F}$.

For every $U \in \mathcal{F}$, let $\Lambda(U)$ denote the subset of Φ defined by

$$\Lambda(U) = \{\lambda_V | U \supset V \in \mathcal{F}\}.$$

Since $\lambda_U \in \Lambda(U)$, $\Lambda(U)$ is nonempty. for any finite subfamily

$$\{U_1, \cdots, U_n\}$$

of $\mathcal{F}$, the intersection

$$U = U_1 \cap \cdots \cap U_n$$

is in $\mathcal{F}$. This implies

$$\Lambda(U) \subset \Lambda(U_1) \cap \cdots \cap \Lambda(U_n)$$

and hence the family $\{\Lambda(U)|U \in \mathcal{F}\}$ has the finite intersection property as defined in Chapter 7. Since Φ is a compact space, it follows from Chapter 7.1 that the intersection

$$\Lambda = \bigcap_{U \in \mathcal{F}} C1[\Lambda(U)]$$

is nonempty. Choose an arbitrary member λ of Λ. Then λ is a function

$$\lambda : \mathcal{C} \to \mathbb{R}.$$

The properties of $\lambda : \mathcal{C} \to \mathbb{R}$ are given by the following lemma.

6.2 Lemma. *Every function $\lambda : \mathcal{C} \to \mathbb{R}$ in A has the following properties*

(a) $\lambda(\phi) = 0$ *and* $\lambda(A) = 1$.

(b) $0 \leq \lambda(K) \leq K : A$ *for every* $K \in \mathcal{C}$.

(c) $\lambda(xK) = \lambda(K)$ *for every* $x \in X$ *and for every* $K \in \mathcal{C}$.

(d) For any two members K *and* L *of* $\mathcal{C}$, $K \subset L$ *implies* $\lambda(K) \leq \lambda(L)$.

(e) For any two members K *and* L *of* $\mathcal{C}$, *we have*

$$\lambda(K \cup L) \leq \lambda(K) + \lambda(L)$$

(f) For any two disjoint members K *and* L *of* $\mathcal{C}$, *we have*

$$\lambda(K \cup L) = \lambda(K) \cup \lambda(L)$$

Proof. For each $K \in \mathcal{C}$, let $p_k : \Phi \to I_K$ denote the natural projection of the topological product Φ onto the coordinate space I_K defined by

$$p_k(\phi) = \phi(K)$$

for every $\phi \in \Phi$. Hence p_k is continuous.

Hence, for every $K \in \mathcal{C}$ and every $U \in \mathcal{F}$, and we have

$$\lambda(K) \in p_k(\Lambda) \subset p_k\{C1[\Lambda(U)]\} \subset C1\{\lambda_V(K)|U \supset V \in \mathcal{F}\}.$$

To establish (a), we take $K = \phi$, or $K = A$. Since $\lambda_V(\phi) = 0$ and $\lambda_V(A) = 1$ for every $V \in \mathcal{F}$, the preceding inclusion implies $\lambda(\phi) = 0$ and $\lambda(A) = 1$.

Since $\lambda \in \Phi$, (b) is obvious.

To establish (c), let $x \in X$ and $K \in \mathcal{C}$ be arbitrarily given. From the continuity of p_K and p_{xK}, it follows that the subset

$$F = \{\phi \in \Phi|\phi(xK) = \phi(K)\}$$

of Φ is closed. Let $U \in \mathcal{F}$. By (c) of (6.1), we have $\Lambda(U) \subset F$. The closedness of F implies

$$\lambda \in C1[\Lambda(U)] \subset F.$$

Hence we obtain $\lambda(xK) = \lambda(K)$.

The proofs of (d) and (e) are similar to that of (c) and hence omitted.

To establish (f), let K and L denote any two disjoint members of $\mathcal{C}$. Form the compactness of K and L, it follows easily that there exists a member $U \in \mathcal{F}$ satisfying

$$KU^{-1} \cap LU^{-1} = \phi.$$

According to (f) in (6.1), we have

$$\lambda_V(K \cup L) = \lambda_V(K) + \lambda_V(L)$$

for every $V \in \mathcal{F}$ with $V \subset U$. This implies that $\Lambda(U)$ is contained in the subset

$$H = \{\phi \in \Phi|\phi(K \cup L) = \phi(K) + \phi(L)\}$$

of Φ. Since H is a closed subset of Φ, we have

$$\lambda \in C1[\Lambda(U)] \subset H$$

and hence $\lambda(K \cup L) = \lambda(K) + \lambda(L)$. □

Throughout the remainder of the section, let $\lambda : \mathcal{C} \to \mathbb{R}$ denote an arbitrtily selected function in the subset Λ of Φ.

Let τ denote the topology of X, that is the collection of all open sets of X. Define a function $\lambda_* : \tau \to \mathbb{R}^*$ by taking

$$\lambda_*(U) = \sup\{\lambda(K) | U \supset K \in \mathcal{C}\}$$

for every $U \in \tau$.

6.3 Lemma. *The function $\lambda_* : \tau \to \mathbb{R}^*$ has the following properties;*

(a) $\lambda_*(\phi) = 0$.

(b) $\lambda_*(xU) = \lambda_*(U)$ *for every* $x \in X$ *and every* $U \in \tau$.

(c) For any two members U and V of τ, $U \subset V$ implies $\lambda_*(U) \leq \lambda_*(V)$.

(d) λ_* *is countably subadditive.*

(e) λ_* *is countably additive.*

Proof. (a) is obvious from the definition of λ_*.

To establish (b), let $x \in X$ and $U \in \tau$ be arbitrarily given. For every $K \in \mathcal{C}$ with $K \subset U$, we have $xK \subset xU$. Since $\lambda(xK) = \lambda(K)$ according to (c) of (6.2), we obtain

$$\{\lambda(K) | U \supset K \in \mathcal{C}\} \subset \{\lambda(K) | xU \supset K \in \mathcal{C}\}.$$

This implies $\lambda_*(U) \leq \lambda_*(xU)$. Since $U = x^{-1}(xU)$, we also have

$$\lambda_*(xU) \leq \lambda_*(U).$$

This proves (b).

To establish (c), let U and V denote any two members of τ with $U \subset V$. For every $K \in \mathcal{C}$ with $K \subset U$, we also have $K \subset U$. Thus

$$\{\lambda(K) | U \supset K \in \mathcal{C}\} \subset \{\lambda(K) | V \supset K \in \mathcal{C}\}.$$

This proves $\lambda_*(U) \leq \lambda_*(V)$.

To establish (d), let us first prove that λ_* is subadditive. For this purpose, let U and V denote any two members of τ. Consider any $K \in \mathcal{C}$ with $K \subset U \cup V$. There exist members F and G of $\mathcal{C}$ with $F \subset U$, $G \subset V$ and $K = F \cup G$. By (e) of (6.2), we have

$$\lambda(k) \leq \lambda(F) + \lambda(G) \leq \lambda_*(U) + \lambda_*(V).$$

Since this is true for all $K \in \mathcal{C}$ satisfying $K \subset U \cup V$, we obtain

$$\lambda_*(U \cup V) \leq \lambda_*(U) + \lambda_*(V).$$

This proves that λ_* is subadditive and hence finitely subadditive.

Next, let $\{U_i | i \in \mathbb{N}\}$ denote any sequence in τ and let K denote any member of $\mathcal{C}$ contained in the union

$$U = \bigcup_{i=1}^{\infty} U_i.$$

Since K is compact, there exists a natural number n with

$$K \subset \bigcup_{i=1}^{n} U_i.$$

Hence we obtain

$$\lambda(K) \leq \lambda_* \left(\bigcup_{i=1}^{n} U_i \right) \leq \sum_{i=1}^{n} \lambda_*(U_i) \leq \sum_{i=1}^{\infty} \lambda_*(U_i).$$

Since this is true for every member K of $\mathcal{C}$ satisfying $K \subset U$, we must have

$$\lambda_*(U) \leq \sum_{i=1}^{\infty} \lambda_*(U_i).$$

This comlpetes the proof of (d).

To establish (e), let us first prove that λ_* is additive. For this purpose, let U and V denote any two disjoint members of τ. Consider arbitrary members F and G of $\mathcal{C}$ with $F \subset U$ and $G \subset V$. Then F and G are disjoint. By (f) of (6.2), this implies

$$\lambda(F) + \lambda(G) = \lambda(F \cup G) \leq \lambda_*(U \cup V).$$

Since $F \subset U$ and $G \subset V$ are arbitrary, we have

$$\lambda_*(U) + \lambda_*(V) \leq \lambda_*(U \cup V).$$

Because of (d), this implies

$$\lambda_*(U) + \lambda_*(V) = \lambda_*(U \cup V).$$

This proves that λ_* is additive and hence finitely additive.

Next, let $\{U_i | i \in \mathbb{N}\}$ denote any disjoint sequence in τ. Consider the union

$$U = \bigcup_{i=1}^{\infty} U_i.$$

By (c) and the finite additivity, we have

$$\lambda_*(U) \geq \lambda_*\left(\bigcup_{i=1}^{n} U_i\right) = \sum_{i=1}^{n} \lambda_*(U_i)$$

for every natural number n. This implies

$$\lambda_*(U) \geq \sum_{i=1}^{\infty} \lambda_*(U_i).$$

Because of (d), this implies

$$\lambda_*(U) = \sum_{i=1}^{\infty} \lambda_*(U_i).$$

This completes the proof of (6.3). □

Now define a function $\mu^* : \wp^X \to \mathbb{R}^*$ on the set $\wp^X$ of all subsets of X by taking

$$\mu^*(E) = \inf\{\lambda_*(U) | E \subset U \in \tau\}$$

for every $E \in \wp(X)$.

6.4 Lemma. *The function $\mu^* : \wp^X \to \mathbb{R}^*$ is an outer measure in X satisfying $\mu^*(xE) = \mu^*(E)$ for every $x \in X$ and every $E \in \wp^X$.*

Proof. To prove that μ^* is an outer measure in X, we have to verify the conditions (OM1-3) in §6.1.

To verify (OM1), we observe $\phi \subset \phi \in \tau$. Since $\lambda_*(\phi) = 0$ ane $\lambda_*(U) \geq 0$ for every $U \in \tau$, we obtain $\mu^*(\phi) = 0$.

To verify (OM2), let D and E denote any two subsets of X with $D \subset E$. For every $U \in \tau$ with $E \subset U$, we have $D \subset U$. This implies $\mu^*(D) \leq \lambda_*(U)$. Since this is true for every $U \in \tau$ with $E \subset U$, we have $\mu^*(D) \leq \mu^*(E)$. Hence μ^* is monotonic.

To verify (OM3), let $\{E_n | n \in \mathbb{N}\}$ denote any sequence of subsets of X. Consider an arbitrarily given positive real number ϵ. For every $n \in \mathbb{N}$, there exists a member U_n of τ satisfying $E_n \subset U_n$ and

$$\lambda_*(U_n) \leq \mu^*(E_n) + 2^{-n}\epsilon.$$

Since λ_* is countably subadditive according to (d) of 6.3, we have

$$\begin{aligned} \mu^*\left(\bigcup_{n=1}^{\infty} E_i\right) &\leq \lambda_*\left(\bigcup_{n=1}^{\infty} U_n\right) \leq \sum_{n=1}^{\infty} \lambda_*(U_n) \\ &\leq \sum_{n=1}^{\infty}[\mu^*(E_n) + 2^{-n}\epsilon] = \sum_{n=1}^{\infty} \mu^*(E_n) + \epsilon. \end{aligned}$$

Since this is true for every positive real number ϵ, we must have

$$\mu^*\left(\bigcup_{n=1}^{\infty} E_i\right) \leq \sum_{n=1}^{\infty} \mu^*(E_n)$$

This completes the proof that μ^* is an outer measure in X.

It remains to prove the left invariance of μ^*. For this purpose, let $x \in X$ and $E \in \wp^X$ be arbitratily given. For every $U \in \tau$ with $E \subset U$, we have $xE \subset xU$. Since $\lambda_*(xU) = \lambda_*(U)$ according to 6.3, we obtain

$$\{\lambda_*(U) | E \subset U \in \tau\} \subset \{\lambda_*(U) | xE \subset U \in \tau\}.$$

This implies $\mu^*(E) \geq \mu^*(xE)$. Since $E = x^{-1}(xE)$, we also have

$$\mu^*(xE) \geq \mu^*(E).$$

This proves

$$\mu^*(xE) = \mu^*(E)$$

and completes the proof of 6.4. □

Let $\mathfrak{M}$ denote the family of all subsets of X which are μ^*-measurable as defined in §3. According to 3.14, $\mathfrak{M}$ is a Boolean σ-algebra and the restriction

$$\mu = \mu^*|\mathfrak{M} : \mathfrak{M} \to \mathbb{R}^*$$

is a measure on $\mathfrak{M}$.

In order to prove that the measure $\mu : \mathfrak{M} \to \mathbb{R}^*$ is a Haar measure in X, we have to use the following lemma.

6.5 Lemma. *The functions λ,λ_*, and μ^*, constructed above have the following two properties;*

(a) For every open subset U of X, we have

$$\mu^*(U) = \lambda_*(U).$$

(b) For every compact subset K of X, we have

$$\mu^*[Int(K)] \leq \lambda(K) \leq \mu^*(K).$$

Proof. To establish (a), we observe $U \subset U \in \tau$. This implies

$$\mu^*(U) \leq \lambda_*(U).$$

On the other hand, we have $\lambda_*(U) \leq \lambda_*(V)$ for every $V \in \tau$ sith $U \subset V$. This implies $\lambda_*(U) \leq \mu^*(U)$. Hence we obtain $\mu^*(U) = \lambda_*(U)$.

To establish (b), let $K \in \mathfrak{C}$ be arbitrarily given. For every $U \in \tau$ with $K \subset U$, we have $\lambda(K) \leq \lambda_*(U)$. This implies $\lambda(K) \leq \mu^*(K)$. On the other hand, for every $L \in \mathfrak{C}$ with $L \subset \text{Int}(K)$. Then we have $L \subset K$ and hence $\lambda(L) \leq \lambda(K)$. Hence we obtain

$$\mu^*[\text{Int}(K)] = \lambda_*[\text{Int}(K)] \leq \lambda(K)$$

as a consequence of (a) and the definition of λ_*. □

6.6 Lemma. *The measure $\mu : \mathfrak{M} \to \mathbb{R}^*$ is a Haar measure in X.*

Proof. We have to verify the three conditions (HM1)-(HM3).

To verify (HM1), let K denote any compact subset of X. To prove $K \in \mathfrak{M}$, let us first consider an arbitrary open subset U of X. Let F be any compact subset of $U \sim K$ and let G be any compact subset of $U \sim F$. Then F and G are disjoint compact subset of X satisfying $F \cup G \subset U$. By (f) of 6.2 and the definition of λ_*, we have

$$\lambda(F) + \lambda(G) = \lambda(F \cup G) \le \lambda_*(U) = \mu^*(U)$$

because of (a) in (6.5). Since this is true for every $G \in \mathfrak{C}$ with $G \subset U \sim F$ and since $U \cap K$ is contained in $U \sim F$, we have

$$\begin{aligned} \lambda(F) + \mu^*(U \cap K) &\le \lambda(F) + \mu^*(U \sim F) \\ &= \lambda(F) + \lambda_*(U \sim F) \le \mu^*(U). \end{aligned}$$

Since this is true for every $F \in \mathfrak{C}$ with $F \subset U \sim K$, we obtain

$$\mu^*(U) \ge \mu^*(U \cap K) + \lambda_*(U \sim K) = \mu^*(U \cap K) + \mu^*(U \sim K)$$

according to the definition of λ_* as well as (a) of 6.5.

Next let S denote any subset of X. Let U be any open subset of X with $S \subset U$. Then we have

$$\mu^*(U) \ge \mu^*(U \cap K) + \mu^*(U \sim K) \ge \mu^*(S \cap K) + \mu^*(S \sim K).$$

Since $\lambda_*(U) = \mu^*(U)$ by (a) of (6.5), it follows from the definition of $\mu^*(S)$ that

$$\mu^*(S) \ge \mu^*(S \cap K) + \mu^*(S \sim K).$$

Since this is true for every subset S of X, K is μ^*-measurable and hence is in $\mathfrak{M}$.

Now let us prove $\mu(K) < \infty$. Since X is locally compact and K is compact, it follows that there exists a compact subset L of X with

$$K \subset \mathrm{Int}(L).$$

Hence by (b) of (6.5), we have

$$\mu(K) = \mu^*(K) \le \mu^*[\mathrm{Int}(L)] \le \lambda(L) < \infty.$$

This completes the verification of(HM1).

To verify (HM2), we observe

$$\mu^*(A) \geq \lambda(A) = 1$$

according to (a) of 6.2 and (b) of (6.5) This completes the verification of (HM2).

Finally since μ is a restriction of μ^*, (HM3) is a consequence of (6.4).

The following corollary is a consequence of (1.5) of sec 6.1

6.7 Corollary. *The measure spece $(X, \mathfrak{M}, \mu)$ is complete.*

combining (6.6) and (6.7) , we obtain the following existence theorem.

6.8 Theorem. *Every locally compact topological group X has a Haar measure $\mu : \mathfrak{M} \to \mathbb{R}^*$ such that $(X, \mathfrak{M}, \mu)$ is a complete measure space.*

PROBLEMS [10.6]

1. Prove that, for every Haar measure $\mu : \mathfrak{M} \to \mathbb{R}^*$ in X and every positive real number c, the function $c\mu : \mathfrak{M} \to \mathbb{R}^*$ is also a Haar measure in X.

2. Prove that X is compact iff it has a finite Haar measure.

3. The Haar measures defined in the text might be called "left Haar measures." Define "right Haar measures" analogously. Let
$$\mu : \mathfrak{M} \to \mathbb{R}^*$$
be any left Haar measure in X. Let
$$\mathfrak{N} = \{E \subset X | E^{-1} \in \mathfrak{M}\}$$
and define $v : \mathfrak{N} \to \mathbb{R}^*$ by taking
$$v(E) = \mu(E^{-1})$$
for every $E \in \mathfrak{N}$. Prove that v is a right Haar measure in X.

Bibliography

[1] Apostol. T. M. "Mathematical Analysis"(2d ed)
Addison-Wesley Reading Mass. (1974)

[2] Bartle. R. G. "The Elements of Real Analysis"
John Wiley & Sons Inc. New York. (1964)

[3] Cohn. D. L. "Measure Theory"
Birkhäuser. Boston (1980)

[4] Dugundji. J. "Topology"
Allyn & Bacon. Boston (1966)

[5] Folland. G. B. "Real Analysis"
John Wiley & Sons. Inc. New York. (1984)

[6] Halmos. D. R. "Measure Theory"
Van Nostrand. Princeton. New York. (1950)

[7] Hasser. N. B. & Sullivan J. A. "Real Analysis"
Van Nostrand Reinhold Company. New York. (1971)

[8] Herstein. I. N. "Topics in Algebra" (2d ed)
Lexington Mass (1975)

[9] Hewitt. E. & Stromberg K. "Real and Abstract Analysis"
Springer-Verlag. New York. (1965)

[10] Hu. S. T. "Elements of Real Analysis"
Holden-Day Inc. San Francisco. (1967)

[11] Jain. P. K. & Gupta. V. P.
"Lebesgue measure and Integration"
A Halsted Press Book. New Delhi. (1986)

[12] Johnsonbaugh. R. "Foundation of Mathematical Analysis"
Marcel Dekker Inc. (1981)

[13] kelly. J. L. "General Topology"
D. Van Nostrand Company Inc. Princeton. New York. (1965)

[14] Kolmogoroff. A. & Fomin. S. "Introductory Real Analysis"
Dover Publications Inc. New York. (1970)

[15] Nicolas. K. A. "Real Analysis"
Southern Illinois Univ. Press. London (1976)

[16] Parzynski. W. R. and Zipse. P. W.
"Introduction to Mathematical Analysis"
International Student Eddition (1982)

[17] Randolph. J. F. "Basic Real and Abstract Real Analysis"
New York Academic Press. (1968)

[18] Royden. H. L. "Real Analysis"
The Macmillan Company. New York. (1970)

[19] Ross. K. A. "Elementary Analysis; The Theory of Calculus"
Springer-Verlag New York. (1980)

[20] Rudin. W. "Principles of Mathematical Analysis"(2d ed)
McGraw-Hill Book Company New York. (1964)

[21] Rudin. W. "Real and Complex Analysis"
McGraw-Hill Book Company New York. (1974)

[22] Spiegel. M. R. "Schaums Outline Series Theory and Problems of Real Variables", Shaums Outline Series. McGraw-Hill Book Company. New York. (1969)

Index

[D]

[E]

[F]

[G]

[H]